AF358824

Power System Analysis, Operation and Control

Power System Analysis, Operation and Control

Editors

Yuan Liao
Ke Xu

Basel • Beijing • Wuhan • Barcelona • Belgrade • Novi Sad • Cluj • Manchester

Editors
Yuan Liao
University of Kentucky
Lexington, KY
USA

Ke Xu
Nanjing Institute of
Technology
Nanjing
China

Editorial Office
MDPI
St. Alban-Anlage 66
4052 Basel, Switzerland

This is a reprint of articles from the Special Issue published online in the open access journal *Energies* (ISSN 1996-1073) (available at: https://www.mdpi.com/journal/energies/special_issues/PSA).

For citation purposes, cite each article independently as indicated on the article page online and as indicated below:

Lastname, A.A.; Lastname, B.B. Article Title. *Journal Name* **Year**, *Volume Number*, Page Range.

ISBN 978-3-7258-1077-2 (Hbk)
ISBN 978-3-7258-1078-9 (PDF)
doi.org/10.3390/books978-3-7258-1078-9

Contents

About the Editors

Yuan Liao

Yuan Liao is currently a professor and holds the endowed Blazie Family professorship with the Department of Electrical and Computer Engineering at the University of Kentucky, Lexington, KY, USA. He is the Director of the Graduate Certificate programs at the Power and Energy Institute of Kentucky. His research interests include power system protection, analysis and planning, smart grids, and renewable energy integration.

Ke Xu

Ke Xu is currently an assistant professor at the School of Electric Power Engineering, Nanjing Institute of Technology, Nanjing, Jiangsu, China. His research interests include power system control, smart grids, power system protection, renewable energy, and AI in power systems.

Preface

To reduce carbon emissions and our reliance on fossil fuels, electric power systems around the world are seeing increasing integration of renewable energies, the adoption of electric vehicles, and the construction of much-needed charging infrastructure. For improved operation and resiliency, more microgrids are emerging and storage devices are being deployed. Many of the distributed energy resources are inverter based, with them having specific short circuit characteristics. The increasing number and unique characteristics of distributed energy resources and controllers, the increasing volume of available measurement data, and environmental constraints pose challenges for analyzing, operating, controlling, and protecting future power systems. New methods and techniques to overcome these challenges and ensure a reliable and resilient power grid are needed.

This Special Issue includes 10 papers from distinguished researchers that present novel techniques and algorithms as well as practical considerations for reliable and efficient power grid performance. We would like to sincerely thank all of the authors for their interest in and outstanding contributions to this Special Issue. We would also like to gratefully acknowledge the excellent support, patience, and valuable time of Ms. Haley Lei, the Managing Editor of this Special Issue.

Yuan Liao and Ke Xu
Editors

Article

CVR Study and Active Power Loss Estimation Based on Analytical and ANN Method

Gaurav Yadav [1], Yuan Liao [1,*], Nicholas Jewell [2] and Dan M. Ionel [1]

[1] Department of Electrical and Computer Engineering, University of Kentucky, Lexington, KY 40506, USA; gya231@uky.edu (G.Y.); dan.ionel@uky.edu (D.M.I.)
[2] Louisville Gas & Electric and Kentucky Utilities (LG&E and KU), Louisville, KY 40202, USA; nicholas.jewell@lge-ku.com
* Correspondence: yuan.liao@uky.edu

Citation: Yadav, G.; Liao, Y.; Jewell, N.; Ionel, D.M. CVR Study and Active Power Loss Estimation Based on Analytical and ANN Method. *Energies* **2022**, *15*, 4689. https://doi.org/10.3390/en15134689

Academic Editor: Abu-Siada Ahmed

Received: 4 June 2022
Accepted: 23 June 2022
Published: 26 June 2022

Publisher's Note: MDPI stays neutral with regard to jurisdictional claims in published maps and institutional affiliations.

Abstract: Conservation through voltage reduction (CVR) aims to reduce the peak load and energy savings in electric power systems and is being deployed at various utilities. The effectiveness of the CVR program depends on the load characteristics, i.e., the sensitivity of the load to voltage variation, and voltage regulation device settings. In the current literature, there is a lack of discussion on the CVR factor calculation using different measurements, and there is a lack of method for active power loss estimation using substation measurements. This paper provides insights into CVR factor calculation based on the measurements captured at the substation and those at the load location. This paper also proposes a new method based on curve fitting and artificial neural network to estimate the active power loss using input active power, input reactive power and input voltage at the substation. The CVR comparison study conducted in this paper helps in understanding the factors affecting CVR factor and may provide guidance in CVR implementation and impact assessment. The proposed loss estimation method sheds light on the impacts of CVR in terms of load and loss reduction. The results based on simulation studies using the IEEE 13-bus and 34-bus systems are reported in this paper, noting that the proposed methods are applicable to larger systems, as long as the required measurements at the substation are available. Future research includes testing and refining the methods using large IEEE and utility distribution systems and considering the stochastic nature of the CVR factor with changing load and voltage regulator control schemes.

Keywords: artificial neural network; conservation through voltage reduction; curve fitting; electric power systems; network power loss; ZIP load model

1. Introduction

The world is currently experiencing a massive surge in energy demand, due to various demographic, environmental and economic reasons. When facing high demand, the electric distribution companies have two options to manage the demand. The first option is to arrange supply for the demand, which may not be available each time. The second option is to curtail the peak load, which may be a more viable method. This reduction may be achieved by decreasing the voltage level on the distribution feeder line. The decrease in the bus voltage leads to a decline in the voltage-dependent load demand, which may also reduce the total energy consumption during a specified time duration. This strategy is called conservation through voltage reduction (CVR) [1]. The objective of CVR is to reduce the peak load or reduce the energy consumption for energy savings, while regulating the voltage within the limit set by the American National Standards Institute (ANSI). ANSI C84.1 standard sets the range for voltages at the secondary terminals of transformer at 120 volts, which can vary by 5% of the aforesaid limit, resulting in a voltage level between 114 and 126 volts [2]. Over the years, numerous CVR-related operations have been conducted in the country, with the earliest being performed by the American Electric

Power system (AEP) in 1973 [3]. Afterwards, various utilities, such as Northeast Utilities (NU) [4], Southern California Edison (SCE) [5], and Dominion Virginia Power [6] have performed CVR tests. These tests resulted in a reduction in the load by 0.3% to 1% per 1% voltage reduction. Ref. [7] conducted a study that estimated a 3.04% reduction in annual energy consumption in the case of deployment of CVR on all distribution feeders of the United States. In addition to the USA, other countries, such as Australia [8] and Ireland [9], have also conducted CVR, resulting in 1% and 1.7% energy savings, respectively. CVR is regarded as an efficient method to manage the load demand in distribution power systems [10].

CVR is one of the most viable methods towards ensuring an economic and environmentally sustainable power network, as it provides numerous benefits to the electric utilities and end-use consumers. CVR can reduce carbon emissions and can extend the life of a transformer by reducing its iron losses by decreasing the voltage [11]. Moreover, increase in energy savings will also lead to less maintenance cost, due to the improvement in the life of substation devices and will also increase the lifespan of consumer appliances, such as incandescent lamps and electric water heaters [12]. The aforesaid benefits of CVR can also ensure improvement in the utilization of various assets linked to the power network, reduce energy demand, and increase the energy efficiency [13].

CVR can be implemented using the following two techniques. The first technique is to reduce the voltage by adding capacitors, line-drop-compensators, and load-tap-changers. The second technique involves decreasing the voltage using a supervisory control and data acquisition (SCADA) system and the advance metering infrastructure (AMI) [6,14]. While applying either of the techniques, it is desired to keep the voltage profile as flat as possible. The aforesaid can be achieved by adding capacitors at different sites to increase the power factor and reduce the power loss. The assessment of CVR program performance is crucial, which can be performed using four assessment methods that are simulation-based, comparison-based, synthesis based, and regression-based methods [15,16]. Various studies have been carried out on the assessment of CVR performance. Dominion Virginia Power developed its own method for CVR calculation by aggregating available AMI data with the existing data management system (DMS) controls over a time duration into a single summary record per meter, with analytical support from the geographic information system (GIS) and planning data [6]. Ref. [17] used the adaptive voltage control technique to control the substation voltage by using automatic control and communications. Both constant or time varying impedance, current and power model, i.e., ZIP model, have been used in CVR assessment studies, as well as in demand and loss reduction optimization [18]. Ref. [19] used a regression and simulation method to calculate the CVR factor for a time-varying ZIP load model. A two-stage recursive least squares algorithm under the regression method was used to calculate the CVR factor and a simulation method was used to determine the accuracy of the CVR calculation method. Ref. [20] used the Kalman filtering method to calculate time-varying ZIP parameters. Ref. [21] devised a coordinated voltage control scheme for voltage control for CVR by varying the voltage (near the operating lower voltage limit) at the on-load tap changer (OLTC) at the substation, in coordination with step voltage regulator (SVR) or shunt capacitor banks at feeders. Ref. [22] presented a multi-objective Volt-Var optimization problem. Different target functions were used to express the system state defined by voltage, active power, reactive power, direct and indirect Volt-Var control methods. Ref. [23] performed a case study on the usage of the CVR/VVO method by various electric utilities across USA with specific focus on the utilities, discussing their methodology for CVR factor calculation. They presented a measurement and verification benchmark of CVR/VVO used by several electric utilities in the U.S. Ref. [24] presented an ANN-based model to accurately calculate the measurements for active power, reactive power, and voltage to obtain optimal output values including power loss. Ref. [25] proposed an algorithm based on the Markov chain and Monte Carlo (MCMC) method to accurately identify the parameters including head and end voltages and active and reactive power on the low voltage side of a power distribution network. Ref. [26] presented a selective particle swarm optimization (SPSO) algorithm to solve an optimization model for optimal placement and sizing of distributed generators to reduce the distribution power losses, keeping technical constraints intact.

However, these studies do not provide a detailed CVR analysis considering both the utility and the consumer perspective, while studying the relationship between active power loss, and input voltage, input active and reactive power affecting the CVR factor. Moreover, they do not depict the relationship between active power loss, active power, reactive power, and voltage. This is the research gap that needs to be addressed in this paper. This paper focuses on the calculation of the CVR factor for the IEEE 13-bus and 34-bus systems, using a simulation-based method from the substation and the load perspective. It also attempts to identify the relationship between active power loss and the input quantities affecting the CVR factor calculation and studies the effect of voltage regulator on the CVR factor.

The paper is structured as follows: Section 2 describes the substation and the load perspective used to calculate the CVR factor. Section 3 focuses on the function methodology used to calculate the active power loss (P_{loss}). It also explains the least squares method used to calculate the constants of the function. Section 4 discusses the artificial neural network (ANN) method used to calculate the (P_{loss}). Section 5 explains the findings achieved for the work carried out in Sections 2–4. Section 6 concludes the methodologies and the results obtained from them. It also explains the future scope of this research.

2. CVR Factor Estimation Based on Simulation Studies

This section describes the CVR comparison study conducted for the IEEE-13 and IEEE-34 bus systems, using the feeder correlated comparison-based method. These two systems are shown in Figures 1 and 2. Two feeders, i.e., feeder 1 and feeder 2, have been selected for CVR-off case and CVR-on case, respectively. The CVR-off case is the scenario when the CVR is not being implemented, i.e., we have selected a base voltage regulator setting for the system, whereas in the CVR-on case, the CVR is being implemented by reducing the base voltage regulator setting to different voltage levels. The voltage and power level between both the feeders have been compared to calculate the CVR factor. The CVR factor for active and reactive power has been calculated using the following formulas:

$$CVR\ factor\ for\ P = \frac{P_{off} - P_{on}}{P_{off}} \bigg/ \frac{V_{off} - V_{on}}{V_{off}} \tag{1}$$

$$CVR\ factor\ for\ Q = \frac{Q_{off} - Q_{on}}{Q_{off}} \bigg/ \frac{V_{off} - V_{on}}{V_{off}} \tag{2}$$

We have used the average voltage magnitude (V), total active power P and reactive power Q of all the phases in the calculation. The aforesaid quantities have been calculated as shown in the following equations:

$$Average\ voltage\ magnitude = \frac{V_A + V_B + V_C}{3}$$
$$where\ V_A,\ V_B\ and\ V_C\ are\ the\ voltages\ of\ phase\ A,\ B\ and\ C,\ respectively. \tag{3}$$

$$Total\ Active\ Power = P_A + P_B + P_C$$
$$where\ P_A,\ P_B\ and\ P_C\ are\ the\ active\ power\ of\ phase\ A,\ B\ and\ C,\ respectively. \tag{4}$$

$$Total\ Reactive\ Power = Q_A + Q_B + Q_C$$
$$where\ Q_A,\ Q_B\ and\ Q_C\ are\ the\ reactive\ power\ of\ phase\ A,\ B\ and\ C,\ respectively. \tag{5}$$

The P, Q and V quantities for the 13-bus and 34-bus systems have been obtained from their model simulated in OpenDSS. For both the systems, the models have been simulated for different voltage levels. For the 13-bus system, simulation has been carried out for different voltage regulator settings levels. The voltage regulator setting of 124 V has been taken as the voltage in the CVR-off scenario, whereas for the CVR-on scenario, the voltage has been reduced to three different levels, i.e., 118 V, 120 V and 122 V. The CVR factor has been calculated for the three voltage levels, where V_{off}, P_{off} and Q_{off} are the input quantities obtained at the voltage regulator setting of 124 V, i.e., the CVR-off case. V_{on}, P_{on} and Q_{on} are the input quantities obtained for the voltage regulator settings considered in the CVR-on case.

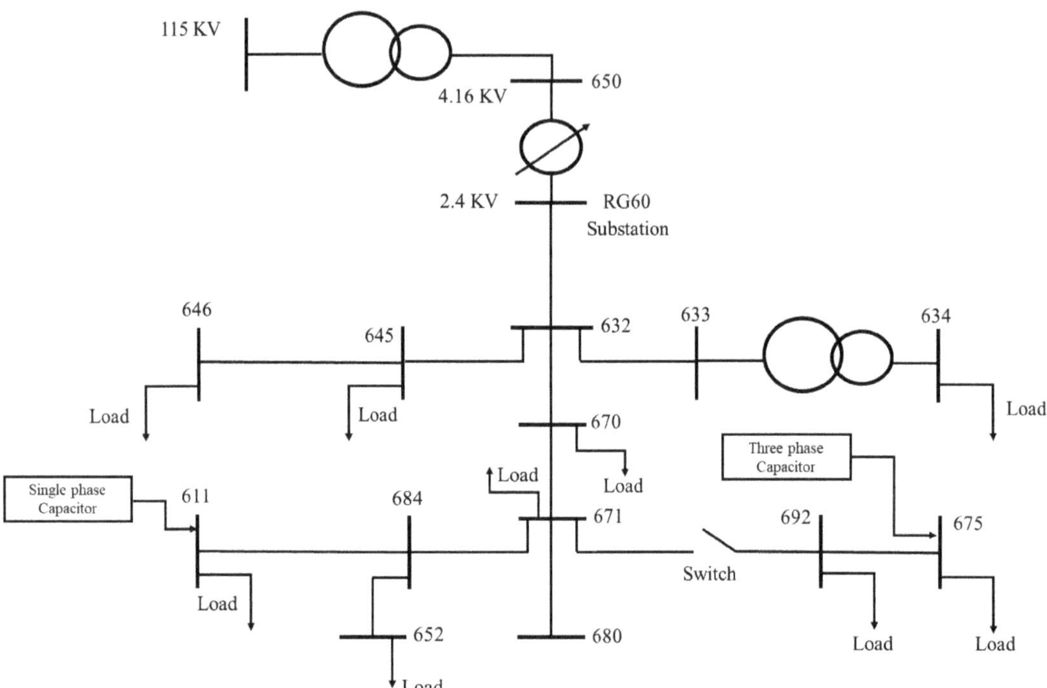

Figure 1. Single line diagram of IEEE 13-bus system.

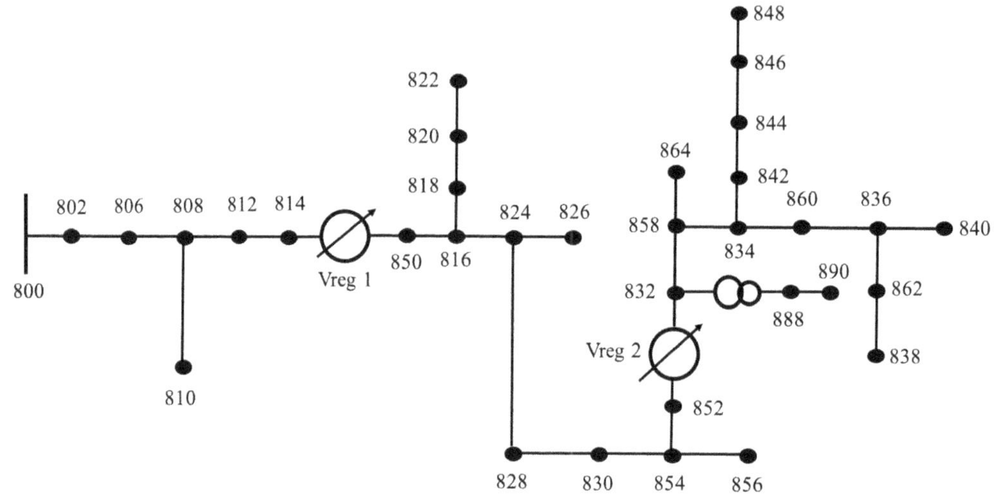

Figure 2. Single line diagram of IEEE 34-bus system.

In the 34-bus system, three cases have been considered, i.e., Case 1, Case 2 and Case 3. Case 1 is the scenario where the substation voltage is being varied when the voltage regulators are present in the system. In Case 2, the voltage regulators are disabled by replacing them with a short transmission line and then the substation voltage is varied.

Case 3 involves proportional changes in the substation voltage and voltage regulator setting. In Case 1 and case 2, the substation voltage of 1.05 per unit (p.u.) has been considered as the voltage in the CVR-off case. For the CVR-on case, the substation voltages considered are 0.95, 1.00, 1.02 and 1.03 p.u. There are two voltage regulators in the 34-bus system whose settings are 122 V and 124 V, which remain the same in Case 1 and Case 2 for the CVR-off and CVR-on case. However, in Case 3, if the voltage from 1.05 p.u. is decreased to 1.02 p.u., then the voltage setting of both the regulators will change to (122 * (1.02/1.05)) V and (124 * (1.02/1.05)) V, respectively. This ratio will apply to other voltage settings as well. The only parameter that will vary in this ratio will be the substation voltages in the CVR-on case. These different scenarios intend to illustrate the impact of voltage regulator settings on CVR results.

For both the systems, eight ZIP load models have been considered in the study for all the loads of the 13-bus and 34-bus systems. The models considered are (0.2 0.2 0.6), (0.2 0.6 0.2), (0.6 0.2 0.2), (0.1 0.2 0.7), (0.7 0.1 0.2), (0 0 1), (0 1 0) and (1 0 0).

Two approaches were used to calculate the CVR factor for the 13-bus and 34-bus systems. The first approach involved utilizing the quantities of interest, i.e., P, Q and V from the input substation bus of the system. The second approach derived P and Q from the sum of all loads connected to the system, and the voltage was selected from the input bus of the system. The first approach will consider system losses and any reactive power compensating devices, as well as load characteristics, while the second approach specifically accounts for the characteristics of the loads themselves.

2.1. CVR Estimation Using Power Captured at Load

The IEEE-13 and IEEE-34 bus systems have 9 and 25 loads, respectively. For the second approach, the summation of P and Q of all the loads is taken as the input quantity and V is the input bus voltage at node RG60 for the 13-bus system. A similar approach for P and Q is followed for the 34-bus system and V is the input bus voltage at bus 800.

2.2. CVR Estimation Using Power Captured at Substation

In the IEEE 13-bus system, when using approach 1, P and Q are the power flow between bus RG60 and 632, and V is the voltage at bus RG60, as shown in Figure 1. For the IEEE-34 bus system, P and Q are the power flow between bus 800 and 802, and V is the voltage at bus 800, as shown in Figure 2.

3. Proposed New Method for Active Power Loss Estimation Based on Least Squares Method

When the voltage is lowered due to the CVR scheme, the system loss will usually be reduced together with the load. Calculating the respective value for the system loss reduction and load reduction is desirable. This section presents a new method to estimate the system loss based on substation voltage and substation active and reactive power input. This method uses two functions, i.e., Function 1 and Function 2, to calculate $P_{loss}^{estimated}$. The constants of the functions will be estimated using the least squares method. This approach was applied to both the 13-bus and 34-bus system.

3.1. Function 1

Function 1 for calculation of the power loss is given as

$$P_{loss} = \frac{K\left(P_1^2 + Q_1^2\right)}{|V_1|^2} + C \tag{6}$$

where K and C are the constants.

The P_{loss} is calculated for each hour over the span of 48 h. P_1, Q_1 and V_1 are the active power, reactive power, and input voltage, respectively, at the substation. All the three quantities are a column vector of order 48 by 1. The aforesaid constants are calculated using the following formula:

$$\begin{bmatrix} K \\ C \end{bmatrix} = [H^t H]^{-1} [H^t Z] \tag{7}$$

where

$H = \begin{bmatrix} a & j \end{bmatrix}$

$a = \dfrac{(P_1^2 + Q_1^2)}{|V_1|^2}$ of order 48×1

$j = [1]_{48 \times 1}$

Z: is the 48×1 P_{loss}.

3.2. Function 2

Function 2 for calculation of the power loss is given as

$$P_{loss} = \frac{K_1 \left(P_1^2 + Q_1^2\right)}{|V_1|^2} + \frac{K_2 \sqrt{\left(P_1^2 + Q_1^2\right)}}{|V_1|} + C \tag{8}$$

Function 2 is a quadratic equation, where P_1, Q_1 and V_1 are a column vector of order 48×1. The constants K_1, K_2 and C are calculated using the following formula:

$$\begin{bmatrix} K_1 \\ K_2 \\ C \end{bmatrix} = [H^t H]^{-1} [H^t Z] \tag{9}$$

where

$H = \begin{bmatrix} a & \sqrt{a} & j \end{bmatrix}$

$a = \dfrac{(P_1^2 + Q_1^2)}{|V_1|^2}$ of order 48×1

$j = [1]_{48 \times 1}$.

After the constants of the functions are obtained, these two functions can be used to estimate the loss based on real time measurements of the voltage, real and reactive power at the substation. The power loss obtained from the functions is termed as "estimated power loss ($P_{loss}^{estimated}$)". The power loss obtained from the model simulated in OpenDSS is called "simulated power loss ($P_{loss}^{simulated}$)". The error in the power loss calculation for each hour has been calculated using the difference between $P_{loss}^{simulated}$ and $P_{loss}^{estimated}$. Using these error values, RMS error has been obtained for each ZIP load model and voltage setting. In addition to individually calculating the errors for each ZIP model and voltage setting, the errors have also been calculated by combining all the cases together. The total number of samples is 1152 (48×24) and 1920 (48×40) for the 13-bus and 34-bus systems, respectively. In the 13-bus system, 24 is the total number of voltage settings, i.e., 3 multiplied by the total number of ZIP models, i.e., 8. In the 34-bus system, 40 is the total number of substation voltage levels, i.e., 5 multiplied by the total number of ZIP models, i.e., 8. Forty-eight is the total number of hours in each case.

4. Proposed New Method for Active Power Loss Estimation Based on ANN Method

This section describes the proposed new ANN-based active power loss estimation method. This method uses feedforward neural network models with one, two or three hidden layers. As an example, Figure 3 illustrates a two hidden layer ANN. These models take P_1, Q_1 and V_1 as inputs and P_{loss} as the target data to train the network. The input quantities are taken from the substation measurement values. The $P_{loss}^{simulated}$ is the loss obtained in simulation and is calculated as the difference between the active power obtained at the substation and load. The feedforward model follows supervised learning training using the input quantities and the simulated loss quantities. Once the network is trained, supplying inputs will lead to estimated loss $P_{loss}^{estimated}$. The difference between $P_{loss}^{estimated}$ and $P_{loss}^{simulated}$ has been used to calculate the RMS value for the 13-bus and 34-bus systems.

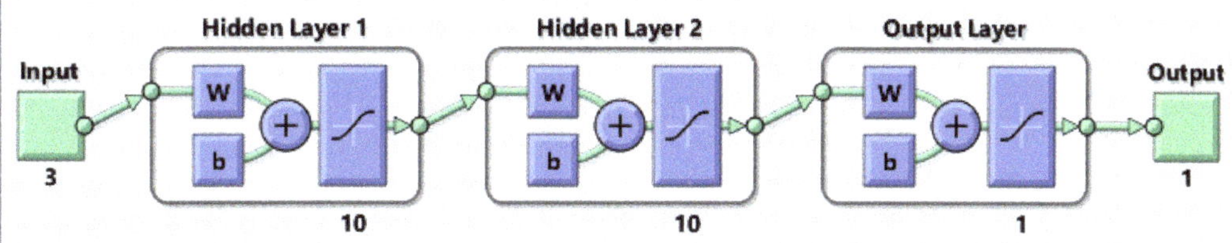

Figure 3. Artificial neural network model consisting of two hidden layers.

The hidden layers shown in Figure 3 contain 10 neurons. The output layer contains only one neuron. The number of neurons in the hidden layer(s) of the above models have been varied from 5 to 16 and P_{loss} has been obtained for each case. The aforesaid variation has been carried out to study the effect of number of neurons and hidden layers on loss estimation accuracy. The variables P_1, Q_1, V_1 and P_{loss} are a matrix combining the respective data for all the cases for all the ZIP load models.

5. Case Studies

Section 5 presents the results obtained for the CVR comparison study using the approaches discussed in Section 2. It also discusses the loss relationship study performed using the function and ANN method discussed in Section 4.

5.1. CVR Studies

The CVR factors for both 13-bus and 34 bus systems are highest when the impedance component is high in the ZIP load, being the highest in the case of the constant impedance load. This is consistent with the theoretical CVR factor being 2.0, 1.0 and 0 for the constant impedance load, constant current load and constant power load, respectively.

For the 34-bus system, a voltage regulator plays a crucial role in CVR factor calculation. In case 1, i.e., when the voltage regulators are enabled, the CVR factor for P is quite low and even negative in some cases. This is due to the restriction in the change in voltage, as both the voltage regulators are only permitted to vary the voltage within the specified bandwidth. In case 2, when we disable the voltage regulators by replacing them with a short transmission line, it can be observed that the CVR factor increases. In case 3, where we vary the voltage regulator setting and substation voltage level in a proportionate manner, it can be observed that the CVR factor increases to a level consistent with the load characteristics.

Analysis of the CVR factor for reactive power is more complicated. When CVR for Q is calculated using the sum of load reactive power consumption, the value is consistent with the load characteristics. However, when the CVR for Q is calculated using the measurements at the substation, i.e., RG60 in the 13-bus system and bus 800 in the 34-bus system, the measurements reflect not only the load characteristics but also the capacitor compensation.

The following sections show the average CVR factor calculated for both the systems for all the eight ZIP load models over the duration of 48 h.

5.1.1. IEEE-13 Bus System

Tables 1 and 2 show the average CVR factor for the IEEE-13 bus system obtained by using the power captured at the loads and substation, respectively.

Table 1. Average CVR factor obtained using power captured at loads.

ZIP Load	Average CVR Factor for P			Average CVR Factor for Q		
	118 V	120 V	122 V	118 V	120 V	122 V
(0.2 0.6 0.2)	1.06	1.06	1.06	1.06	1.06	1.06
(0.2 0.2 0.6)	0.65	0.66	0.66	0.65	0.66	0.66
(0.6 0.2 0.2)	1.39	1.37	1.29	1.39	1.37	1.30
(0.1 0.2 0.7)	0.44	0.45	0.45	0.44	0.45	0.45
(0.7 0.1 0.2)	1.52	1.53	1.54	1.52	1.53	1.54
(0 0 1)	0.00	0.00	0.00	0.00	0.00	0.00
(0 1 0)	1.06	1.07	1.06	1.06	1.06	1.06
(1 0 0)	1.96	1.97	1.98	1.96	1.97	1.98

Table 2. Average CVR factor obtained using power captured at substation.

ZIP Load	Average CVR Factor for P			Average CVR Factor for Q		
	118 V	120 V	122 V	118 V	120 V	122 V
(0.2 0.6 0.2)	1.03	1.04	1.03	0.21	0.20	0.20
(0.2 0.2 0.6)	0.61	0.62	0.62	−0.60	−0.60	−0.60
(0.6 0.2 0.2)	1.42	1.43	1.43	0.94	0.96	0.95
(0.1 0.2 0.7)	0.39	0.40	0.40	−1.02	−1.03	−1.02
(0.7 0.1 0.2)	1.51	1.52	1.53	1.12	1.13	1.13
(0 0 1)	−0.07	−0.07	−0.06	−1.93	−1.93	−1.92
(0 1 0)	1.04	1.04	1.03	0.22	0.21	0.19
(1 0 0)	1.96	1.97	1.98	1.96	1.97	1.98

Table 1 clearly reveals that the CVR factors calculated using the power measurements at the load location reflect the load characteristics. In contrast, it can be observed from Table 2 that some CVR values for Q are negative due to capacitor bank compensation. When the substation voltage reduces, the load reactive power reduces, but the reactive power injection from the capacitor bank also reduces and in addition, the reactive power loss on the lines increases, so the reactive power input at the substation increases, leading to a negative CVR value for Q. A numerical example is provided below.

If one takes one hour power flow as an example, in the base case with 124 V, the reactive power from RG60 to 632 is 677.87 kVar. The total load reactive power consumption is 1298.74 kVar. The capacitor bank reactive power injection is 730.02 kVar. The line reactive power loss is 109.15 kVar. In the case with 118 V, the reactive power from RG60 to 632 is 715.41 kVar. The total load reactive power consumption is 1260.27 kVar. The capacitor bank reactive power injection is 659.85 kVar. The line reactive power loss is 114.99 kVar. The CVR for Q calculated using load measurements would be 0.61, while the value calculated using substation measurements would be −1.14.

5.1.2. IEEE-34 Bus System

Tables 3 and 4 show the average CVR factor obtained for case 1 using both the CVR calculation approaches discussed in Section 2. Tables 5 and 6 show the average CVR factor obtained for case 2 using the aforesaid methods. Tables 7 and 8 show the results for case 3.

Table 3. Average CVR factor obtained using power captured at load for case 1.

ZIP Load	Average CVR Factor for P				Average CVR Factor for Q			
	0.95	**1**	**1.02**	**1.03**	**0.95**	**1**	**1.02**	**1.03**
(0.2 0.6 0.2)	0.88	0.83	0.83	0.83	0.93	0.87	0.87	0.87
(0.2 0.2 0.6)	0.62	0.51	0.50	0.50	0.66	0.53	0.53	0.53
(0.6 0.2 0.2)	1.12	1.11	1.11	1.11	1.18	1.16	1.17	1.17
(0.1 0.2 0.7)	0.51	0.35	0.34	0.34	0.54	0.37	0.36	0.36
(0.7 0.1 0.2)	1.19	1.18	1.18	1.18	1.24	1.24	1.24	1.24
(0 0 1)	0.30	0.04	0.01	0.00	0.33	0.05	0.01	0.00
(0 1 0)	0.91	0.85	0.84	0.84	0.95	0.89	0.89	0.89
(1 0 0)	1.53	1.53	1.53	1.53	1.59	1.60	1.61	1.61

Table 4. Average CVR factor using power captured at substation for case 1.

ZIP Load	Average CVR Factor for P				Average CVR Factor for Q			
	0.95	**1**	**1.02**	**1.03**	**0.95**	**1**	**1.02**	**1.03**
(0.2 0.6 0.2)	0.14	−0.04	−0.09	−0.09	−3.56	−2.87	−3.73	−4.76
(0.2 0.2 0.6)	0.02	−0.07	−0.10	−0.10	6.68	4.53	4.83	5.74
(0.6 0.2 0.2)	0.26	0.01	−0.08	−0.08	−3.39	−2.59	−3.09	−3.28
(0.1 0.2 0.7)	−0.05	−0.09	−0.11	−0.10	3.47	2.36	2.52	2.80
(0.7 0.1 0.2)	0.28	0.02	−0.09	−0.10	−0.89	−0.52	−1.06	−1.20
(0 0 1)	−0.20	−0.15	−0.13	−0.12	−0.75	−0.13	0.32	0.26
(0 1 0)	0.15	−0.04	−0.09	−0.09	−6.91	−5.27	−6.29	−8.27
(1 0 0)	0.39	0.05	−0.09	−0.09	2.22	2.49	2.53	2.74

Table 5. Average CVR factor obtained using power captured at load for case 2.

ZIP Load	Average CVR Factor for P				Average CVR Factor for Q			
	0.95	**1**	**1.02**	**1.03**	**0.95**	**1**	**1.02**	**1.03**
(0.2 0.6 0.2)	0.25	0.09	0.05	0.04	0.24	0.09	0.04	0.03
(0.2 0.2 0.6)	0.16	0.06	0.03	0.03	0.15	0.05	0.03	0.02
(0.6 0.2 0.2)	0.33	0.13	0.06	0.05	0.32	0.12	0.05	0.04
(0.1 0.2 0.7)	0.11	0.04	0.02	0.02	0.11	0.04	0.02	0.02
(0.7 0.1 0.2)	0.35	0.14	0.06	0.04	0.34	0.13	0.05	0.03
(0 0 1)	0.00	0.00	0.00	0.00	0.00	0.00	0.00	0.00
(0 1 0)	0.25	0.09	0.05	0.04	0.24	0.08	0.04	0.03
(1 0 0)	0.44	0.18	0.07	0.06	0.43	0.16	0.05	0.04

Table 6. Average CVR factor using power captured at substation for case 2.

ZIP Load	Average CVR Factor for P				Average CVR Factor for Q			
	0.95	1	1.02	1.03	0.95	1	1.02	1.03
(0.2 0.6 0.2)	1.29	1.21	1.18	1.16	21.52	25.21	28.48	29.66
(0.2 0.2 0.6)	1.05	0.90	0.85	0.82	−6.63	−8.16	−9.01	−9.97
(0.6 0.2 0.2)	1.51	1.49	1.48	1.48	3.94	4.45	4.65	4.63
(0.1 0.2 0.7)	0.96	0.76	0.70	0.67	1.20	0.90	1.43	1.34
(0.7 0.1 0.2)	1.57	1.56	1.56	1.56	−0.60	−0.90	−1.30	−1.55
(0 0 1)	0.79	0.48	0.40	0.37	7.86	8.41	9.94	11.24
(0 1 0)	1.31	1.23	1.20	1.18	3.57	4.00	4.38	4.36
(1 0 0)	1.91	1.95	1.97	1.98	1.91	1.95	1.97	1.98

Table 7. Average CVR factor obtained using power captured at load for case 3.

ZIP Load	Average CVR Factor for P				Average CVR Factor for Q			
	0.95	1	1.02	1.03	0.95	1	1.02	1.03
(0.2 0.6 0.2)	0.78	0.77	0.78	0.76	0.82	0.81	0.82	0.81
(0.2 0.2 0.6)	0.46	0.46	0.47	0.47	0.48	0.49	0.49	0.50
(0.6 0.2 0.2)	1.07	1.08	1.08	1.08	1.12	1.14	1.14	1.14
(0.1 0.2 0.7)	0.31	0.31	0.31	0.31	0.33	0.32	0.33	0.33
(0.7 0.1 0.2)	1.15	1.16	1.16	1.15	1.20	1.22	1.22	1.21
(0 0 1)	0.00	0.00	0.00	0.00	0.00	0.00	0.00	0.00
(0 1 0)	0.78	0.77	0.77	0.76	0.82	0.81	0.81	0.80
(1 0 0)	1.52	1.53	1.53	1.53	1.59	1.60	1.60	1.61

Table 8. Average CVR factor using power captured at substation for case 3.

ZIP Load	Average CVR Factor for P				Average CVR Factor for Q			
	0.95	1	1.02	1.03	0.95	1	1.02	1.03
(0.2 0.6 0.2)	0.94	0.91	0.93	0.92	−5.40	−6.54	−7.36	−10.50
(0.2 0.2 0.6)	0.49	0.44	0.46	0.47	24.06	24.77	26.15	24.72
(0.6 0.2 0.2)	1.31	1.35	1.36	1.37	−5.70	−5.67	−5.99	−5.53
(0.1 0.2 0.7)	0.28	0.21	0.22	0.23	16.23	17.28	17.91	18.16
(0.7 0.1 0.2)	1.42	1.46	1.47	1.46	−0.29	−0.37	−0.31	−0.21
(0 0 1)	−0.14	−0.28	−0.27	−0.26	6.93	6.25	7.32	8.06
(0 1 0)	0.94	0.92	0.93	0.92	−14.09	−16.07	−17.13	−21.26
(1 0 0)	1.91	1.96	1.98	1.99	1.93	2.01	2.06	2.11

5.2. Active Power Loss Estimation Using Least Squares Method

There is no direct relationship between active power loss and CVR factor. The active power loss is obtained as a function of substation voltage and, active and reactive power inputs. This section presents the active power loss estimation using the least squares method for the IEEE-13 and 34 bus system. For both the systems, the value of $P_{loss}^{estimated}$ obtained using curve fitting approach is very similar to $P_{loss}^{simulated}$, which can be observed in the following figures.

5.2.1. IEEE-13 Bus System

For the IEEE-13 bus system, the coefficients for function 1 were obtained as K = 45.1501 and $C = -0.076$. The coefficients for function 2 were $K_1 = 44.1627$, $K_2 = 1.2666$ and $C = -0.0018$.

Figures 4 and 5 depict the active power loss estimation obtained from function 1 and 2, respectively.

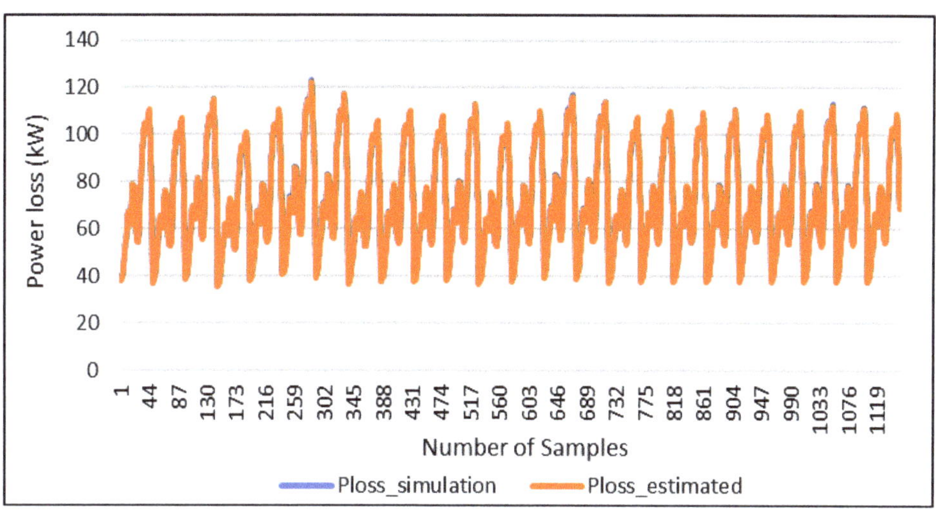

Figure 4. Active power loss obtained from function 1.

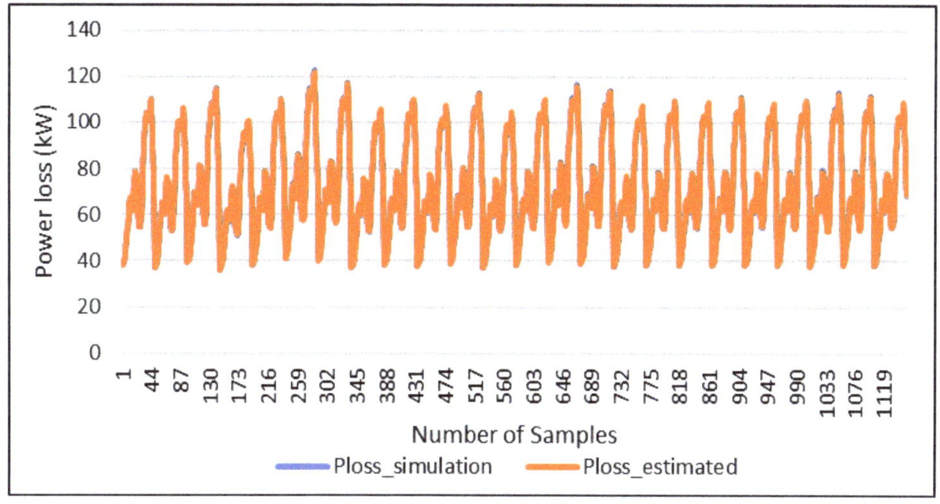

Figure 5. Active power loss obtained from function 2.

In the 13-bus system, it can be observed from the graphs that the estimated power loss obtained from both functions is very close to the simulated power loss obtained from OpenDSS. The RMS error obtained using function 2 (0.47 kW) was lower than function 1 (0.53 kW). This was true for combined cases, where all ZIP cases are combined, as well as for individual cases, where each ZIP case is dealt with individually. Table 9 presents the RMS error in kW for active power loss estimation for function 1 and 2 for each ZIP case.

Table 9. RMS error for active power loss estimation for function 1 and 2 for IEEE-13 bus system.

ZIP Parameter	Voltage Setting	Function 1	Function 2
(0.2 0.6 0.2)	118 V	0.31	0.08
	120 V	0.30	0.08
	122 V	0.33	0.09
(0.6 0.2 0.2)	118 V	0.34	0.06
	120 V	0.35	0.06
	122 V	0.36	0.07
(0.2 0.2 0.6)	118 V	0.28	0.10
	120 V	0.26	0.10
	122 V	0.31	0.11
(0.1 0.2 0.7)	118 V	0.82	0.41
	120 V	0.83	0.47
	122 V	0.88	0.41
(0.7 0.1 0.2)	118 V	0.89	0.44
	120 V	0.91	0.42
	122 V	0.88	0.41
(1 0 0)	118 V	0.38	0.04
	120 V	0.40	0.04
	122 V	0.41	0.06
(0 1 0)	118 V	0.31	0.08
	120 V	0.30	0.08
	122 V	0.33	0.09
(0 0 1)	118 V	0.21	0.13
	120 V	0.19	0.13
	122 V	0.24	0.09

5.2.2. IEEE-34 Bus System

Figures 6–11 depict the power loss obtained after calculating the constants for function 1 and 2, respectively, for all three cases by combining all the voltage levels and ZIP load models in a single matrix. The difference between $P_{loss}^{simulated}$ and $P_{loss}^{estimated}$ provides the RMS error for function 1 and function 2. The RMS error obtained for the aforesaid cases indicate function 2 as the most efficient relationship between power loss and substation voltage, active and reactive power, because of its low RMS error in comparison to function 1.

Function 1

For the IEEE-34 bus system, the coefficients for function 1 were obtained as $K = 15{,}043$ and $C = 32$ for case 1, $K = 16{,}874$ and $C = 21$ for case 2 and, $K = 15{,}593$ and $C = 31$ for case 3. The RMS error is 5.80 kW for case 1, 7.78 kW for case 2 and 5.31 kW for case 3.

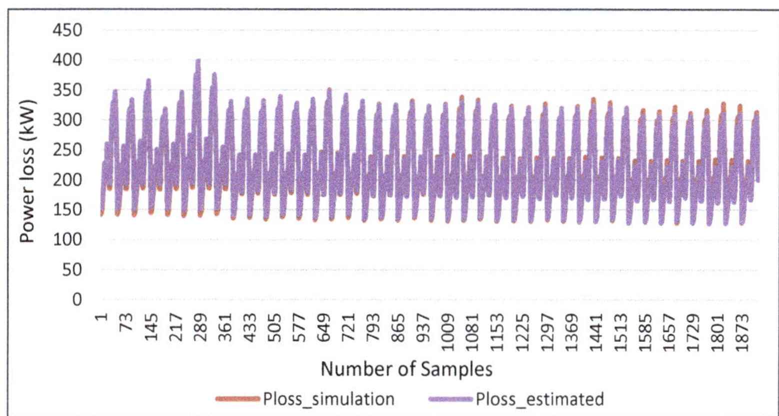

Figure 6. Active power loss obtained for case 1 for function 1.

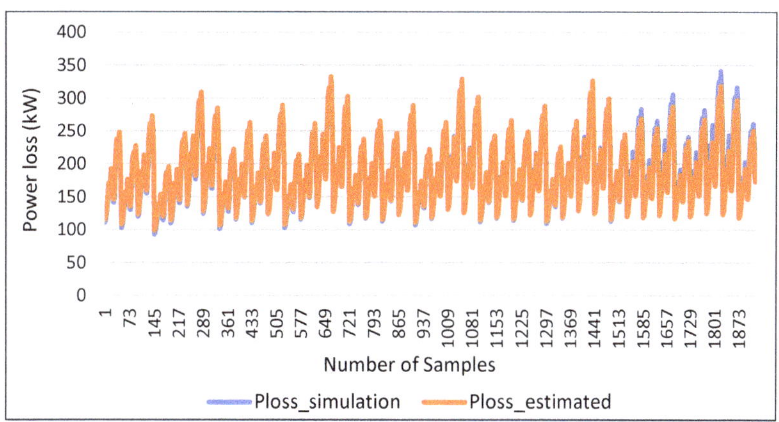

Figure 7. Active power loss obtained for case 2 for function 1.

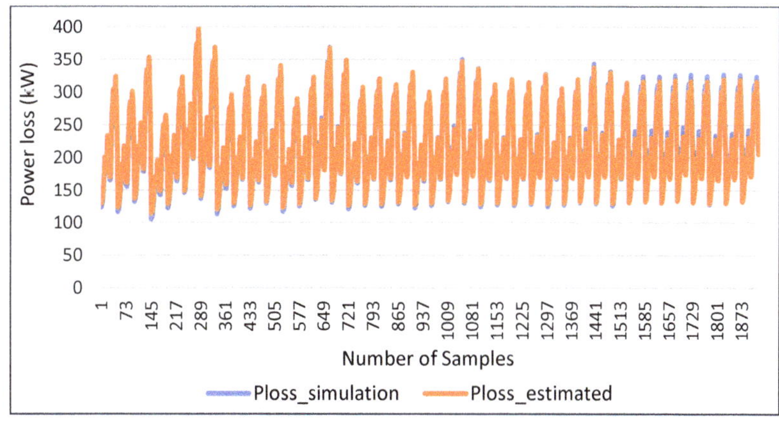

Figure 8. Active power loss obtained for case 3 for function 1.

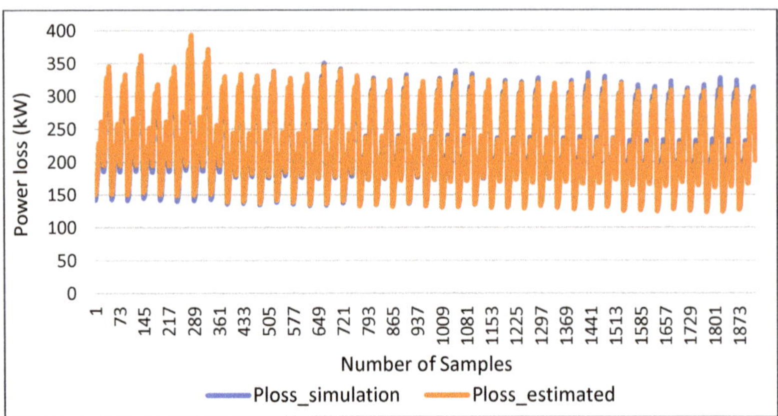

Figure 9. Active power loss obtained for case 1 for function 2.

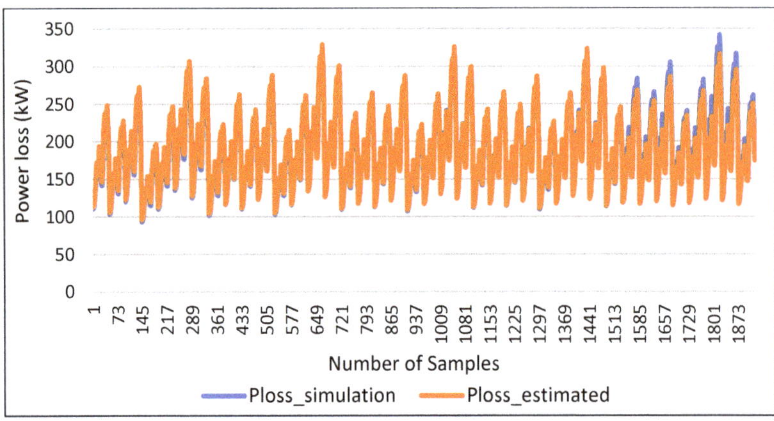

Figure 10. Active power loss obtained for case 2 for function 2.

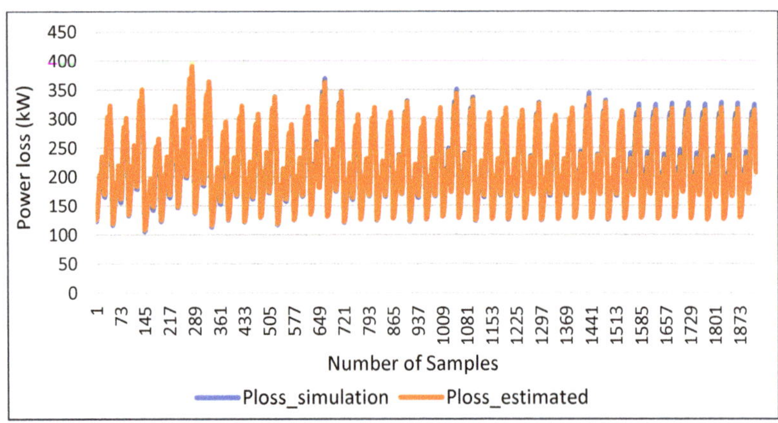

Figure 11. Active power loss obtained for case 3 for function 2.

Table 10 presents the RMS error for active power loss estimation for function 1 for each ZIP case.

Table 10. RMS error for active power loss estimation for function 1 for IEEE-34 bus system.

ZIP Parameter	Voltage Setting	Case 1	Case 2	Case 3
(0.2 0.6 0.2)	0.95	0.79	0.78	1.35
	1.00	1.01	0.93	1.28
	1.02	1.04	1.11	1.24
	1.03	1.18	1.15	1.28
(0.6 0.2 0.2)	0.95	0.79	0.69	1.25
	1.00	1.00	0.86	1.21
	1.02	1.03	0.90	1.27
	1.03	1.18	0.90	1.33
(0.2 0.2 0.6)	0.95	0.79	0.72	1.49
	1.00	0.96	1.03	1.33
	1.02	1.05	1.38	1.29
	1.03	1.16	1.48	1.32
(0.1 0.2 0.7)	0.95	0.82	0.70	1.48
	1.00	0.95	1.06	1.36
	1.02	1.08	1.45	1.31
	1.03	1.14	1.62	1.30
(0.7 0.1 0.2)	0.95	0.79	0.67	1.20
	1.00	1.00	0.83	1.17
	1.02	1.02	0.84	1.29
	1.03	1.18	0.84	1.33
(1 0 0)	0.95	0.77	0.55	1.12
	1.00	0.97	0.57	1.21
	1.02	1.03	0.60	1.23
	1.03	1.15	0.61	1.25
(0 1 0)	0.95	0.82	0.76	1.39
	1.00	1.01	0.92	1.28
	1.02	1.04	1.09	1.24
	1.03	1.17	1.13	1.28
(0 0 1)	0.95	1.00	0.59	1.57
	1.00	0.97	1.24	1.45
	1.02	1.06	1.55	1.40
	1.03	1.13	1.85	1.28

Function 2

For the IEEE-34 bus system, the coefficients for function 2 were obtained as $K_1 = 11,218$, $K_2 = 872$ and $C = -17$ for case 1, $K_1 = 12,838$, $K_2 = 818$ and $C = -19$ for case 2 and, $K_1 = 10,680$, $K_2 = 1087$ and $C = -28$ for case 3. The RMS error is 5.69 kW for case 1, 7.73 kW for case 2 and 5.09 kW for case 3.

Table 11 presents the RMS error for active power loss estimation for function 2 for each ZIP case.

It is noted that the RMS error for the combined cases is generally larger than that for each ZIP case. In practical applications, the ZIP parameters can be estimated using feeder head measurements including voltage and power. Then, the function coefficients can be estimated using that specific ZIP case for more accurate active power loss estimation.

Table 11. RMS error for active power loss estimation for function 2 for IEEE-34 bus system.

ZIP Parameter	Voltage Setting	Case 1	Case 2	Case 3
(0.2 0.6 0.2)	0.95	0.12	0.12	0.51
	1.00	0.37	0.11	0.55
	1.02	0.45	0.08	0.50
	1.03	0.48	0.06	0.52
(0.6 0.2 0.2)	0.95	0.09	0.07	0.54
	1.00	0.40	0.10	0.51
	1.02	0.44	0.04	0.51
	1.03	0.51	0.03	0.56
(0.2 0.2 0.6)	0.95	0.15	0.16	0.47
	1.00	0.37	0.11	0.55
	1.02	0.45	0.16	0.53
	1.03	0.47	0.12	0.55
(0.1 0.2 0.7)	0.95	0.18	0.16	0.44
	1.00	0.37	0.12	0.59
	1.02	0.46	0.20	0.52
	1.03	0.45	0.14	0.52
(0.7 0.1 0.2)	0.95	0.09	0.07	0.49
	1.00	0.40	0.07	0.49
	1.02	0.45	0.03	0.53
	1.03	0.51	0.03	0.57
(1 0 0)	0.95	0.07	0.12	0.46
	1.00	0.36	0.03	0.49
	1.02	0.44	0.04	0.51
	1.03	0.52	0.04	0.52
(0 1 0)	0.95	0.12	0.13	0.51
	1.00	0.36	0.10	0.51
	1.02	0.45	0.07	0.50
	1.03	0.48	0.06	0.51
(0 0 1)	0.95	0.28	0.11	0.42
	1.00	0.41	0.13	0.61
	1.02	0.47	0.22	0.53
	1.03	0.45	0.25	0.48

5.3. Active Power Loss Estimation Using ANN Method

Table 12 show the RMS error for loss estimation versus the number of neurons in the hidden layer for a given number of hidden layers for the IEEE-13 bus system. Tables 13–15 list the RMS error for loss estimation for the IEEE-34 bus system for case 1, 2 and 3, respectively.

The results indicate that the number of neurons and number of hidden layers do not have a substantial effect on the estimation accuracy. Thus, for simplicity, one hidden layer network with five neurons in the hidden layer will suffice.

Table 12. RMS error for active power loss estimation for IEEE-13 bus system.

RMS Error		Number of Neurons in Hidden Layer											
		5	6	7	8	9	10	11	12	13	14	15	16
Number of hidden layers	1	0.4806	0.4807	0.4808	0.4808	0.4808	0.4809	0.4808	0.4808	0.4808	0.4808	0.4808	0.4807
	2	0.4807	0.4807	0.4807	0.4809	0.4807	0.4808	0.4808	0.4808	0.4808	0.4808	0.4808	0.4812
	3	0.4808	0.4808	0.4808	0.4808	0.4808	0.4808	0.4808	0.4808	0.4808	0.4808	0.4812	0.4808

Table 13. RMS error for active power loss estimation for IEEE-34 bus system (case 1).

RMS Error		Number of Neurons in Hidden Layer											
		5	6	7	8	9	10	11	12	13	14	15	16
Number of hidden layers	1	5.6860	5.6861	5.6873	5.6875	5.6876	5.6877	5.6873	5.6872	5.6872	5.6874	5.6872	5.6873
	2	5.6872	5.6869	5.6873	5.6874	5.6872	5.6872	5.6872	5.6872	5.6872	5.6872	5.6872	5.6872
	3	5.6873	5.6873	5.6871	5.6869	5.6872	5.6872	5.6878	5.6872	5.6872	5.6872	5.6872	5.6872

Table 14. RMS error for active power loss estimation for IEEE-34 bus system (case 2).

RMS Error		Number of Neurons in Hidden Layer											
		5	6	7	8	9	10	11	12	13	14	15	16
Number of hidden layers	1	7.7288	7.7314	7.7298	7.7295	7.7298	7.7298	7.7298	7.7298	7.7299	7.7298	7.7298	7.7298
	2	7.7297	7.7299	7.7297	7.7298	7.7297	7.7298	7.7298	7.7298	7.7298	7.7298	7.7298	7.7298
	3	7.7299	7.8642	7.7298	7.7297	7.7298	7.7298	7.7301	7.7298	7.7299	7.7299	7.7298	7.7298

Table 15. RMS error for active power loss estimation for IEEE-34 bus system (case 3).

RMS Error		Number of Neurons in Hidden Layer											
		5	6	7	8	9	10	11	12	13	14	15	16
Number of hidden layers	1	5.6860	5.6861	5.6873	5.6875	5.6876	5.6877	5.6873	5.6872	5.6872	5.6874	5.6872	5.6873
	2	5.6872	5.6869	5.6873	5.6874	5.6872	5.6872	5.6872	5.6872	5.6872	5.6872	5.6872	5.6872
	3	5.6873	5.6873	5.6871	5.6867	5.6872	5.6872	5.6878	5.6872	5.6872	5.6872	5.6872	5.6872

6. Conclusions

This paper presents new insights about CVR factor calculation and factors affecting CVR implementation. This paper also proposes two new methods, a curve fitting method and ANN-based method, for estimating system active power loss by utilizing real and reactive power and voltage measurements at the substation. The presented CVR studies and methods are demonstrated through the IEEE 13-bus and 34-bus systems.

The CVR factor calculation under the CVR comparison study was carried out using the power captured at the load and the substation. For the IEEE 13-bus system, three different voltage regulator settings were considered. The study indicated that for the 13-bus system, the CVR factor increased with the increase in impedance component of the ZIP load, being the highest in the case of constant impedance load. For the IEEE 34-bus system, three different cases were created to calculate the CVR factor for different substation voltage levels and to study the effect of voltage regulator settings on the CVR factor. In case 1, the CVR factor was very low, as the voltage regulator restricted the variation in the voltage level at load locations by keeping it inside the defined bandwidth. The second case focused on CVR calculation without voltage regulators, by replacing them with a short line. The CVR factor improved considerably. In case 3, the voltage regulator setting and the substation voltage were proportionally, leading to an improved CVR factor. The results indicate that effective CVR implementation would require careful voltage regulator settings and that CVR effect assessment needs careful selection of measurements.

The active power loss was estimated using two different approaches, i.e., curve fitting and ANN. For the curve fitting approach, two predefined functions were considered. The lower value of rms error for function 2 indicated that it is the better function to define the relationship between the active power loss and the substation active and reactive power inputs and voltage. In the ANN approach, a neural network model with three different hidden layers, with neurons ranging from 5 to 16, was considered. The input data included the values of P, Q and V combined for all the ZIP models, and the target data included the active power loss. Simulation data obtained through OpenDSS for the 13-bus and 34 bus systems were used to calculate the function coefficients and train the ANN model. It was found that the ANN approach is a feasible option for modeling the loss relationship. It is noted that the proposed method is not used to identify control actions to minimize system losses, but instead to estimate system losses, given the substation measurements.

The methods presented in this paper for CVR study and active power loss estimation are applicable to larger systems. The studies help us to understand the factors that affect the CVR factor and its implementation. The active power loss calculation provides us with a better picture of CVR benefits in terms of load reduction and loss reduction.

To show the efficacy and scalability of our proposed method, a larger system (IEEE 8500 node system) will be studied and field data from distribution networks of LG&E and KU will be harnessed in the research. The stochastic nature of the CVR factor when loads and voltage regulator control schemes vary will be researched.

Author Contributions: Conceptualization, Y.L.; Formal analysis, G.Y. and Y.L.; Funding acquisition, D.M.I.; Methodology, G.Y. and Y.L.; Supervision, Y.L., N.J. and D.M.I.; Validation, G.Y.; Writing—original draft, G.Y. and Y.L.; Writing—review and editing, Y.L., N.J. and D.M.I. All authors have read and agreed to the published version of the manuscript.

Funding: This research was funded by LG&E and KU, Louisville, Kentucky, USA.

Data Availability Statement: Not applicable.

Conflicts of Interest: The authors declare no conflict of interest.

References

1. Fan, W.; Hossan, M.S.; Zheng, H.; Cook, A.; Zaid, S.; Fard, S.A.; Khodaei, A.; Paaso, A. A CVR On/Off Status Detection Algorithm for Measurement and Verification. In Proceedings of the 2021 IEEE Power & Energy Society Innovative Smart Grid Technologies Conference (ISGT), Washington, DC, USA, 16–18 February 2021; pp. 1–5. [CrossRef]
2. ANSI. ANSI C84.1-2020: Electric Power Systems Voltage Ratings (60 Hz)-ANSI Blog. 2020. Available online: https://blog.ansi.org/2020/10/ansi-c84-1-2020-electric-voltage-ratings-60/?_ga=2.222408667.1147844074.1647109334-405374262.1647109334 (accessed on 7 April 2022).
3. William, A.; Feb, J.; Hua, E.; Roland, S.; Oct, S.; Max, D.; Henry, J.; Hutchison, M.A.; Wilson, W.J.; Kusano, S.; et al. 1979 Index IEEE Transactions on Power Apparatus and Systems. *IEEE Trans. Power Appar. Syst.* **2008**, *PAS-98*, 2354a–2354al. [CrossRef]
4. Lauria, D.M. Conservation Voltage Reduction (CVR) at Northeast Utilities. *IEEE Trans. Power Deliv.* **1987**, *2*, 1186–1191. [CrossRef]
5. Williams, B. Distribution capacitor automation provides integrated control of customer voltage levels and distribution reactive power flow. In Proceedings of the Power Industry Computer Applications Conference, Salt Lake City, UT, USA, 7–12 May 1995; pp. 215–220. [CrossRef]
6. Peskin, M.A.; Powell, P.W.; Hall, E.J. Conservation Voltage Reduction with feedback from Advanced Metering Infrastructure. In Proceedings of the IEEE Power Engineering Society Transmission and Distribution Conference, Orlando, FL, USA, 7–10 May 2012; pp. 1–8. [CrossRef]
7. Pacific Northwest National Laboratory, & PNNL. Evaluation of CVR on a National Level. Report, July 2010. Available online: http://www.pnl.gov/main/publications/external/technical_reports/PNNL-19596.pdf (accessed on 7 April 2022).
8. Ellens, W.; Berry, A.; West, S. A quantification of the energy savings by Conservation Voltage Reduction. In Proceedings of the 2012 IEEE International Conference on Power System Technology, POWERCON, Auckland, New Zealand, 30 October–2 November 2012. [CrossRef]
9. Diskin, E.; Fallon, T.; O'Mahony, G.; Power, C. Conservation voltage reduction and voltage optimisation on Irish distribution networks. In *CIRED 2012 Workshop: Integration of Renewables into the Distribution Grid*; IET: London, UK, 2012; p. 264. [CrossRef]
10. El-Shahat, A.; Haddad, R.J.; Alba-Flores, R.; Rios, F.; Helton, Z. Conservation voltage reduction case study. *IEEE Access* **2020**, *8*, 55383–55397. [CrossRef]
11. Diaz-Aguilo, M.; Sandraz, J.; Macwan, R.; de Leon, F.; Czarkowski, D.; Comack, C.; Wang, D. Field-Validated Load Model for the Analysis of CVR in Distribution Secondary Networks: Energy Conservation. *IEEE Trans. Power Deliv.* **2013**, *28*, 2428–2436. [CrossRef]
12. Wilson, T.L. Energy conservation with voltage reduction-fact or fantasy. In Proceedings of the 2002 Rural Electric Power Conference. Paper Presented at the 46th Annual Conference (Cat. No. 02CH37360), Colorado Springs, CO, USA, 5–7 May 2002. [CrossRef]
13. Singh, R.; Tuffner, F.; Fuller, J.; Schneider, K. Effects of distributed energy resources on conservation voltage reduction (CVR). In Proceedings of the 2011 IEEE Power and Energy Society General Meeting, Detroit, MI, USA, 24–28 July 2011; pp. 1–7. [CrossRef]
14. O'Connell, A.; Keane, A. Volt-var curves for photovoltaic inverters in distribution systems. *IET Gener. Transm. Distrib.* **2017**, *11*, 730–739. [CrossRef]
15. Wang, Z.; Begovic, M.; Wang, J. Analysis of conservation voltage reduction effects based on multistage SVR and stochastic process. *IEEE Trans. Smart Grid* **2014**, *5*, 431–439. [CrossRef]

16. Wang, Z.; Wang, J. Review on implementation and assessment of conservation voltage reduction. *IEEE Trans. Power Syst.* **2014**, *29*, 1306–1315. [CrossRef]
17. Wilson, T.L. Measurement and verification of distribution voltage optimization results for the IEEE power & energy society. In Proceedings of the IEEE PES General Meeting, Minneapolis, MN, USA, 25–29 July 2010; pp. 1–9. [CrossRef]
18. Wen, F.; Liao, Y.; See, J.; Goins, B.; Gill, C.; Petreshock, J.; Bridges, J.D. Distribution system voltage and var optimization. In Proceedings of the 2012 IEEE Power and Energy Society General Meeting, San Diego, CA, USA, 22–26 July 2012; pp. 1–8. [CrossRef]
19. Hossan, M.S.; Maruf, H.M.M.; Chowdhury, B. Comparison of the ZIP load model and the exponential load model for CVR factor evaluation. In Proceedings of the 2017 IEEE Power & Energy Society General Meeting, Chicago, IL, USA, 16–20 July 2017; pp. 1–5. [CrossRef]
20. Zhang, Y.; Liao, Y.; Jones, E.; Jewell, N.; Ionel, D. ZIP load modeling for single and aggregate loads and CVR factor estimation. In Proceedings of the PAC World Conference, Glasgow, UK, 31 August–1 September 2021.
21. Go, S.I.; Yun, S.Y.; Ahn, S.J.; Kim, H.W.; Choi, J.H. Heuristic coordinated voltage control schemes in distribution network with distributed generations. *Energies* **2020**, *13*, 2849. [CrossRef]
22. Raz, D.; Beck, Y. An operational approach to multi-objective optimization for Volt-VAr control. *Energies* **2020**, *13*, 5871. [CrossRef]
23. Hossein, Z.S.; Khodaei, A.; Fan, W.; Hossan, M.S.; Zheng, H.; Fard, S.A.; Paaso, A.; Bahramirad, S. Conservation voltage reduction and volt-VAR optimization: Measurement and verification benchmarking. *IEEE Access* **2020**, *8*, 50755–50770. [CrossRef]
24. Biserica, M.; Besanger, Y.; Caire, R.; Chilard, O.; Deschamps, P. Neural networks to improve distribution state estimation-Volt var control performances. *IEEE Trans. Smart Grid* **2012**, *3*, 1137–1144. [CrossRef]
25. Wang, H.; Jiao, H.; Chen, J.; Liu, W. Parameter Identification for a Power Distribution Network Based on MCMC Algorithm. *IEEE Access* **2021**, *9*, 104154–104161. [CrossRef]
26. Rizwan, M.; Waseem, M.; Liaqat, R.; Sajjad, I.A.; Dampage, U.; Salmen, S.H.; Obaid, S.; Al Mohamed, M.A.; Annuk, A. Spso based optimal integration of dgs in local distribution systems under extreme load growth for smart cities. *Electronics* **2021**, *10*, 2542. [CrossRef]

Article

Nonlinear Self-Synchronizing Current Control for Grid-Connected Photovoltaic Inverters [†]

Moath Alqatamin and Michael L. McIntyre *

Electrical and Computer Engineering Department, Speed School of Engineering, University of Louisville, Louisville, KY 40292, USA; moath.alqatamin@louisville.edu
* Correspondence: michael.mcintyre@louisville.edu
† This paper is an extended version of our paper published in 2020 American Control Conference (ACC), Denver, CO, USA, 1–3 July 2020; pp. 192–197, doi:10.23919/ACC45564.2020.9147600.

Abstract: Three-phase inverters for photovoltaic grid-connected applications typically require some form of grid voltage phase-angle detection in order to properly synchronize to the grid and control real and reactive power generation. Typically, a phase-locked loop scheme is used to determine this real-time phase angle information. However, in the present work, a novel method is proposed whereby the phase angle of the grid can be accurately identified solely via the grid current feedback. This phase-angle observer is incorporated into a current controller which can manage the real and reactive power of the grid-connected PV inverter system. Moreover, the maximum power point of the photovoltaic arrays is achieved without using a DC–DC converter. The proposed method achieves the grid current and DC-link voltage control objectives without the knowledge of the grid information and without the need for a cascaded control scheme. The design of this combined observer/controller scheme is motivated and validated via a Lyapunov stability analysis. The experimental setup is prototyped utilizing a real-time Typhoon HIL 603 and National Instrument cRIO embedded controller in order to validate the proposed observer/controller scheme under different operation scenarios such as irradiation changes, frequency changes, reactive power injection, and operation with a distorted grid. The results show that the DC-link voltage and the active and reactive powers are well regulated from the proposed control scheme without the measurement of the grid phase and frequency.

Keywords: photovoltaic system; grid angle estimation; self-synchronization current control

Citation: Alqatamin, M.; McIntyre, M.L. Nonlinear Self-Synchronizing Current Control for Grid-Connected Photovoltaic Inverters. *Energies* **2022**, *15*, 4855. https://doi.org/10.3390/en15134855

Academic Editors: Yuan Liao and Ke Xu

Received: 1 June 2022
Accepted: 29 June 2022
Published: 2 July 2022

1. Introduction

Solar energy is the most abundant and the cleanest form of renewable energy sources available in the world, which is used for a variety of applications, such as generation of electricity for domestic, commercial, or industrial applications [1,2]. Nowadays, photovoltaic (PV) generation is becoming a vital part of modern energy systems due to the cost reduction of manufacturing PV modules. Additionally, improvements in the energy extraction from solar panels has made PV generation more efficient and reliable than other generating sources [2,3]. Electrical power generated by PV array depends on the temperature and the solar irradiation, which has a strong nonlinear effect. Recently, the proliferation of installed grid-connected PV systems have begun to impact the stability of the power grid [4,5]. Thus, real-time control of grid-connected inverter systems has drawn recent attention in the literature [6,7]. The affects seen on the power system due to the stochastic nature of PV power outputs has drawn attention to the control system aspect of PV systems [8]. In the literature, connecting PV arrays to the utility grid has been accomplished in many forms and configurations [8,9]. Traditional PV systems utilize a two-stage power converter system with closed loop control schemes utilized throughout. A DC–DC converter is typically utilized as a first stage to regulate the voltage of the DC-link capacitor, thereby ensuring maximum power point operation for the PV arrays. The second stage is the DC–AC power converter, which injects AC power into the grid and/or supports a local load [10]. Using a two stage topology results in a reduction of the system efficiency as well as an increase in the system cost [2]. For these reasons, a single-stage, three-phase, grid-connected PV system with maximum power point tracking (MPPT) and DC–AC converter will be used in this work.

Unintentional connection of the PV-based grid-connected inverter systems to the utility grid is undesirable as this could lead to system to instability [11]. Thus, in order to perform safe connection between the distribution generation unit (such as the PV-based inverter) and the grid, a synchronization unit is necessary to obtain the grid's phase angle and frequency before connection to avoid an instantaneous out of phase connection. The main method that is used in the literature for this purpose is a Phase Locked Loop (PLL). Moreover, the grid's information that is extracted from PLLs is used in the coordinate transformation in three-phase systems [12–14]. A PLL system is mainly integrated with the primary current control scheme as an outer loop, thus creating a cascaded control system [13,15]. In such cascaded schemes, the dynamics of the outer loop (in this case the PLL) should be slower (bandwidth) than the inner loop (current control loop) to maintain the stability of the system. On the other hand, the slow dynamic of a PLL-loop means delays in the detection of the grid's phase and frequency, both of which are necessary for the current control loop. Moreover, adding a PLL loop to the main control loop will add more complexity to the system [16]. Many approaches have been proposed in the literature to improving the performance of the PLL schemes to address any potential instability issues resulting from the interaction between the PLL schemes and current control loop [17,18]. These challenges have motivated researchers to develop control schemes for grid-connected inverters that do not require a synchronization system such as PLLs [19–22]. For instance, the authors in [23] proposed a direct power controller for a single-phase grid-connected inverter without a PLL scheme. Moreover, we proposed in our work in [24] a method for a PLL-less current control scheme for three-phase, grid-connected inverters. Although these methods do not need a PLL scheme, they do not consider a grid-connected PV system or the additional dynamics this would entail. Previous studies [19–24] have considered a fixed dc voltage as an input. Hence, the dynamic of the PV panels, dc voltage regulator, and maximum power point tracking algorithm are not included in their control schemes.

A few works have investigated a PLL-less synchronization procedure for PV inverter systems. For example, in [25], the authors propose a current control scheme for single-phase PV inverter systems without a PLL scheme. The problem with this approach is that the control scheme has multiple current loops, which is problematic in managing the bandwidths with the various loops to insure the stability of the whole system. In [26], we propose a PLL-less current control scheme for PV inverter systems. Simulation results demonstrated the effectiveness of this approach to simultaneously control the grid current, DC-link voltage, and estimate the grid phase and frequency.

In this paper, a single-stage, three-phase, grid-connected photovoltaic system is controlled by a novel nonlinear current controller. The main contribution of the proposed controller is that the current tracking objective is achieved without knowledge of the grid phase angle and frequency. To do so, an estimated rotating reference frame ($\gamma\delta$-frame) is utilized [27]. Moreover, the proposed control scheme ensures that tracking of the DC-link capacitor reference voltage-generated from MPPT algorithm is achieved. This algorithm ensures the PV array(s) achieves their maximum power operating point. Within the control scheme, adaptive compensation terms facilitate the voltage and current tracking objectives and simultaneously account for the unavailable grid frequency and phase, hence eliminating the need for an additional feedback system for synchronization, such as a PLL. The voltage sensors required for PLL-type systems are expensive and introduce electrical noise and dc offsets, which must be compensated for in other ways.

The main contributions of the present work over our previous published work [26] and other works in the literature are:

1. The proposed controller/observer scheme eliminates the need for a cascaded control scheme. A cascaded approach is necessary in traditional control schemes for such an application to regulate the DC-link capacitor voltage, control the grid current, and determine the grid information extraction loop (PLL).
2. In terms of the control design, a new term has been added to both the DC-link voltage dynamics and grid current dynamics which accounts for un-modeled dynamics terms. Inclusion of these terms was motivated by validation experiments. Then, during the control design process, adaptive compensation terms have been included in the control to compensate for these uncertainties.
3. In terms of validation, an experimental setup has been prototyped to further validate the proposed scheme, including both control and observer aspects. A combination between Typhoon HIL [28–31] and National Instrument real time compact RIO has been used for this purpose. Furthermore, the proposed controller/observer scheme has been tested in different operating scenarios including sudden change to solar irradiance, grid frequency fluctuation, reactive power injection, and distorted grid utility.

2. System Model

A single-stage, three-phase, PV grid-connected inverter system with an inductive filter is shown in Figure 1. By applying Kirchhoff's voltage law, the dynamic mathematical model for this system can be represented as

$$v_{abc} \triangleq L\dot{I}_{abc} + RI_{abc} + e_{abc}, \tag{1}$$

where $v_{abc}(t) \in \mathbb{R}^{3 \times 1}$ are the phase voltages, $I_{abc}(t) \in \mathbb{R}^{3 \times 1}$ are the phase currents, and $e_{abc} \in \mathbb{R}^{3 \times 1}$ are the voltages of the point of common coupling (PCC) at which the system is connected to the public utility grid. Assuming a switching average model, the three-phase inverter output voltages as the product to the DC-link voltage V_{dc} and the duty cycles of their respective switching legs can be represented as

$$v_{abc} \triangleq V_{dc}D_{abc} \tag{2}$$

where the duty cycles, $D_{abc}(t) \in [0,1]$. Moreover, assuming a balanced utility grid, the three-phase PCC voltages can be rewritten as

$$e_{abc} \triangleq E_m \cos(\theta \pm \phi) \tag{3}$$

where E_m is the amplitude of the PCC voltage, $\theta(t)$ is the phase angle of the same voltage which will be donated as a grid phase angle, and $\phi = 0$ and 120 to represent balanced three-phase voltages. By substituting (2) and (3) into (1), the three-phase grid current dynamics can be obtained as

$$L\dot{I}_{abc} = V_{dc}D_{abc} - RI_{abc} - E_m \cos(\theta \pm \phi). \tag{4}$$

From Figure 1, the dynamic equation of the DC-link capacitor of the PV System can be obtained by applying Kirchhoff's current law (KCL) at the capacitor node as follows:

$$C\dot{V}_{dc} = I_{pv} - I_{dc} \tag{5}$$

where I_{pv} is the output current from PV panels, and I_{dc} is the DC input current to the inverter.

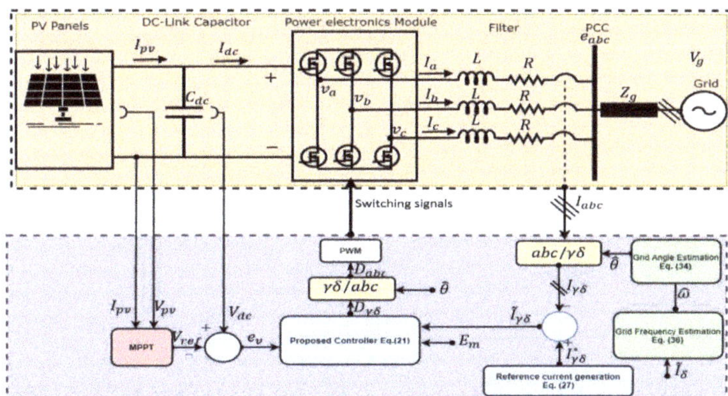

Figure 1. Three-Phase Grid Connected PV System and proposed control scheme.

By using the standard Clarke and Park transformations, the dq-reference frame model for the system dynamics is obtained as

$$L\begin{bmatrix} \dot{I}_d \\ \dot{I}_q \end{bmatrix} = V_{dc}\begin{bmatrix} D_d \\ D_q \end{bmatrix} - \begin{bmatrix} R & -\omega L \\ \omega L & R \end{bmatrix}\begin{bmatrix} I_d \\ I_q \end{bmatrix} - E_m\begin{bmatrix} 1 \\ 0 \end{bmatrix} \tag{6}$$

where $\omega(t) \triangleq \dot{\theta}(t)$ is the grid frequency.

In this work, the grid angle $\theta(t)$ is an unmeasurable and unknown parameter. Therefore, designing an observer for the grid angle $\hat{\theta}(t)$ is required to perform the dq transformations in the estimated reference frame [27]. This results in a rotating orthogonal axis system denoted as $\gamma\delta$. After performing the transformation in the $\gamma\delta$ frame, the system shown in (6) can be written as

$$L\begin{bmatrix} \dot{I}_\gamma \\ \dot{I}_\delta \end{bmatrix} = V_{dc}\begin{bmatrix} D_\gamma \\ D_\delta \end{bmatrix} - \begin{bmatrix} R & -\dot{\hat{\theta}}L \\ \dot{\hat{\theta}}L & R \end{bmatrix}\begin{bmatrix} I_\gamma \\ I_\delta \end{bmatrix} - E_m\begin{bmatrix} \cos\tilde{\theta} \\ \sin\tilde{\theta} \end{bmatrix} + d_i\begin{bmatrix} 1 \\ 0 \end{bmatrix} \tag{7}$$

where $\tilde{\theta}(t)$ is the error between the actual grid phase angle, and the observed phase angle will be defined later. In (7), $d_i(t) \in \mathbb{R}$ is an unknown, slowly time-varying term added to compensate for any unknown disturbance and imperfect inverter modelling.

The dynamic equation of the DC-link capacitor in (5) can be modified by using the active power balance principle between the dc and ac system sides as follows:

$$P_{dc} \triangleq P_{ac} \tag{8}$$

where

$$P_{dc} = V_{dc}I_{dc} \tag{9}$$

and

$$P_{ac} = e_{abc}i_{abc} = E_m\cos(\theta \pm \phi)I_{abc} \tag{10}$$

Based on the Clarke transformation, (10) in the $\alpha\beta$-frame can be written after some mathematical simplifications as

$$P_{ac} = \frac{3}{2}E_m[\cos\theta \quad \sin\theta]\begin{bmatrix} I_\alpha \\ I_\beta \end{bmatrix} \tag{11}$$

By performing the transformation of $\alpha\beta$ to $\gamma\delta$ on (11), and after some mathematical simplifications, the power equation can be obtained in $\gamma\delta$- estimated frame as follows:

$$P_{ac} = \frac{3}{2}E_m\left(I_\gamma \cos\tilde{\theta} + I_\delta \sin\tilde{\theta}\right). \tag{12}$$

Using (8), (9), and (12), the I_{dc} equation can be written as

$$I_{dc} = \frac{3}{2}\frac{E_m}{V_{dc}}\left(I_\gamma \cos\tilde{\theta} + I_\delta \sin\tilde{\theta}\right) \tag{13}$$

Finally, the dynamic equation of the DC-link capacitor voltage for the PV inverter system in the $\gamma\delta$ estimated reference frame is obtained by substituting (13) into (5) as shown in (14):

$$C\dot{V}_{dc} = I_{pv} - \frac{3}{2}\frac{E_m}{V_{dc}}\left(I_\gamma \cos\tilde{\theta} + I_\delta \sin\tilde{\theta}\right) + d_v \tag{14}$$

where $d_v(t) \in \mathbb{R}$ is a term added to compensate for any slowly time-varying, external, unknown disturbance.

3. Control System Development

The main objective of the proposed controller/observer scheme is designing duty cycle control signals D_γ and D_δ of the PV single-stage, three-phase, grid-connected inverter system in the absence of the grid angle $\theta(t)$ and frequency $\omega(t)$ measurements. The $I_\gamma(t)$ and $I_\delta(t)$ can follow their respective reference currents $I_\gamma^*(t)$ and $I_\delta^*(t)$, hence, $I_\gamma(t) \to I_\gamma^*(t)$ and $I_\delta(t) \to I_\delta^*(t)$. as $t \to \infty$. Additionally, the PV array must operate at the maximum power point. To ensure this, the reference active current $I_\gamma^*(t)$ is designed in order to make sure that the DC-link capacitor voltage tracks its reference trajectory. Hence, $V_{dc}(t) \to V_{ref}(t)$ as $t \to \infty$ where the reference voltage trajectory $V_{ref}(t)$ is obtained from a standard MPPT algorithm. The subsequent stability analysis will prove that when the grid angle observer converges, $\tilde{\theta}(t) = 0$, then $I_\gamma(t) = I_d(t)$, $I_\delta(t) = I_q(t)$, which means that the active and reactive power that will be injected into the grid are controllable. The following assumptions are made to facilitate this control development.

Assumption 1. *$\alpha\beta$-frame currents $I_\alpha(t), I_\beta(t)$ are measurable. $V_{pcc}(t), V_{dc}(t), I_{PV}(t)$ are also measurable.*

Assumption 2. *The nominal system parameters R, L are known a priori.*

Assumption 3. *The grid frequency $\omega(t)$ is unknown but is assumed to be a positive constant. Additionally, the grid phase angle $\theta(t)$ is unknown.*

Assumption 4. *The disturbance terms $d_i(t)$, $d_v(t)$ are assumed to be slowly time-varying; hence, $\dot{d}_i(t), \dot{d}_v(t) \cong 0$.*

3.1. Error System Development

To facilitate the subsequent control development, the following current tracking errors are defined as

$$\begin{bmatrix} \tilde{I}_\gamma \\ \tilde{I}_\delta \end{bmatrix} \triangleq \begin{bmatrix} I_\gamma^* \\ I_\delta^* \end{bmatrix} - \begin{bmatrix} I_\gamma \\ I_\delta \end{bmatrix} \tag{15}$$

The tracking error dynamics of the DC-link voltage for the PV inverter system are defined as

$$e_v \triangleq V_{dc} - V_{ref}. \tag{16}$$

Since the grid phase angle and frequency are unknown, the estimation error signals are defined as

$$\tilde{\omega} \triangleq \omega - \hat{\omega} \tag{17}$$

$$\tilde{\theta} \triangleq \theta - \hat{\theta} \tag{18}$$

Taking the time derivative of (15), pre-multiplying by L, and utilizing the system dynamics from (7), the open loop error dynamics for the current are obtained as follows:

$$L\begin{bmatrix}\dot{\tilde{I}}_\gamma \\ \dot{\tilde{I}}_\delta\end{bmatrix} = L\begin{bmatrix}\dot{I}^*_\gamma \\ \dot{I}^*_\delta\end{bmatrix} + \begin{bmatrix}R & -\hat{\dot{\theta}}L \\ \hat{\dot{\theta}}L & R\end{bmatrix}\begin{bmatrix}I_\gamma \\ I_\delta\end{bmatrix} + E_m\begin{bmatrix}\cos\tilde{\theta} \\ \sin\tilde{\theta}\end{bmatrix} - V_{dc}\begin{bmatrix}D_\gamma \\ D_\delta\end{bmatrix} - d_i\begin{bmatrix}1 \\ 0\end{bmatrix}. \tag{19}$$

To simplify the analysis, assume that $\tilde{\theta}(t)$ is small and centered about 0; hence, $\cos\tilde{\theta} \approx 1$. Based on this assumption, (19) can be simplified in the following form:

$$L\begin{bmatrix}\dot{\tilde{I}}_\gamma \\ \dot{\tilde{I}}_\delta\end{bmatrix} = L\begin{bmatrix}\dot{I}^*_\gamma \\ \dot{I}^*_\delta\end{bmatrix} + \begin{bmatrix}R & -\hat{\dot{\theta}}L \\ \hat{\dot{\theta}}L & R\end{bmatrix}\begin{bmatrix}I_\gamma \\ I_\delta\end{bmatrix} + E_m\begin{bmatrix}1 \\ \sin\tilde{\theta}\end{bmatrix} - V_{dc}\begin{bmatrix}D_\gamma \\ D_\delta\end{bmatrix} - d_i\begin{bmatrix}1 \\ 0\end{bmatrix}. \tag{20}$$

3.2. Control Design

Motivated by the subsequent stability analysis and using the open loop error dynamics for the current in (20), the following duty cycle control inputs are defined as follows:

$$\begin{bmatrix}D_\gamma \\ D_\delta\end{bmatrix} \triangleq \frac{1}{V_{dc}}\left(L\begin{bmatrix}\dot{I}^*_\gamma \\ \dot{I}^*_\delta\end{bmatrix} + \begin{bmatrix}R & -\hat{\dot{\theta}}L \\ \hat{\dot{\theta}}L & R\end{bmatrix}\begin{bmatrix}I_\gamma \\ I_\delta\end{bmatrix} + \begin{bmatrix}E_m \\ 0\end{bmatrix} + k_1\begin{bmatrix}\tilde{I}_\gamma \\ \tilde{I}_\delta\end{bmatrix} - \hat{d}_i\begin{bmatrix}1 \\ 0\end{bmatrix}\right) \tag{21}$$

where $k_1 \in \mathbb{R}^+$ is a positive control gain and $\hat{d}_i(t)$ is the estimation of the unknown disturbance $d_i(t)$. The update law of $d_i(t)$ is designed based on the subsequent stability analysis and found to be as follows:

$$\dot{\hat{d}}_i = -k_{di}\tilde{I}_\gamma \tag{22}$$

where $k_{d_i} \in \mathbb{R}^+$ is a positive estimation gain. The closed loop error dynamics of the current are obtained by substituting (21) into the open loop error dynamics in (20) as follows:

$$L\begin{bmatrix}\dot{\tilde{I}}_\gamma \\ \dot{\tilde{I}}_\delta\end{bmatrix} = -k_1\begin{bmatrix}\tilde{I}_\gamma \\ \tilde{I}_\delta\end{bmatrix} + \begin{bmatrix}0 \\ E_m\sin\tilde{\theta}\end{bmatrix} - \tilde{d}_i\begin{bmatrix}1 \\ 0\end{bmatrix} \tag{23}$$

where $\tilde{d}_i(t)$ is the estimation error defined as:

$$\tilde{d}_i \triangleq d_i - \hat{d}_i \tag{24}$$

Taking the time derivative of (16), and pre-multiplying it by C, and using the DC-link dynamic from (14) as well as the assumption $\cos\tilde{\theta} \approx 1$, the open loop error dynamic of the DC- link voltage can be obtained as

$$C\dot{e}_v = I_{pv} - \frac{3}{2}\frac{E_m}{V_{dc}}\left(I_\gamma + I_\delta\sin\tilde{\theta}\right) + d_v - C\dot{V}_{ref} \tag{25}$$

Utilizing the tracking error of the current $I_\gamma(t)$ from (15), the above equation in (25) can be rewritten as

$$C\dot{e}_v = I_{pv} - \frac{3}{2}\frac{E_m}{V_{dc}}\left(I^*_\gamma - \tilde{I}_\gamma + I_\delta\sin\tilde{\theta}\right) + d_v - C\dot{V}_{ref}. \tag{26}$$

Motivated by subsequent stability analysis, $I^*_\gamma(t)$ can be designed based on the Equation (26) as follows:

$$I^*_\gamma \triangleq \frac{2}{3}\frac{V_{dc}}{E_m}\left[I_{pv} - C\dot{V}_{ref} + k_v e_v + \hat{d}_v\right] \tag{27}$$

where $K_v \in \mathbb{R}^+$ is a positive control gain and $\hat{d}_v(t)$ is the estimation of the unknown disturbance with the following updating law, which is obtained based on the subsequent stability analysis:

$$\dot{\hat{d}}_v = -k_{dv} e_v \tag{28}$$

where $k_{d_v} \in \mathbb{R}^+$ is a positive estimation gain.

Remark 1. *$I_\gamma^*(t)$ is designed to stabilize the DC-link voltage and to make sure that the PV arrays are operating at their maximum power point. Moreover, $I_\delta^*(t)$ could be selected to control the amount of the injected reactive power to the grid.*

The closed loop error dynamics for $e_v(t)$ can be obtained by substituting (27) into (25) as follows:

$$C\dot{e}_v = \frac{3}{2}\frac{E_m}{V_{dc}}\left(\tilde{I}_\gamma - I_\delta \sin\tilde{\theta}\right) - k_v e_v - \tilde{d}_v \tag{29}$$

where $\tilde{d}_v(t)$ is the estimation error of the unknown disturbance defined as

$$\tilde{d}_v \triangleq d_v - \hat{d}_v. \tag{30}$$

In order to design the grid phase-angle observer, the derivative of (18) is taken to obtain

$$\dot{\tilde{\theta}} = \omega - \dot{\hat{\theta}}. \tag{31}$$

Then, motivated by the subsequent stability analysis and the above equation in (31), the angle observer can be obtained after some mathematical operations as follows:

$$\dot{\hat{\theta}} \triangleq \hat{\omega} + \tilde{I}_\delta + E_m \sin\tilde{\theta} - \frac{3}{2}\frac{e_v}{V_{dc}}I_\delta \tag{32}$$

It is notable that the above update law for the grid angle observer is unrealizable due to the unknown and unmeasurable signals such as $\tilde{\theta}(t)$. To make this observer realizable and implementable, substitute $E_m \sin\tilde{\theta}$ term from the closed loop error dynamics of $\tilde{I}_\delta(t)$ from (23) into (32) to obtain

$$\dot{\hat{\theta}} \triangleq \hat{\omega} + (k_1 + 1)\tilde{I}_\delta + L\dot{\tilde{I}}_\delta - \frac{3}{2}\frac{e_v}{V_{dc}}I_\delta. \tag{33}$$

Integrating both sides of (33) yields the following realizable form:

$$\hat{\theta} = L\tilde{I}_\delta + \int \left[\hat{\omega} + (k_1 + 1)\tilde{I}_\delta - \frac{3}{2}\frac{e_v}{V_{dc}}I_\delta\right]d\sigma \tag{34}$$

where $\hat{\omega}(t) \in \mathbb{R}$ is the subsequently designed update law for the unknown grid frequency. Note that a realizable form of $\dot{\hat{\theta}}(t)$ is still undefined, which is necessary for generating the duty cycle control inputs $D_\gamma(t), D_\delta(t)$ as shown in (21). For this reason, $\hat{\omega}(t)$ is subsequently designed. Given the relationship between these variables, this substitution is easily justified. Motivated by the stability analysis, the unrealizable design form for the grid frequency estimator is obtained as follows:

$$\dot{\hat{\omega}} = k_\omega E_m \sin\tilde{\theta} \tag{35}$$

where $k_\omega \in \mathbb{R}^+$ is a positive estimator gain. To obtain a realizable form of this estimator, again the term $E_m \sin\tilde{\theta}$ from (23) should be used in (35) and then the integration of both sides is taken to obtain

$$\hat{\omega} = k_\omega \left(L\tilde{I}_\delta + k_1 \int \tilde{I}_\delta(\sigma)\right)d\sigma \tag{36}$$

3.3. Stability Analysis

Theorem 1. *The closed loop error dynamics from (23) and (29) ensure that the error signals defined in (15), (16), and (18) are regulated as $\tilde{I}_\gamma(t)$, $\tilde{I}_\delta(t)$, $e_v(t), \tilde{\theta}(t) \to 0$ as $t \to \infty$.*

Proof of Theorem 1. To analyze the stability of the system, a Lyapunov function is chosen as follows:

$$V = \frac{C}{2}e_v^2 + \frac{1}{2}L\tilde{I}_\gamma^2 + \frac{1}{2}L\tilde{I}_\delta^2 + E_m(1 - \cos\tilde{\theta}) + \frac{1}{2k_\omega}\tilde{\omega}^2 + \frac{1}{2k_{di}}\tilde{d}_i^2 + \frac{1}{2k_{dv}}\tilde{d}_v^2. \tag{37}$$

Assumption 5. $V(t)$ in (37) is positive definite in the local region $\tilde{\theta}(t) \in (-2\pi, 2\pi)$. Assuming that $\tilde{\theta}$ is wrapped such that its effective domain is $\tilde{\theta}(t) \in [-\pi, \pi)$, the function is effectively globally positive definite.

The time derivative of $V(t)$ is taken as follows:

$$\dot{V} = Ce_v\dot{e}_v + L\tilde{I}_\gamma\dot{\tilde{I}}_\gamma + L\tilde{I}_\delta\dot{\tilde{I}}_\delta + E_m\dot{\tilde{\theta}}\sin\tilde{\theta} - \frac{1}{k_\omega}\tilde{\omega}\dot{\hat{\omega}} - \frac{1}{k_{di}}\tilde{d}_i\dot{\hat{d}}_i - \frac{1}{k_{dv}}\tilde{d}_v\dot{\hat{d}}_v \tag{38}$$

Substituting the closed loop error dynamics from (23) and (29) as well as the update laws from (22), (28), (32), and (35) a long (31) into the above equation and simplifying, $\dot{V}(t)$ can be upper bounded as

$$\dot{V} \le -k_v e_v^2 - k_1\tilde{I}_\gamma^2 - k_1\tilde{I}_\delta^2 - E_m^2\sin^2\tilde{\theta} + \frac{3}{2}\frac{E_m}{V_{dc}}\left|e_v\right|\left|\tilde{I}_\gamma\right|. \tag{39}$$

By using the inequality $|e_v||\tilde{I}_\gamma| \le \frac{1}{2}|e_v|^2 + \frac{1}{2}|\tilde{I}_\gamma|^2$, $\dot{V}(t)$ can be further upper bounded as

$$\dot{V} \le -k_v e_v^2 - k_1\tilde{I}_\gamma^2 - k_1\tilde{I}_\delta^2 - E_m^2\sin^2\tilde{\theta} + \frac{3E_m}{4V_{dc}}|e_v|^2 + \frac{3E_m}{4V_{dc}}|\tilde{I}_\gamma|^2 \tag{40}$$

$$\dot{V} \le -\left(k_v - \frac{3E_m}{4V_{dc}}\right)e_v^2 - \left(k_1 - \frac{3E_m}{4V_{dc}}\right)\tilde{I}_\gamma^2 - k_1\tilde{I}_\delta^2 - E_m^2\sin^2\tilde{\theta}. \tag{41}$$

It is clear that $\dot{V}(t)$ is a negative semi-definite function if $k_v > \frac{3E_m}{4V_{dc}}$ and $k_1 > \frac{3E_m}{4V_{dc}}$. From $V(t)$, $\dot{V}(t)$ it is clear that $e_v, \tilde{I}_\gamma, \tilde{I}_\delta, \sin^2\tilde{\theta}, \tilde{\omega}, \tilde{d}_i, \tilde{d}_v \in \mathcal{L}_\infty$. Since $\tilde{I}_\gamma, \tilde{I}_\delta, \tilde{\theta} \in \mathcal{L}_\infty$, from closed loop error dynamics for $\tilde{I}_\gamma, \tilde{I}_\delta$ and e_v it is shown that $\dot{\tilde{I}}_\gamma, \dot{\tilde{I}}_\delta, \dot{e}_v \in \mathcal{L}_\infty$. Moreover, $\tilde{\omega} \in \mathcal{L}_\infty$ and from Assumption 3, $\omega \in \mathcal{L}_\infty$ and therefore $\hat{\omega} \in \mathcal{L}_\infty$. Since d_i and d_v are bounded by assumption, then from (24) and (30) one can conclude that $\hat{d}_i, \hat{d}_v \in \mathcal{L}_\infty$. Since $\tilde{I}_\gamma, \tilde{I}_\delta, \tilde{\theta}, \hat{\omega} \in \mathcal{L}_\infty$, from definition of $\dot{\hat{\theta}}$ it is clear that $\dot{\hat{\theta}} \in \mathcal{L}_\infty$. Now, $\omega, \dot{\hat{\theta}} \in \mathcal{L}_\infty$ and from definition of $\dot{\tilde{\theta}}$, $\dot{\tilde{\theta}} \in \mathcal{L}_\infty$. Finally $\tilde{I}_\gamma, \tilde{I}_\delta, \dot{e}_v, \dot{\tilde{\theta}} \in \mathcal{L}_\infty$ and $\ddot{V}(t) \in \mathcal{L}_\infty$, by Barbalat's Lemma it is clear that $\dot{V}(t) \to 0$ as $t \to \infty$ and thus $\tilde{I}_\gamma, \tilde{I}_\delta, e_v, \tilde{\theta} \to 0$ as $t \to \infty$. $\square$

4. Experimental Results

To validate the proposed controller, a hardware-in-the-loop setup has been implemented. The PV panels and the grid-connected inverter system shown in Figure 1 were emulated in Typhoon HIL603 hardware using the small-time step library. The control algorithm was implemented inside LabVIEW FPGA and then programmed into the National Instruments CompactRIO (cRIO) 9032. A digital output module NI 9401 on the cRIO is used to send the switching signals generated by the controller to the HIL603 via a digital input module. Moreover, the analog output module of the HIL603 is used to measure the emulated currents and the voltages of the inverter model and sends them to the controller via the analog input module NI 9222 of the cRIO. Analog input module NI 9015 is used to measure the DC-link voltage and PV current. Figure 2 shows the experimental setup components that are used in this C-HIL validation setup. Table 1 details the PV panels parameters that are used in Typhoon PV model generation. Table 2 summarizes the system parameters and the controller gains that are used throughout the following experiments. The controller gains have been found to meet the stability analysis conditions and have the best performance.

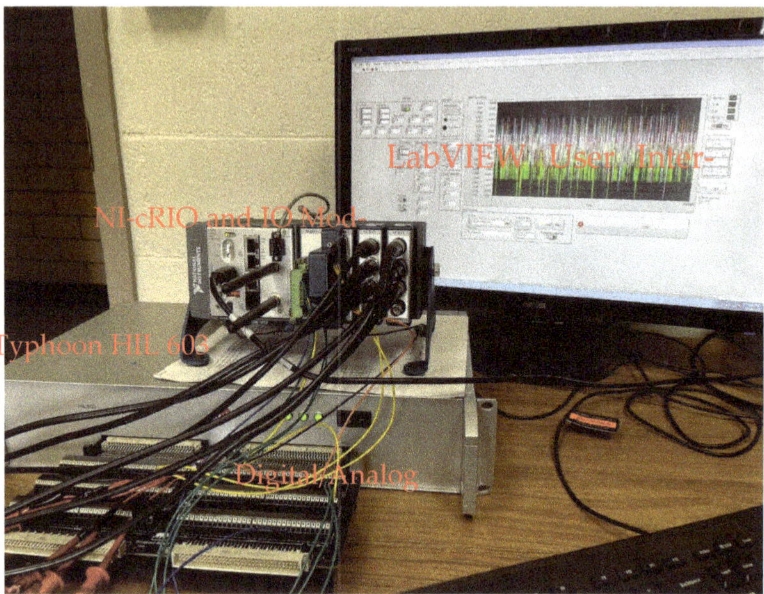

Figure 2. C-HIL Experimental setup used for control system validation.

Table 1. Photovoltaic Panels, System and Controller Parameters.

Parameters	Value
Nominal open circuit voltage of PV array, V_{oc}	32.9 V
Nominal short circuit current of PV array, I_{sc}	8.21 A
Panels connected in series	20
Solar cells in a string	54
Solar cells strings connected in parallel	3

Table 2. System Parameters and Controller gains.

Parameters	Value
Nominal Line Inductance, L	10 mH
Nominal Line Resistance, R	0.1 Ω
Switching Frequency, f_{sw}	10 kHz
Grid Voltage, V_g	110 V_{rms}
Control Gain, k_1	20
Control Gain, k_ω	0.001
Control Gain, k_v	1
Estimation Gains, k_{di}, k_{dv}	10
Nominal System Frequency, ω	377 rad/s

4.1. Tracking Performance of Proposed Controller

In the first scenario, the tracking performance of the proposed controller/observer scheme, the PV system experiences standard atmospheric conditions for which the irradiation and the temperature are 1000 W/m^2 and 25 °C, respectively. The $\gamma\delta$-axis currents with their respective reference currents are shown in Figures 3 and 4. It can be seen that $\gamma\delta$-axis currents successfully follow the reference currents. The root mean square (RMS) of the steady-state errors for active and reactive currents $\tilde{I}_\gamma(t)$ and $\tilde{I}_\delta(t)$ are calculated to be 0.45 A and 0.06 A, respectively. Figure 5 shows the voltage of the DC-link capacitor and the reference voltage generated by the MPPT algorithm (the MPPT algorithm will not be studied in this work). It is clear that the DC-link voltage tracks its reference voltage with the RMS steady-state error, which is around 0.1 V. The controller output (Duty cycle) that is needed for the proposed controller/observer scheme for the PV single-stage, three-phase, grid-connected inverter is shown in Figure 6. The observer of the grid angle follows the angle generated from PLL as shown in Figure 7. The PLL algorithm is used here only for comparison and is never used in the control scheme.

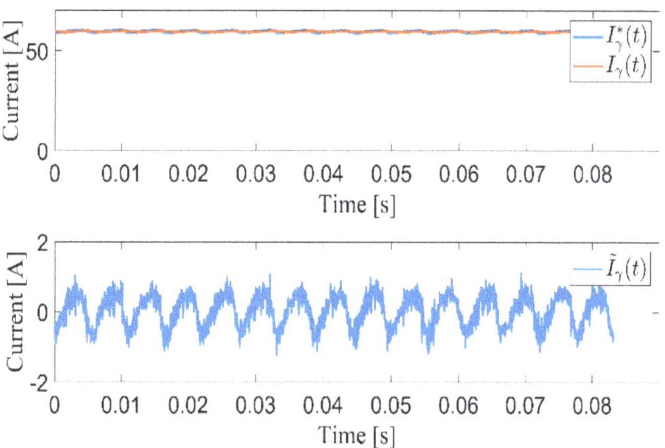

Figure 3. Injected grid current in γ-axis for steady-state operation.

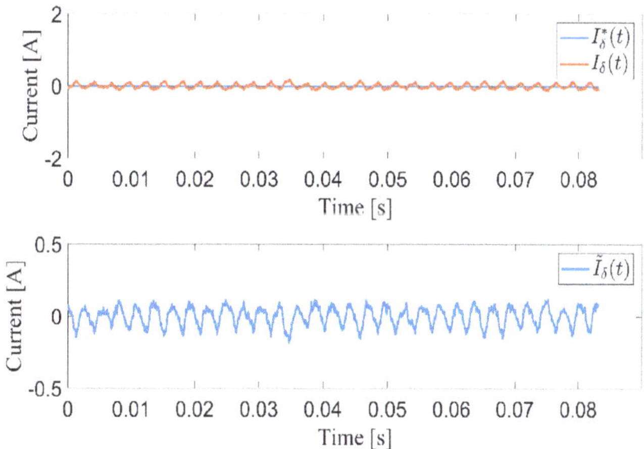

Figure 4. Injected grid current in δ-axis for steady-state operation.

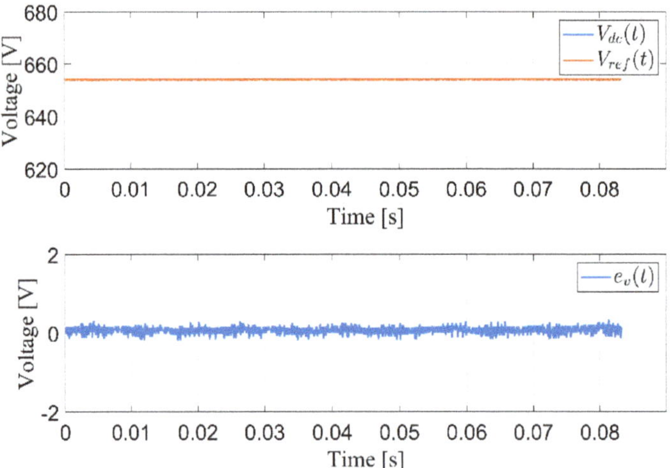

Figure 5. DC-link bus performance in steady state operation.

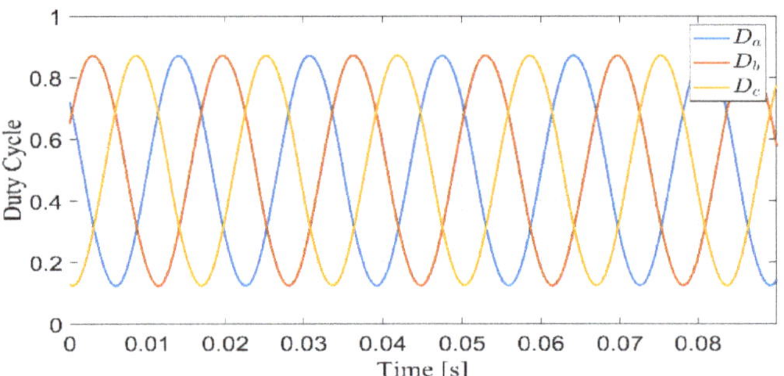

Figure 6. Duty cycle for the inverter.

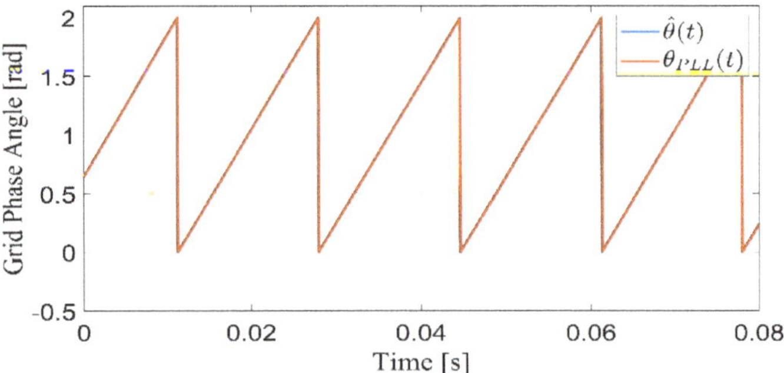

Figure 7. Performance of the proposed observer and PLL-based grid phase angle.

Under real-world conditions, the irradiation of the sun is continuously changing, which affects the output power of the PV panels. For this reason, the controller for such a system should be tested in the presence of irradiation changes. In this test, a sudden change in irradiation occurs at 3.6 s. from 1000 W/m^2 to 800 W/m^2. It can be seen that the γ-axis currents follow the reference currents smoothly and very quickly, as shown in Figure 8. Moreover, the DC-link voltage $V_{dc}(t)$ follows the reference voltage $V_{ref}(t)$, which is generated from the MPPT algorithm, as shown in Figure 9. It is clear from this figure that the voltage control works very well. Figure 10 shows the output active power changes based on the irradiation step change. It is clear that the active power is reduced according to irradiation drop with smooth performance. Figure 11 demonstrates the tracking performance of the proposed grid phase angle and frequency observer scheme. The grid frequency is changed from 60 Hz to 58 Hz and then back to 60 Hz. It is seen that the proposed observer estimates the grid frequency without any over or undershoot.

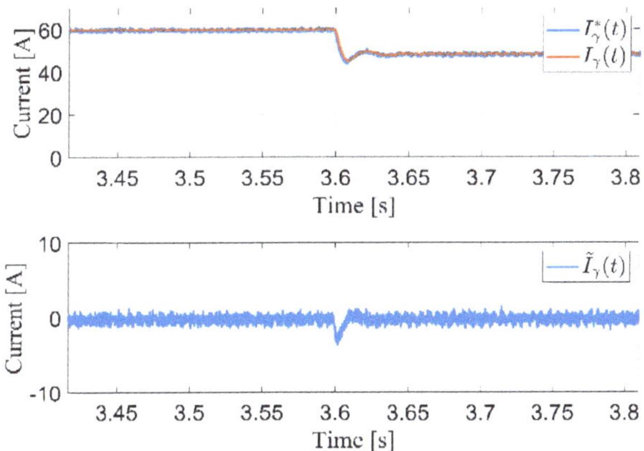

Figure 8. Current tracking performance for irradiation change from 1000 W/m^2 to 800 W/m^2.

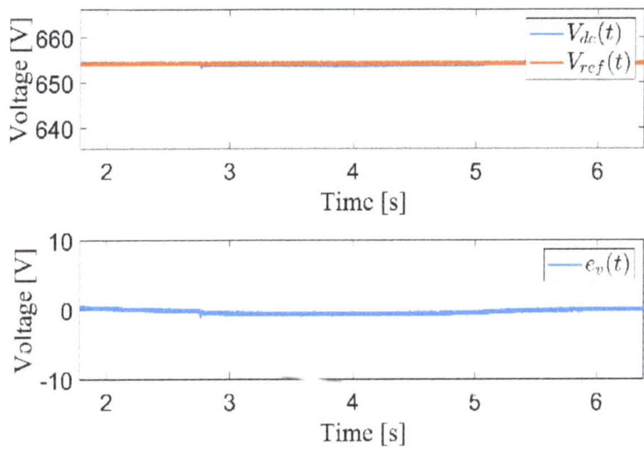

Figure 9. DC-link bus performance for irradiation changes from 1000 W/m^2 to 800 W/m^2.

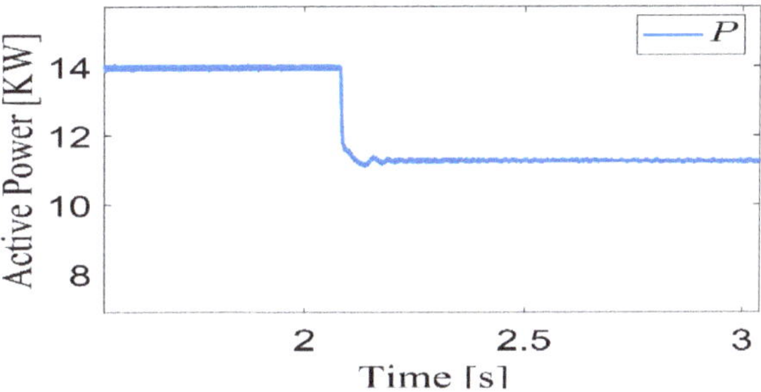

Figure 10. Active power performance during the irradiance changes.

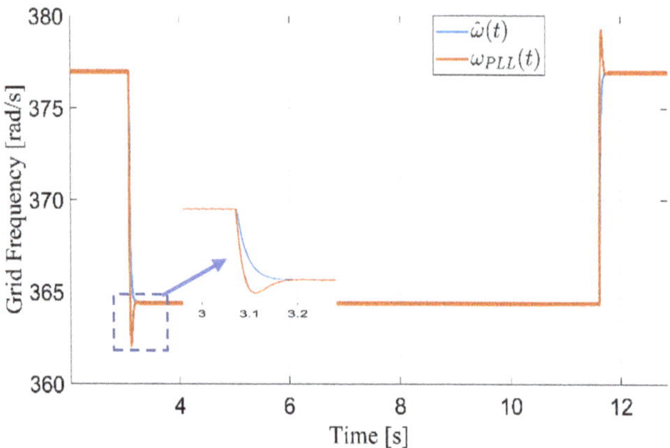

Figure 11. Tracking performance of the estimation of the grid frequency.

4.2. Proposed Controller Performance under Low- Voltage Ride-Through (LVRT) Scenario

Generally, LVRT is a common disturbance in the utility grid voltage which creates a severe operating condition for PV grid-connected inverters. An LVRT scenario has been employed to test the robustness of our proposed controller/observer scheme against this type of grid disturbance. This scenario is emulated by dropping the magnitude of the grid voltage from 110 V_{rms} to 80 V_{rms} for 150 ms. Figure 12 shows the active power injected into the grid during this kind of disturbance along with the grid voltage profile. As can be seen, the controller maintains constant power output to the grid. Since the grid voltage decreases, the only way to keep the power constant is to increase the current supplied from the PV system. In this case, the current limiter is added to the circuit in order to keep the current within the PV panel-rated current if the voltage drops to severe levels. Figure 13 shows the tracking performance of the injected current in this operating point.

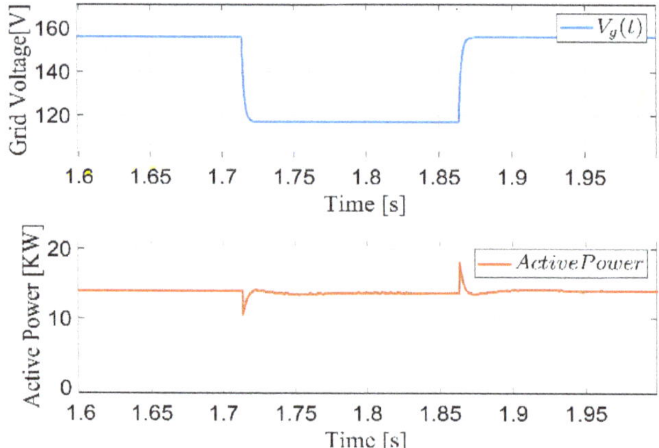

Figure 12. System performance due to LVRT scenario (Voltage drop from 110 V_{rms} to 80 V_{rms}).

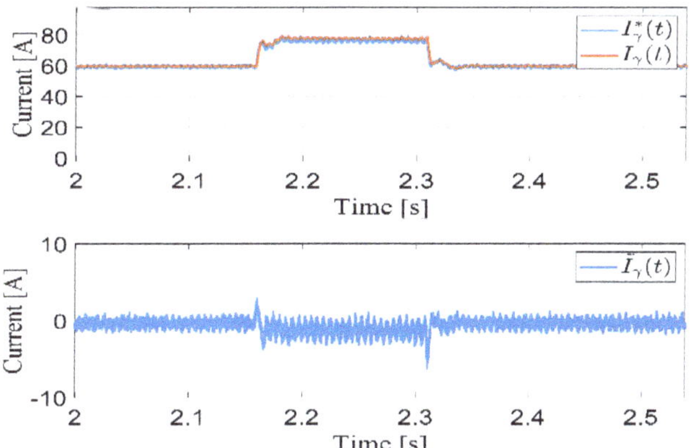

Figure 13. Current tracking performance corresponding to LVRT condition.

4.3. Proposed Controller Performance in the Presence of Harmonic Rich Grid

Harmonics are very common in grid voltage. The effect of these harmonics on renewable based power systems must be tested. In this scenario, the gird voltage with a Total Harmonics Distortion (THD) of 16% is used in order to demonstrate the capability of the proposed controller to harvest the maximum power from the PV panels and to inject power into the grid. Figure 14 shows the active power that is injected into the grid. Figure 15 shows the controller performance—in this case, in terms of the error signals for the injected currents and DC-Link voltage. Moreover, Figure 16 shows the proposed observer performance for the phase angle and frequency of the grid voltage.

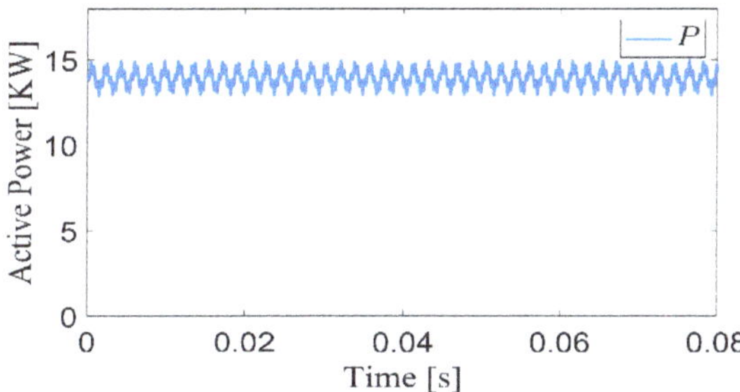

Figure 14. Active power injected into the grid under distorted grid.

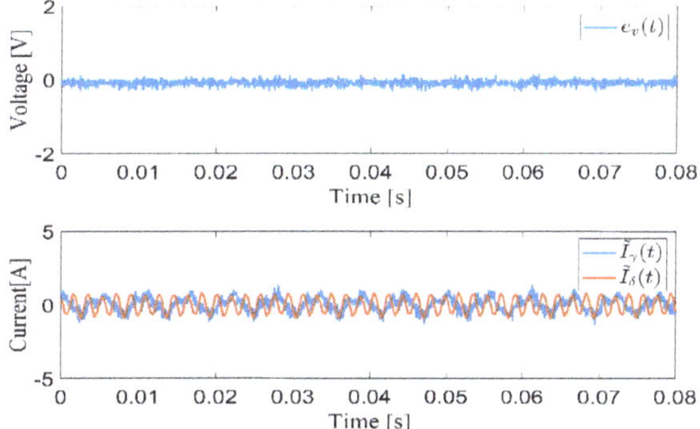

Figure 15. Steady-state error signals under distorted grid conditions.

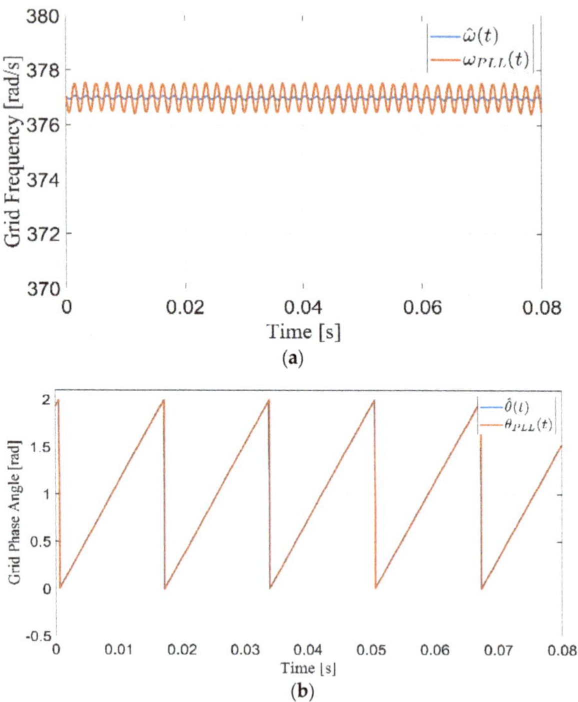

Figure 16. Steady-state observer performance under distorted grid (**a**) grid frequency [rad/s] (**b**) grid phase angle [rad].

4.4. Proposed Controller Performance under Reactive Power Injection Condition

The proposed controller has been designed to handle both active power harvesting and reactive power injection operating conditions. In this test, the reactive power injection condition is applied by increasing the q-axis reference current to inject 7.5 KVAR into the grid. Figure 17 shows the performance during the injection condition. It is seen that the reactive power is injected with a small percentage overshoot of less than 5%. Moreover, Figure 17 shows the power factor at the grid side before and after reactive power injection. The power factor is unity before the injection instant and is calculated to be 0.86 after the injection of reactive power since the phase angle between the voltage and current is around 30° .

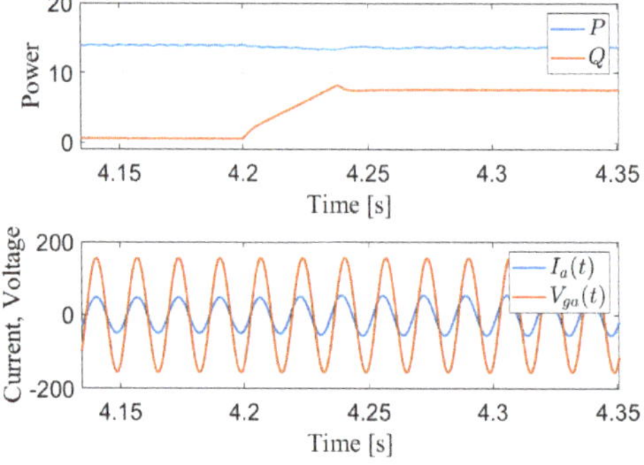

Figure 17. Active and reactive power performance during reactive power injection condition and power factor change due to the injection.

5. Conclusions

A current and voltage control scheme for a single-stage, PV grid-connected, three-phase inverter has been proposed. The proposed controller/observer scheme achieves the grid current and DC-link voltage control objectives without the knowledge of the grid information and without the need for a cascaded control scheme. Moreover, self-synchronization between the PV inverter and the grid has been guaranteed without using a PLL scheme or a synchronization unit. The proposed scheme adaptively estimated the grid phase angle and frequency. The controller and estimation schemes are validated by a Lyapunov stability analysis, and the closed loop stability has been proven. From the hardware-in-the loop experimental results, it is clear that the proposed scheme has acceptable performance for both transient and steady-state operations. Many other real-world test cases have also been evaluated. In all cases, the proposed scheme demonstrated acceptable operations in the presence of missing system information.

Author Contributions: Conceptualization, M.L.M. and M.A.; methodology, M.L.M. and M.A.; software, M.A.; validation, M.A.; formal analysis, M.A.; investigation, M.A. and M.L.M.; resources, M.L.M.; data curation, M.A.; writing—original draft preparation, M.A.; writing—review and editing, M.L.M.; visualization, M.A.; supervision, M.L.M.; project administration, M.L.M.; funding acquisition, M.L.M. All authors have read and agreed to the published version of the manuscript.

Funding: This research received no external funding.

Institutional Review Board Statement: Not applicable.

Informed Consent Statement: Not applicable.

Data Availability Statement: Not applicable.

Conflicts of Interest: The authors declare no conflict of interest.

References

1. Rashid, M.H. *Alternative Energy in Power Electronics*; Butterworth-Heinemann: Oxford, UK, 2014; pp. 1–363.
2. Kouro, S.; Leon, J.I.; Vinnikov, D.; Franquelo, L.G. Grid-Connected Photovoltaic Systems: An Overview of Recent Research and Emerging PV Converter Technology. *IEEE Ind. Electron. Mag.* **2015**, *9*, 47–61. [CrossRef]
3. Hernández-Callejo, L.; Gallardo-Saavedra, S.; Alonso-Gómez, V. A review of photovoltaic systems: Design, operation and maintenance. *Solar Energy* **2019**, *188*, 426–440. [CrossRef]
4. Rahman, S.; Saha, S.; Islam, S.N.; Arif, M.T.; Mosadeghy, M.; Haque, M.E.; Oo, A.M.T. Analysis of Power Grid Voltage Stability With High Penetration of Solar PV Systems. *IEEE Trans. Ind. Appl.* **2021**, *57*, 2245–2257. [CrossRef]
5. You, S.; Kou, G.; Liu, Y.; Zhang, X.; Cui, Y.; Till, M.J.; Yao, W.; Liu, Y. Impact of High PV Penetration on the Inter-Area Oscillations in the U.S. Eastern Interconnection. *IEEE Access* **2017**, *5*, 4361–4369. [CrossRef]
6. Joshi, J.; Swami, A.K.; Jately, V.; Azzopardi, B. A Comprehensive Review of Control Strategies to Overcome Challenges During LVRT in PV Systems. *IEEE Access* **2021**, *9*, 121804–121834. [CrossRef]
7. Singhal, A.; Ajjarapu, V.; Fuller, J.; Hansen, J. Real-Time Local Volt/Var Control Under External Disturbances With High PV Penetration. *IEEE Trans. Smart Grid* **2019**, *10*, 3849–3859. [CrossRef]
8. Romero-Cadaval, E.; Francois, B.; Malinowski, M.; Zhong, Q. Grid-Connected Photovoltaic Plants: An Alternative Energy Source, Replacing Conventional Sources. *IEEE Ind. Electron. Mag.* **2015**, *9*, 18–32. [CrossRef]
9. Basu, T.S.; Maiti, S. A Hybrid Modular Multilevel Converter for Solar Power Integration. *IEEE Trans. Ind. Appl.* **2019**, *55*, 5166–5177. [CrossRef]
10. Chen, L.; Amirahmadi, A.; Zhang, Q.; Kutkut, N.; Batarseh, I. Design and Implementation of Three-Phase Two-Stage Grid-Connected Module Integrated Converter. *IEEE Trans. Power Electron.* **2014**, *29*, 3881–3892. [CrossRef]
11. Dhlamini, N.; Chowdhury, S.P.D. Solar Photovoltaic Generation and its Integration Impact on the Existing Power Grid. In Proceedings of the 2018 IEEE PES/IAS PowerAfrica, Cape Town, South Africa, 28–29 June 2018; pp. 710–715.
12. Golestan, S.; Guerrero, J.M.; Musavi, F.; Vasquez, J.C. Single-Phase Frequency-Locked Loops: A Comprehensive Review. *IEEE Trans. Power Electron.* **2019**, *34*, 11791–11812. [CrossRef]
13. Zou, Z.; Liserre, M. Modeling Phase-Locked Loop-Based Synchronization in Grid-Interfaced Converters. *IEEE Trans. Energy Convers.* **2020**, *35*, 394–404. [CrossRef]
14. Hui, N.; Feng, Y.; Han, X. Design of a High Performance Phase-Locked Loop With DC Offset Rejection Capability Under Adverse Grid Condition. *IEEE Access* **2020**, *8*, 6827–6838. [CrossRef]
15. Huang, L.; Xin, H.; Li, Z.; Ju, P.; Yuan, H.; Lan, Z.; Wang, Z. Grid-Synchronization Stability Analysis and Loop Shaping for PLL-Based Power Converters With Different Reactive Power Control. *IEEE Trans. Smart Grid* **2020**, *11*, 501–516. [CrossRef]

16. Sun, Y.; de Jong, E.C.W.; Wang, X.; Yang, D.; Blaabjerg, F.; Cuk, V.; Cobben, J.F.G. The Impact of PLL Dynamics on the Low Inertia Power Grid: A Case Study of Bonaire Island Power System. *Energies* **2019**, *12*, 1259. Available online: https://www.mdpi.com/1996-1073/12/7/1259 (accessed on 2 April 2019). [CrossRef]
17. Šimek, P.; Valouch, V. Cascaded Delayed Signal Cancellation Based Pre-Filtering Technique to Improve Frequency Locked Loop for Grid Synchronization. In Proceedings of the 2019 International Conference on Electrical Drives & Power Electronics (EDPE), The High Tatras, Slovakia, 24–26 September 2019; pp. 391–396.
18. Hu, W.; Wu, Y.; Shen, Y.; Yang, F.; Quan, X.; Deng, F.; Zou, Z. One-Step-Prediction Discrete Observer Based Frequency-Locked-Loop Technique for Three-Phase System. *IEEE Access* **2021**, *9*, 95401–95411. [CrossRef]
19. Zhong, Q.; Nguyen, P.; Ma, Z.; Sheng, W. Self-Synchronized Synchronverters: Inverters Without a Dedicated Synchronization Unit. *IEEE Trans. Power Electron.* **2014**, *29*, 617–630. [CrossRef]
20. Perenyi, C.; Alqatamin, M.; Harzig, T.; McIntyre, M.; Grainger, B.M. System Frequency Dynamic Response of a Novel, Self-Synchronizing Inverter in a High Renewable Penetration Grid. In Proceedings of the 2020 22nd European Conference on Power Electronics and Applications (EPE'20 ECCE Europe), Lyon, France, 7–11 September 2020; pp. 1–10.
21. Konstantopoulos, G.C.; Zhong, Q.; Ming, W. PLL-Less Nonlinear Current-Limiting Controller for Single-Phase Grid-Tied Inverters: Design, Stability Analysis, and Operation Under Grid Faults. *IEEE Trans. Ind. Electron.* **2016**, *63*, 5582–5591. [CrossRef]
22. Guo, X.; Liu, W.; Zhang, X.; Sun, X.; Lu, Z.; Guerrero, J.M. Flexible Control Strategy for Grid-Connected Inverter Under Unbalanced Grid Faults Without PLL. *IEEE Trans. Power Electron.* **2015**, *30*, 1773–1778. [CrossRef]
23. Ma, J.; Song, W.; Jiao, S.; Zhao, J.; Feng, X. Power Calculation for Direct Power Control of Single-Phase Three-Level Rectifiers Without Phase-Locked Loop. *IEEE Trans. Ind. Electron.* **2016**, *63*, 2871–2882. [CrossRef]
24. Alqatamin, M.; Latham, J.; Smith, Z.T.; Grainger, B.M.; McIntyre, M.L. Current Control of a Three-Phase, Grid-Connected Inverter in the Presence of Unknown Grid Parameters Without a Phase-Locked Loop. *IEEE J. Emerg. Sel. Top. Power Electron.* **2021**, *9*, 3127–3136. [CrossRef]
25. Tsang, K.M.; Chan, W.L.; Tang, X. PLL-less single stage grid-connected photovoltaic inverter with rapid maximum power point tracking. *Solar Energy* **2013**, *97*, 285–292. [CrossRef]
26. Alqatamin, M.; Bhagwat, B.; Hawkins, N.; Latham, J.; McIntyre, M.L. Self-Synchronizing Current Control for Single-Stage Three-Phase Grid-Connected Photovoltaic Systems. In Proceedings of the 2020 American Control Conference (ACC), Denver, CO, USA, 1–3 July 2020; pp. 192–197.
27. Krause, P.; Wasynczuk, O.; Sudhoff, S.D.; Pekarek, S. Reference-Frame Theory. In *Analysis of Electric Machinery and Drive Systems*; Wiley: Hoboken, NJ, USA, 2013; pp. 86–120. [CrossRef]
28. Goli, C.S.; Manjrekar, M.; Sahu, P.; Chanda, A.; Essakiappan, S. Implementation of Stationary and Synchronous Frame Current Regulators for Grid Tied Inverter using Typhoon Hardware in Loop System. In Proceedings of the 2021 IEEE 12th International Symposium on Power Electronics for Distributed Generation Systems (PEDG), Chicago, IL, USA, 28 June–1 July 2021; pp. 1–8.
29. He, X.; Geng, H. PLL Synchronization Stability of Grid-Connected Multi-Converter Systems. *IEEE Trans. Ind. Appl.* **2021**, *58*, 830–842. [CrossRef]
30. Achlerkar, P.; Panigrahi, B.K. Dynamic Harmonic Domain Modeling and Stability Augmented Design of Inverter Interface to Weak and Unbalanced Grid. *IEEE Trans. Power Deliv.* **2021**, 1. [CrossRef]
31. Gautam, G.; Poddar, S. Real Time Simulation of 3-φ Grid-Connected Converter with Real and Reactive Power Control Under Different Grid Fault Conditions. In Proceedings of the Advances in Smart Grid Automation and Industry 4.0, Singapore, 29 May 2021; Reddy, M.J.B., Mohanta, D.K., Kumar, D., Ghosh, D., Eds.; Springer: Singapore, 2021; pp. 601–612.

Article

Ultra-Short-Term Load Dynamic Forecasting Method Considering Abnormal Data Reconstruction Based on Model Incremental Training

Guangyu Chen [1,*], Yijie Wu [1], Li Yang [2], Ke Xu [1], Gang Lin [3], Yangfei Zhang [1] and Yuzhuo Zhang [1]

[1] School of Electric Power Engineering, Nanjing Institute of Technology, Nanjing 211167, China
[2] State Grid Fujian Electric Power Company Limited, Fuzhou 350001, China
[3] State Grid Fujian Electric Power Company Quanzhou Power Supply Company, Quanzhou 362000, China
* Correspondence: cgyhhu@163.com

Abstract: In order to reduce the influence of abnormal data on load forecasting effects and further improve the training efficiency of forecasting models when adding new samples to historical data set, an ultra-short-term load dynamic forecasting method considering abnormal data reconstruction based on model incremental training is proposed in this paper. Firstly, aiming at the abnormal data in ultra-short-term load forecasting, a load abnormal data processing method based on isolation forests and conditional adversarial generative network (IF-CGAN) is proposed. The isolation forest algorithm is used to accurately eliminate the abnormal data points, and a conditional generative adversarial network (CGAN) is constructed to interpolate the abnormal points. The load-influencing factors are taken as the condition constraints of the CGAN, and the weighted loss function is introduced to improve the reconstruction accuracy of abnormal data. Secondly, aiming at the problem of low model training efficiency caused by the new samples in the historical data set, a model incremental training method based on a bidirectional long short-term memory network (Bi-LSTM) is proposed. The historical data are used to train the Bi-LSTM, and the transfer learning is introduced to process the incremental data set to realize the adaptive and rapid adjustment of the model weight and improve the model training efficiency. Finally, the real power grid load data of a region in eastern China are used for simulation analysis. The calculation results show that the proposed method can reconstruct the abnormal data more accurately and improve the accuracy and efficiency of ultra-short-term load forecasting.

Keywords: ultra-short-term load forecasting; abnormal data reconstruction; isolation forests; conditional generation adversarial network; bi-directional long short-term memory network; transfer learning

Citation: Chen, G.; Wu, Y.; Yang, L.; Xu, K.; Lin, G.; Zhang, Y.; Zhang, Y. Ultra-Short-Term Load Dynamic Forecasting Method Considering Abnormal Data Reconstruction Based on Model Incremental Training. *Energies* **2022**, *15*, 7353. https://doi.org/10.3390/en15197353

Academic Editor: Andrzej Bielecki

Received: 30 August 2022
Accepted: 2 October 2022
Published: 6 October 2022

1. Introduction

Accurate load forecasting provides a basis for power system construction planning, dispatching the decision making and production planning of power generation enterprises [1]. In recent years, with the continuous increase in the scale of new energy grid connection and the increasing popularity of electric vehicles, the power load presents volatility, nonlinearity and randomness, and the difficulty of power grid regulation increases, which puts forward higher requirements for the accuracy of load forecasting.

At present, load forecasting mainly includes statistical analysis methods and artificial intelligence methods [2,3]. Statistical analysis methods mainly include multiple linear regression modeling, autoregressive summation moving average and exponential smoothing [4,5]. Statistical analysis methods are mainly used to deal with linear and stable load data, ignoring the influence of climate, date type and other factors on load prediction, and the accuracy of load prediction is poor.

With the rise of artificial intelligence, machine learning and deep learning have been widely used in power grid fault identification and load forecasting because of their strong nonlinear fitting ability [6,7]. In [8], based on the decision tree classifier (DTC), an enhanced DTC is designed to obtain better prediction accuracy, but the machine learning algorithm often ignores the time-series dependence, and the prediction effect of long-time series is not as good as that of the deep learning algorithm. The deep learning algorithm has unique advantages in the field of load forecasting with its strong time series learning ability. An LSTM network and a GRU network improved with a recurrent neural network can effectively deal with long, high-dimensional time series [9]. Reference [10] proposed convolutional neural networks (CNN) and long short-term memory network (LSTM) fusion network model, which significantly improved the prediction efficiency and accuracy of individual household electric load by using the powerful feature extraction ability of CNN. Reference [11] constructed a load-forecasting model based on GRU network that effectively improved the forecasting accuracy compared with a LSTM network. Reference [12] proposed an integrated load forecasting model based on CNN and LSTM that significantly improved the forecasting efficiency and accuracy of multidimensional characteristic loads. In order to improve the prediction accuracy of power loads in different time ranges, [13] proposed a fusion model of a long-short-term memory network and neural prophet (LSTM-NP). Simulation results show that: compared with traditional prediction methods, LSTM-NP improves the prediction accuracy of three different types of load forecasting. The research in the above literature has achieved good prediction accuracy when dealing with high-dimensional long time series, but most of them are based on the premise of accurate load history data, and the impact of abnormal data on prediction accuracy is not fully considered. Abnormal data destroy the original distribution of the data set and also cause insufficient data redundancy, which hinders the improvement of load forecasting accuracy. Therefore, it is necessary to process the abnormal data and restore the initial data distribution of the historical data set.

Abnormal data processing methods mainly include deletion and filling [14,15]. The deletion method is to directly delete the abnormal data and their associated data, which is simple and easy to perform. However, when the data are abnormal in a large area, it may lead to the loss of important information [16]. Filling methods can be divided into two categories: statistical methods and machine learning methods [17–19]. Statistical methods mainly include mean filling, nearest distance filling, and regression filling. The filling results obtained by these methods are relatively stable, but they are easily affected by other types of data, and the filling accuracy is poor. Machine learning mainly includes K-nearest neighbor filling, missing forest, and K-means clustering filling. Reference [20] introduced linear interpolation, matrix combination, and matrix transfer to improve the random forest; the simulation results show that the improved random forest algorithm has high filling accuracy when filling the power missing data, but the machine learning easily ignores the timing information, and the filling accuracy is low.

The generative adversarial network (GAN) is an unsupervised generative learning model that has been widely used in many areas such as image generation and data filling [21]. On the basis of GAN, reference [22] proposed self-attention based on time-series impaction networks, and it improved the filling accuracy of the missing data. In order to improve the reliability evaluation of transmission gears with insufficient data and imbalance, [23] proposed a conditional generative adversarial network-mean-covariance balancing labeling (CGAN-MBL) model: On the basis of constructing CGAN model, MBL was introduced to improve the authenticity of the CGAN-generated data. The above method could effectively reconstruct the missing data, but it did not fully consider the correlations between other features and missing values. Moreover, JS divergence and Wasserstein distance are mostly used as loss functions, and gradient disappearance is easy to occur in network training.

Most of the traditional modeling methods are based on fixed load data. When new samples are added to the historical data set, it is often necessary to remodel and train the new data set. With increasing amounts of training data, it not only consumes a great deal of training time but also leads to the disappearance of gradient and underfitting phenomenon, which affects the accuracy and efficiency of load forecasting. Therefore, in order to improve the accuracy and efficiency of ultra-short-term load forecasting under abnormal data, the following two problems need to be solved: (1) how to discover the hidden nonlinear relationship between abnormal data and other characteristic data, improve the authenticity of data reconstruction, and restore the integrity of load series; (2) when new samples are added to the load history data set, how to realize the incremental training of the model and improve the efficiency of ultra-short-term load forecasting.

Based on the in-depth analysis of the above literature, this paper proposes an ultra-short-term load dynamic forecasting method based on model incremental training and considering abnormal data reconstruction. Firstly, aiming at the abnormal data in ultra-short-term load forecasting, a load abnormal data processing method based on IF-CGAN is proposed. The isolation forest algorithm is used to accurately eliminate the abnormal data points, and the condition generation countermeasure network (CGAN) is constructed to interpolate the abnormal points. The load-influencing factors are taken as the condition constraints of CGAN, and the weighted loss function is introduced to improve the reconstruction accuracy of abnormal data. Secondly, aiming at the problem of low model training efficiency caused by the new samples in the historical data set, a model incremental training method based on Bi-LSTM is proposed. The historical data are used to train the Bi-LSTM, and transfer learning is introduced to process the incremental data set to realize the adaptive and rapid adjustment of the model weight and improve the model training efficiency. Finally, the simulation analysis is carried out with the real power grid load data in a certain area. The calculation results show that the proposed method can reconstruct the abnormal data more accurately and improve the accuracy and efficiency of ultra-short-term load forecasting.

2. Abnormal Data Reconstruction Method Based on IF-CGAN

The abnormal data in the load data set destroy the integrity of the time series and affect the prediction accuracy of the model. In this chapter, outliers are eliminated through isolation forest model, and then CGAN is constructed to interpolate the missing point data.

2.1. Isolation Forest

Isolation forest algorithms are widely used in the outlier detection of massive data [24]. Its principle is to randomly select a plane to divide the sample data space into two subspaces and then use the same method to divide the two subspaces many times until there is only one data stronghold in the subspace. This segmentation method is similar to the formation method of binary tree: The sample data space is the root of the tree, and the data points are the branches and leaves of the tree. Therefore, the path $L(y)$ from the branch and leaf point y to the root can be calculated to judge whether the data point y is outlier.

An isolation forest is a collection of multiple isolation trees. Suppose that the data space contains m data points, and the anomaly index of data point y can be expressed as:

$$S(y, m) = 2^{-\frac{E(L(y))}{C(m)}} \tag{1}$$

$$C(m) = 2\left[\ln(m - 1) + \xi - \frac{m - 1}{m}\right] \tag{2}$$

where $S(y,m)$ is the abnormal score of data point y, and $E(L(y))$ is the expected value of path $L(y)$ of y in multiple trees. $C(m)$ is the average path length of the isolation tree, ξ is Euler constant with a value of 0.577.

2.2. Conditional Generative Adversarial Network

GAN consists of a generator and a discriminator. The essence of the generator and discriminator is to learn the distribution of real data. The function of the generator is to approximate the real data as much as possible, and the function of the discriminator is to identify and generated data as much as possible. When the game between the generator and the discriminator reaches equilibrium, the output data of generator are infinitely close to the real value. CGAN is improved on the basis of GAN. It applies supervised learning to GAN. CGAN retains the game structure of GAN and adds conditional values to the inputs of generator and discriminator to speed up the convergence of the network. Its basic structure is shown in Figure 1. Circles represent neurons and colored lines represent connections between neurons.

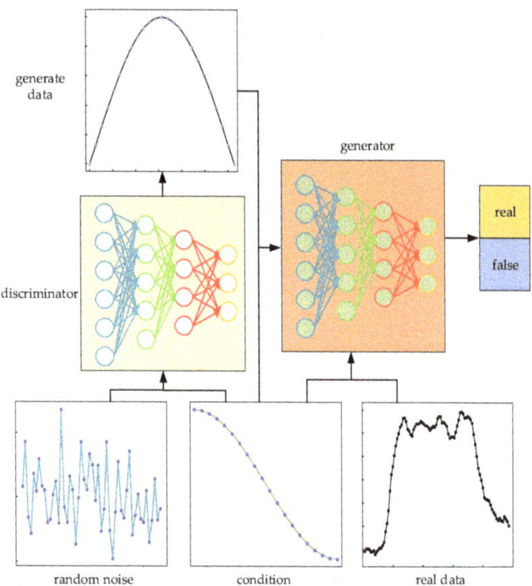

Figure 1. Basic structure of CGAN.

The random noise z and condition c are combined and input into the generator, and the generator outputs the generated sample $G(z|c)$. The input of the discriminator is the combination of the real value t of the load data and the condition c, and the combination of the generated sample $G(z|c)$ and the condition c. the discriminator needs to judge whether the distribution between the generated sample and the real sample is similar and whether the generated sample meets condition c. The generator and discriminator update the parameters according to the discrimination results. The loss functions of generator and discriminator in CGAN are:

$$L_G = -E_{(z,c)}[D(G(z|c)|c)] \tag{3}$$

$$L_D = -E_{(t,c)}[D(t|c)] + E_{(z,c)}[D(G(z|c)|c)] \tag{4}$$

where E represents the expected value of the corresponding distribution and $G(z|c)$ and $D(t|c)$ represent the output of the generator and discriminator respectively. The generator improves the authenticity of the generated sample $G(z|c)$ through continuous iteration, and the discriminator hopes to reduce the authenticity of the generated data and improve the accuracy of distinguishing the real sample. Therefore, CGAN gradually balances in the game between the two, and the objective function can be defined as:

$$\min_{G} \max_{D} L_{CGAN} = E_{(t,c)}[D(t|c)] - E_{(z,c)}[D(G(z|c)|c)] \tag{5}$$

It is easy for the gradient to disappear in the training of the original GAN. This is because when the generator generation effect is too excellent, the loss function is equivalent to JS divergence. When the generated data distribution does not overlap with the real data distribution, JS divergence is constant. Therefore, at the initial stage of model training and when the generator is too excellent, the loss function is constant, which makes network training difficult and convergence slow. Reference [25] proposed a WGAN-GP model, selected Wasserstein distance as the loss function of discriminator, and introduced gradient punishment mechanism to realize the Lipschitz constraint on the concentration area of true and false samples and their cross transition area. The objective function of CGAN is:

$$\min_{G} \max_{D} L_{CGAN} = E_{(t,c)}[D(t|c)] - E_{(z,c)}[D(G(z|c)|c)] + \lambda E[\|\nabla D(\ \sim)\| - 1]^2 \quad (6)$$

where λ is the gradient penalty coefficient.

Smooth $L1$ loss function has strong robustness and stability in the solution process. It is widely used in neural network training, which can effectively avoid gradient explosion and improve the convergence speed of the network. Smooth $L1$ loss function can be defined as:

$$L_{SL1} = \begin{cases} 0.5 \times (t - G(z|c))^2, |t - G(z|c)| < 1 \\ |t - G(z|c)| - 0.5, |t - G(z|c)| \geq 1 \end{cases} \quad (7)$$

Cosine similarity measures the difference between two vectors by calculating the cosine of the angle between vectors. Cosine similarity has high accuracy for vector similarity discrimination and can identify the trajectory change trend of vector [26]. In this paper, cosine similarity loss function is introduced to improve the authenticity of generated samples. Its loss function is defined as:

$$L_{\cos} = 1 - \frac{\sum\limits_{i=1}^{n} t_i G_i}{\sqrt{\sum\limits_{i=1}^{n} t_i^2} \sqrt{\sum\limits_{i=1}^{n} G_i^2}} \quad (8)$$

Therefore, the objective function of CGAN is as follows:

$$L = \min_{G} \max_{D} L_{CGAN} + \lambda_1 L_{SL1} + \lambda_2 L_{\cos} \quad (9)$$

where λ_1, λ_2 is the weighting coefficient of the corresponding loss function.

2.3. Abnormal Data Reconstruction Strategy Based on CGAN

In practical training, when GAN processes high-dimensional data, the convergence speed is slow, and it is prone to problems such as underfitting and gradient disappearance. Compared with fully connected network, convolutional neural network (CNN) has a simple structure with strong feature extraction ability and can process high-dimensional data sets efficiently. Therefore, CNN is used to construct discriminator and generator in this paper.

In order to ensure that the convolution kernel can effectively extract the features of the data, the structure design of CGAN is shown in Figure 2. The blue cuboid represents the output of the hidden layer.

The input of the generator is the 13th-order matrix of the combination of random noise and conditional value. A three-layer CNN is constructed to extract the features of the input matrix, and then the samples are generated from the output of the full connection layer. In order for the generator to efficiently learn the nonlinear relationship between the generated samples and the condition values, the ReLU activation function is used in the convolution layer, and the regularization processing is carried out between the convolution layers. The specific network parameters are shown in Table 1.

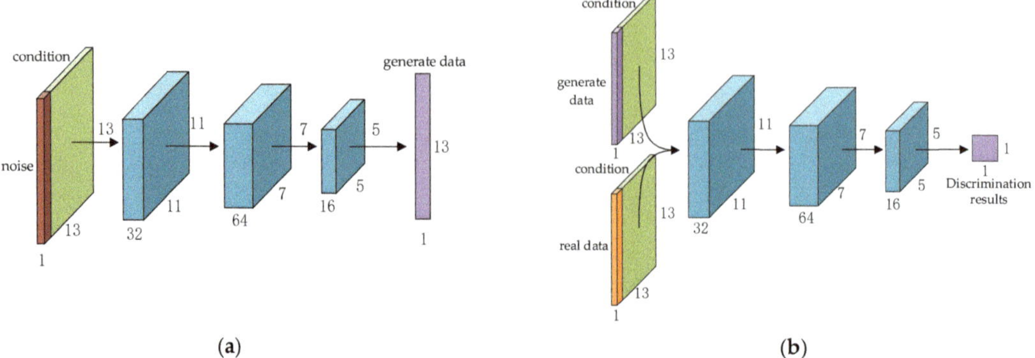

Figure 2. CGAN structure design: (**a**) Generator network structure; (**b**) Discriminator network structure.

Table 1. Generator network parameters.

Model	Parameter Name		Specification
CNN 1	2D convolution	convolution kernel	3×3
		number of filters	32
		step	1
	activation function		ReLU
	regularization		32
CNN 2	2D convolution	convolution kernel	5×5
		number of filters	64
		step	2
	activation function		ReLU
	regularization		64
CNN 3	2D convolution	convolution kernel	3×3
		number of filters	16
		step	1
	activation function		ReLU
	regularization		16
FC	output dimension		13×1

The condition value, the real sample, and the generated sample are combined as the two input matrices of the discriminator. The discriminator needs to extract and identify the features of the input matrix, and its model structure is basically consistent with that of the generator. Different from the generator, because the WGAN-GP model introduces the gradient punishment mechanism, there is no regularization between the convolution layers in the discriminator. The LeakyReLU function is selected as the activation function, and the full connection layer outputs the discrimination results of the input matrix. The specific network parameters are shown in Table 2.

Table 2. Discriminator network parameters.

Model	Parameter Name		Specification
CNN 1	2D convolution	convolution kernel	3×3
		number of filters	32
		step	1
	activation function		LeakyReLU
CNN 2	2D convolution	convolution kernel	5×5
		number of filters	64
		step	2
	activation function		LeakyReLU
CNN 3	2D convolution	convolution kernel	3×3
		number of filters	16
		step	1
	activation function		LeakyReLU
FC	output dimension		1×1

2.4. Abnormal Data Reconstruction Based on IF-CGAN

Figure 3 shows the abnormal data reconstruction process based on IF-CGAN.

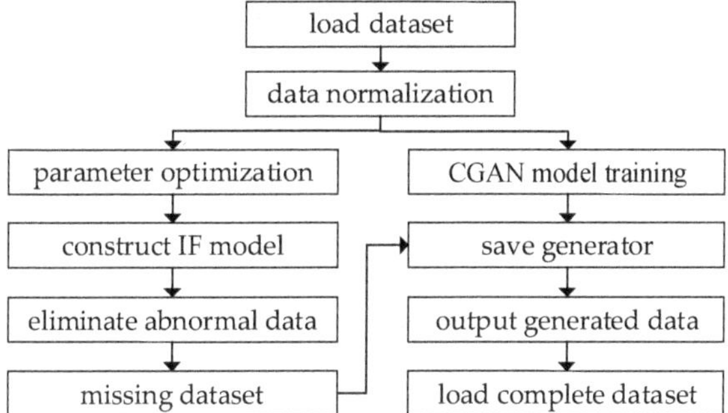

Figure 3. Abnormal data reconstruction process based on IF-CGAN.

3. Dynamic Forecasting of Ultra-Short-Term Load Based on Incremental Model Training

When the load data set changes, in order to obtain the latest load information, it is often necessary to re model and train the new data set, and the modeling method is inefficient. Based on the establishment of the Bi-LSTM model, this chapter introduces migration learning to process incremental data sets to realize the rapid adjustment of model weight and improve modeling efficiency.

3.1. Transfer Learning

Transfer learning [27] applies the knowledge learned from an old task to a different but related new task, avoiding learning new tasks from scratch and shortening the time of learning new tasks. There are two basic concepts of transfer learning: domain and task. The field with sufficient historical sample data is called the source field, and the field with limited data sample size is called the target field. The mathematical model can be expressed as:

$$d_s = \{x_s, y_s\} \tag{10}$$

$$d_t = \{x_t, y_t\} \tag{11}$$

where d_s and d_t respectively represent the source domain and target domain; x_s and y_s respectively represent the source domain samples and their corresponding labels; and x_t

and y_t respectively represent the target domain samples and their corresponding tags. The task mathematical model of the source domain and target domain is expressed as:

$$t_s = \{\, x_s, f_s(\,\sim)\}$$ (12)

$$t_t = \{\, x_t, f_t(\,\sim)\}$$ (13)

where t_s and t_t represent the tasks of source domain and target domain, respectively, and $f(\sim)$ represents the mapping relationship between domain data x and target value y. Migration learning is to migrate the source domain mapping relationship $f_s(\sim)$ to the target domain. When the data distribution of the source domain and the target domain is similar, the target domain model can be modified by fine tuning to obtain the target domain mapping relationship $f_t(\sim)$.

3.2. Bi-LSTM

When dealing with long time series, RNN has problems such as gradient explosion and disappearance. To solve such problems, Hochreiter proposed a LSTM network. A LSTM network controls the preservation and loss of time-series information through a forgetting gate, input gate, and output gate, so as to effectively avoid the disappearance of gradient caused by long series. The structure of LSTM unit is shown in Figure 4.

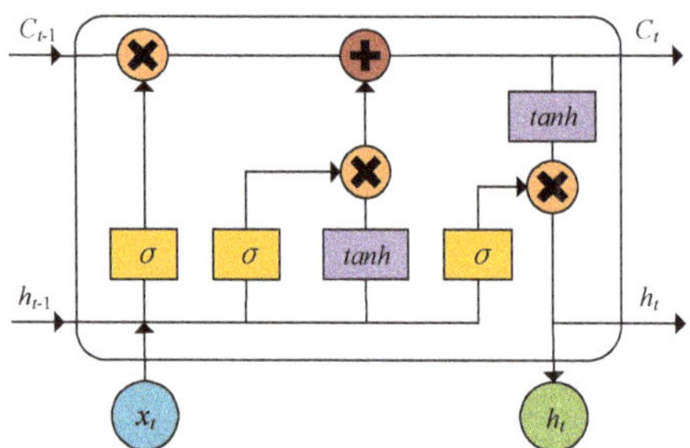

Figure 4. LSTM structure.

The forgetting gate controls the information flow of the previous time to ensure the backward transmission of effective information. The input gate updates the current data to the storage unit. The function of the output gate is to transmit the information of the storage unit to the next time. The calculation process is as follows:

$$f_t = \sigma\left(W_f x_t + U_f h_{t-1} + b_f\right)$$ (14)

$$i_t = \sigma(W_i x_t + U_i h_{t-1} + b_i)$$ (15)

$$\tilde{c}_t = Tanh(W_c x_t + U_c h_{t-1} + b_c)$$ (16)

$$c_t = f_t \odot c_{t-1} + i_t \odot \tilde{c}_t$$ (17)

$$o_t = \sigma(W_0 x_t + U_0 h_{t-1} + b_0)$$ (18)

$$h_t = o_t \odot tanh(c_t)$$ (19)

where W_f, U_f, b_f, W_i, U_i, b_i, W_o, U_o and b_0 are the parameters corresponding to the three gates respectively, W_c, U_c, b_c are the weight parameters corresponding to the cell state, x_t and h_t are the input of LSTM unit and the output of hidden layer at time t, which σ are sigmoid functions, and tanh represents hyperbolic tangent functions.

The Bi-LSTM model is composed of two LSTM networks with opposite directions and shared weights. The principle is shown in Figure 5. The blue circle represents the forward LSTM network output. The green circle represents the reverse LSTM network output. Bi-LSTM can learn load time series information from both positive and negative directions, integrate past and future load information to update weight parameters, and improve the regression accuracy of long time series.

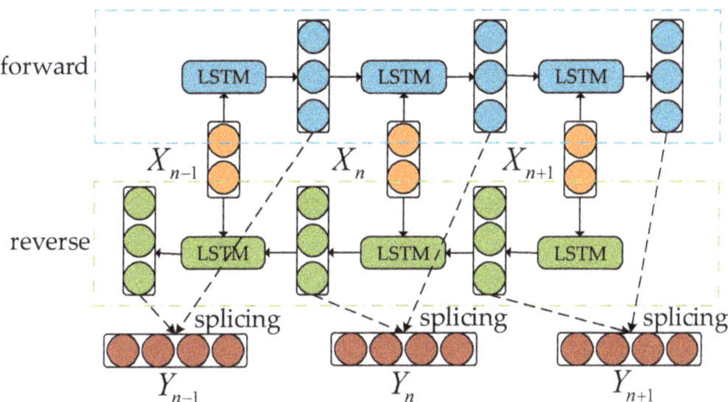

Figure 5. Bi-LSTM structure.

3.3. Incremental Model Training Method Based on Bi-LSTM

Based on the establishment of a load forecasting model, this paper introduces transfer learning to realize the rapid update of model weight when load data changes. The principle is shown in Figure 6.

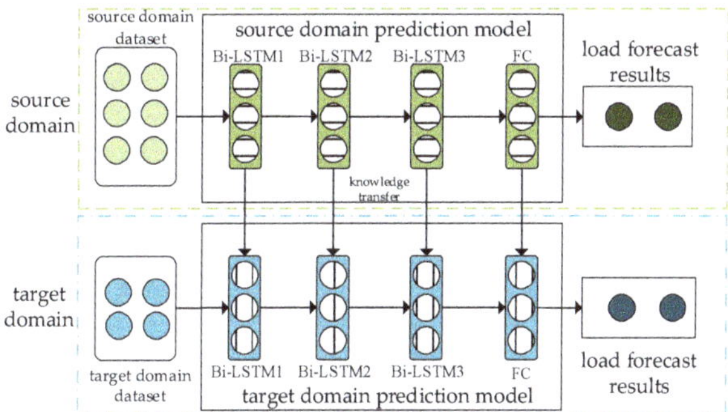

Figure 6. Transfer learning principle.

The maximum mean difference (MMD) is mainly used to measure the distance between two different samples [28]. Therefore, MMD is usually used in transfer learning to evaluate the correlation of data distribution between source domain and target domain. When MMD is small, the data distribution of source domain and target domain is very similar. Adjusting the model structure will reduce the fitting accuracy of the model. The source domain data studied in this paper are the collected historical load samples, and the target domain data are the new load samples. The data distribution difference between the source domain and the target domain is small, so there is no need to adjust the source domain model structure.

The source domain model can be divided into three parts according to its functionality. The shallow network extracts the detailed characteristics of the source domain data, the deep network extracts the overall data information, and the output layer outputs the prediction results. Taking the source domain model parameters as the initial parameters of the target domain model, fixing different Bi-LSTM network layers, and using the target domain data to update the parameters of other network layers, the network parameters can be updated efficiently.

The model incremental training process based on Bi-LSTM is shown in Figure 7, and the specific steps are as follows.

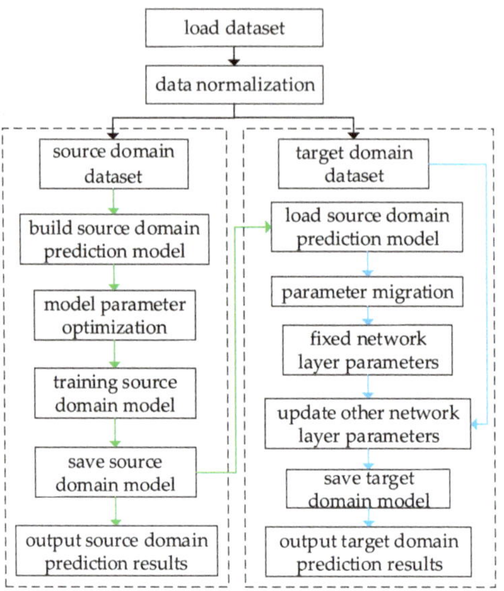

Figure 7. Bi-LSTM network incremental training flow chart.

Step 1: build the source domain model. The method proposed in Section 2.2 is used to establish the source domain model for the historical load data, and the three parameters of network layers, time step, and iteration times are selected to save the source domain model with the highest prediction accuracy.

Step 2: train the target domain model. Import the source domain model, migrate the source domain model parameters to the target domain, fix the different Bi-LSTM network layers, train other network layers with the target domain data, and save the target domain model with the highest prediction accuracy.

Step 3: output the prediction results. Load the target domain model, normalize the load influencing factors and reshape them into a three-dimensional matrix, input the target domain model, and then inverse normalize the output value to obtain the load forecasting value.

4. Ultra-Short-Term Load Dynamic Forecasting Method Considering Abnormal Data Reconstruction Based on Model Incremental Training

A high-quality load data set is the premise of establishing the prediction model. In this paper, isolation forest and conditional generation countermeasure network are used to deal with outliers to improve the quality of the data set. Secondly, the prediction model based on Bi-LSTM network is constructed, and migration learning is introduced to realize the incremental training of prediction model. The basic framework is shown in Figure 8.

(1)   Abnormal data reconstruction based on IF-CGAN.

Firstly, the historical load data are normalized by Equation (20), the isolation forest model is constructed, and the load outliers are screened and deleted. The conditional value of CGAN is the load influencing factor, mainly including time factor (time, rest day), climate factor (wind speed, air temperature, dew point temperature, air pressure, cloud amount) and load factor (the load value is engraved at the same time in the first two days and the load is engraved at the same time in the next two days). The random noise and the conditional value are spliced horizontally and input into the generator. The convolution kernel is used to extract the features and output the generated samples. The generated sample, real load, and condition value are spliced horizontally and input into the discriminator for true and false discrimination. In the process of game between generator and discriminator, the nonlinear relationship between load and influencing factors is excavated. After the game between generator and discriminator reaches balance, the generator is saved and the missing load data are filled:

$$a_n = \frac{a - a_{\min}}{a_{\max} - a_{\min}} \tag{20}$$

where a_n is the result of data normalization, $a_{\max}$ and $a_{\min}$ are the maximum and minimum values respectively, and a is the initial value of data.

(2) Construction of load forecasting model based on Bi-LSTM network.

After obtaining the complete historical load data, the data are normalized and reconstructed into three-dimensional matrix format of sequence length, time step, and number of sequence features. The load forecasting model is built based on a Bi-LSTM network. The optimizer selects Adam, the learning rate is set to 0.001, the loss function is MSE, and the optimal model is saved according to the prediction accuracy evaluation index.

(3) Network incremental training method based on Transfer Learning.

After obtaining the complete target domain data, load the source domain model, fix different network parameters, input the target domain data, train and update the remaining network layer weights, save the optimal target domain model, and output the target domain prediction results.

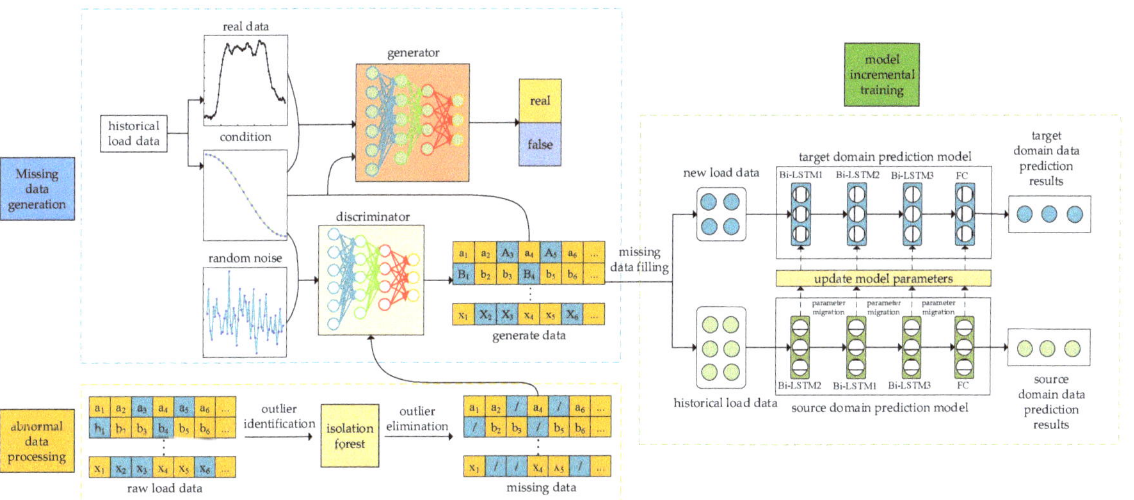

Figure 8. Ultra-short-term load dynamic forecasting framework considering abnormal data reconstruction based on model incremental training.

5. Example Analysis

In order to verify the effectiveness of the method proposed in this paper, the load data from November 2012 to August 2013 in a region of eastern China are used as an example. The sampling frequency of the load data is 15 min, the time span is 10 months, and a total of 28,277 load data are collected. Among them, the load information of 8 months from November 2012 to June 2013 is selected as the source domain data, the load information of July 2013 is selected as the target domain data, and the load information of August are the test set data.

5.1. Evaluation Index Construction

The reconstruction accuracy and R square are selected as the evaluation indexes of data reconstruction. The closer the reconstruction accuracy and R square are to 1, the higher the authenticity of model reconstruction. Root mean square error (RMSE) and mean absolute error (MAE) are used as the evaluation indexes of the prediction model. The smaller RMSE and MAE are, the closer the predicted value of the model is to the real value [13]. The calculation formula of evaluation index is as follows:

$$E_{R^2} = 1 - \frac{\sum\limits_{i=1}^{n} (z_g - z_t)^2}{\sum\limits_{i=1}^{n} (z_m - z_t)^2} \tag{21}$$

$$E_{acc} = \frac{1}{n} \sum\limits_{i=1}^{n} \left(1 - \frac{|z_t - z_g|}{z_t} \right) \times 100\% \tag{22}$$

$$E_{rmse} = \sqrt{\frac{1}{n} \sum\limits_{i=1}^{n} (z_t - z_p)^2} \tag{23}$$

$$E_{mae} = \frac{1}{n} \sum\limits_{i=1}^{n} |z_t - z_p| \tag{24}$$

where z_g, z_t and z_p are the generated value, real value and predicted value of load respectively, and z_m represents the average value of missing data.

5.2. Comparative Analysis of Abnormal Data Reconstruction Results

In the process of power grid data acquisition, the meter flies away, communication faults and other phenomena occur from time to time, and the obtained data sets easily contain missing and abnormal data. Missing values are randomly generated in the original load data set to simulate missing data caused by communication failures. The value of abnormal data ranges from 1.5 to 1.8 times of the real value, isolation forests are used to simulate the data anomalies caused by flight, and labels are set to verify the identification effect of abnormal data. The number of isolation trees is set to 100, and the containment parameter is set according to the proportion of outliers. An isolation forest model is established to detect outliers in the data set, and the detection results are compared with the real labels.

The detection results of abnormal data are shown in Figure 9. The blue and black circles represent the normal data and abnormal data of the load respectively, while the red circles represent the detected values of isolation forests. When the red circles coincide with the black circles, it means that isolation forests identify the outliers correctly. As can be seen from Figure 9, isolation forests can more accurately identify abnormal data caused by meter flight.

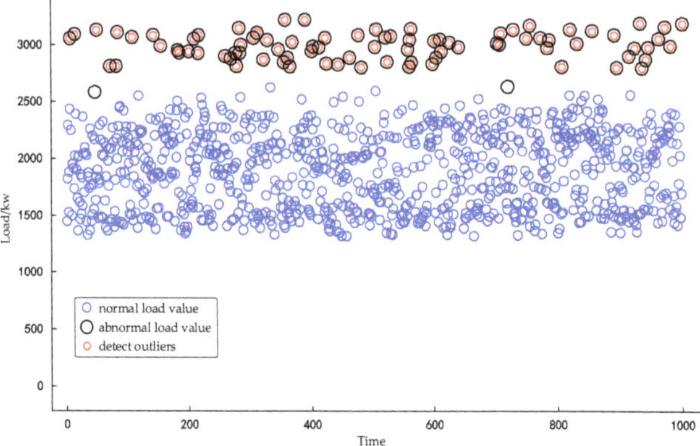

Figure 9. Abnormal data detection results.

The original load data were processed by isolated forest to form missing data sets with missing rates of 10%, 20%, 40%, and 60%, the gradient penalty coefficient λ and the weight of objective function λ_1, λ_2 are 40, 10 and 0.6 respectively. The complete data under different deletion rates are sent to the CGAN model for training. After the game between the generator and the discriminator reaches balance, the generated samples are output. The filling accuracy and R square are selected as the evaluation indexes of the generated samples, and compared with the mean interpolation method, KNN and random forest. The weight of KNN is set to "distance", K is set to 8, the number of decision trees of random forest is 50, and the maximum depth is 20.

Figure 10 shows the data-filling effect of different models with a deletion rate of 40%. It can be seen from the figure that CGAN fully excavates the nonlinear relationship between load and influencing factors and can reconstruct the abnormal data more accurately. The generated load samples are closest to the real value. The filling effect of random forest and KNN is poor when dealing with long time series.

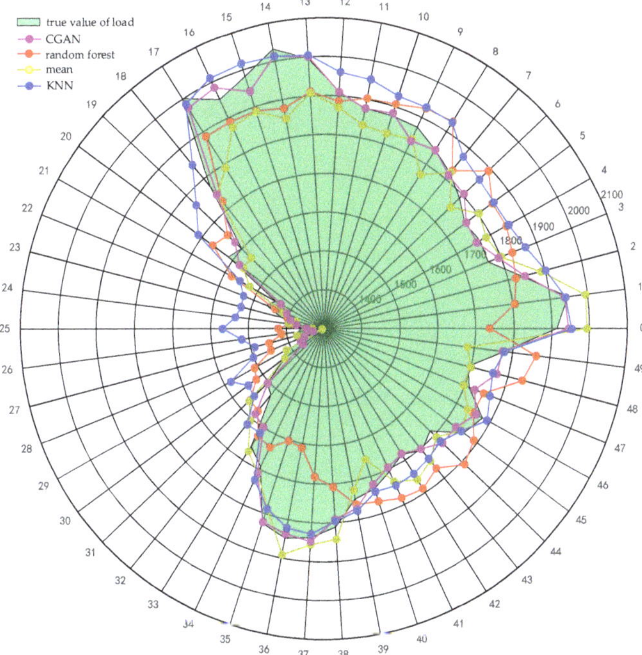

Figure 10. Reconstruction results with a deletion rate of 40%.

From the filling quantitative results of different models in Tables 3 and 4, it can be seen that the R square sum accuracy of reconstructed data of CGAN model under different deletion rates is higher than that of the other three models, which verifies the effectiveness of the reconstructed data in the CGAN model. In different loss rate data sets, the reconstruction index of mean interpolation method is the least ideal, the reconstruction effect of CGAN model is relatively stable, and the reconstruction accuracy of KNN and RF is similar, especially in high loss rate data, the reconstruction accuracy fluctuates greatly. Under different deletion rates, the reconstruction accuracy of the CGAN model constructed in this paper is improved by 8.1% and 3.5% compared with the other three models.

Table 3. *R* square comparison of reconstruction results of different models.

Deletion Rate	R Square			
	CGAN	KNN	RF	Mean
10%	0.9447	0.9255	0.9166	0.8931
20%	0.9248	0.9116	0.8869	0.8745
40%	0.8938	0.8525	0.8674	0.8695
60%	0.8785	0.8193	0.8123	0.8227

Table 4. Comparison of accuracy of reconstruction results of different models.

Deletion Rate	Accuracy/%			
	CGAN	KNN	RF	Mean
10%	97.00	95.84	94.96	94.53
20%	96.35	95.20	93.09	93.86
40%	96.02	94.76	93.47	93.57
60%	95.53	92.82	92.31	92.58

5.3. Analysis on the Influence of Data Processing Methods on Prediction Results

The data generated by the model in Section 5.2 are used to interpolate the load anomaly data set, and the Bi-LSTM network is constructed according to the parameter settings in Section 5.4. Then, the complete data set and the data set obtained by the deletion method are sent to the network to train the Bi-LSTM network to verify the effectiveness of the time series reconstruction of the model proposed in this paper.

After using the above five methods to process the data set with a deletion rate of 40%, the prediction results of the Bi-LSTM network are shown in Figure 11. It can be seen from the figure that the data set reconstructed by CGAN model achieves the best fitting effect, which verifies the effectiveness of the reconstruction timing of CGAN model. The deletion method destroys the integrity of the data sequence, and the prediction effect of the obtained data set is the worst. Mean interpolation, KNN, and random forest restore the integrity of the sequence to a certain extent. The overall prediction effect is better than the deletion method, but there is still a gap with the CGAN model. The quantitative results of network prediction error of the complete data set obtained under different deletion rates are shown in Tables 5 and 6.

Table 5. Influence of different processing methods on root mean square error of prediction.

Deletion Rate	RMSE/KW				
	CGAN	KNN	RF	Mean	Deletion
10%	43.48	48.35	64.41	46.11	44.87
20%	48.87	57.81	62.59	51.45	54.66
40%	58.31	64.58	71.92	65.27	68.31
60%	70.09	80.53	79.61	82.69	84.26

Table 6. Influence of different processing methods on mean absolute error of prediction.

Deletion Rate	MAE/KW				
	CGAN	KNN	RF	Mean	Deletion
10%	34.26	37.71	48.67	35.94	35.69
20%	38.33	45.78	46.44	40.26	42.79
40%	44.18	50.41	49.27	52.03	51.38
60%	51.31	55.08	53.91	57.65	60.81

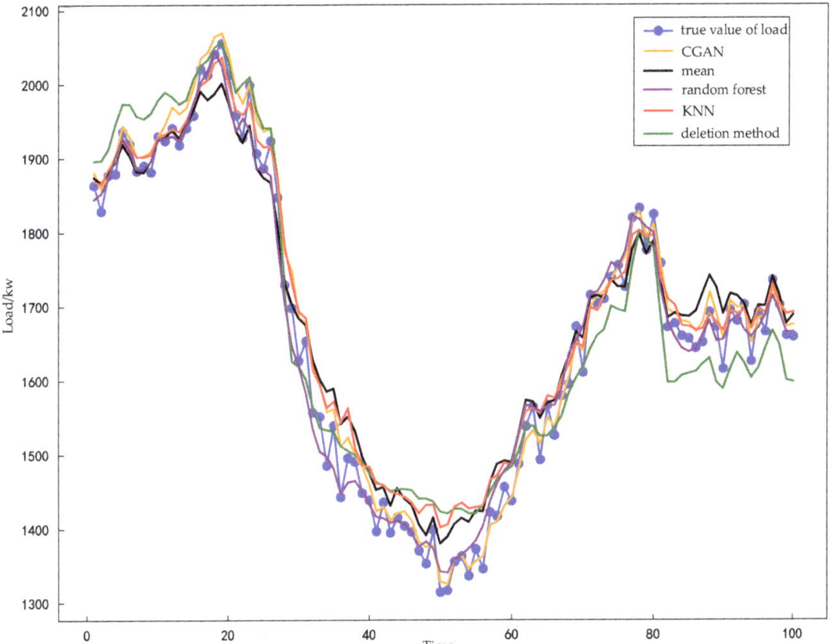

Figure 11. Prediction results under different data processing methods.

It can be seen from Tables 5 and 6 that for the load data set repaired by CGAN, the prediction error of Bi-LSTM network is the smallest, and the prediction accuracy is higher than that of the other interpolation models. When the deletion rate is 10%, the data set obtained by directly deleting abnormal data is the closest to CGAN; with the increase of data missing rate, the prediction accuracy of the data set obtained by deletion method and mean interpolation becomes clearly worse. On the data set reconstructed by CGAN, RMSE and MAE of Bi-LSTM network are reduced by at least 3.1% and 4.8% compared with the other four processing methods.

5.4. Comparative Analysis of Prediction Results of Source Domain Model

The Bi-LSTM network can effectively avoid the phenomenon of gradient disappearance when dealing with long time series. There are many Bi-LSTM network parameters, and different parameter settings have a great impact on the prediction effect of the network. In this paper, three parameters that have a great impact on the network are selected for research. The ultra-short-term load forecasting experiment is carried out on the data set processed by CGAN model. The source domain data are divided according to the ratio of 4:1. 80% of the data is used as the training set, and the other data is the test set. The prediction error of the test set with different parameter combinations is shown in Table 7.

Table 7. Comparison of prediction errors under different parameter combinations.

Epoch	Time Step	Batchsize	RMSE	MAE
200	10	16	91.87	63.20
200	10	32	67.13	49.65
200	10	64	91.60	68.23
200	15	32	48.87	38.33
200	20	32	216.46	159.6
100	15	32	88.56	63.04
300	15	32	170.6	102.4

As can be seen from Table 7, when the number of iterations, time step, and batch size of the Bi-LSTM network are set to 200, 15, and 32, respectively, the load information learned by the network is the most abundant and the prediction error is the smallest. If the number of iterations is increased or reduced, the network will appear over fitting or under fitting, resulting in the reduction of prediction accuracy.

The Bi-LSTM network is constructed according to the above parameters and compared with BP network, RNN, and SVR. The number of neurons in the hidden layer of BP network is set to 32, and the RNN parameter setting is consistent with that of the Bi-LSTM network. The kernel function of SVR model is set as Gaussian kernel function, and the penalty coefficient is 1.5. The load forecasting results of a certain day in the test set are shown in Figure 12.

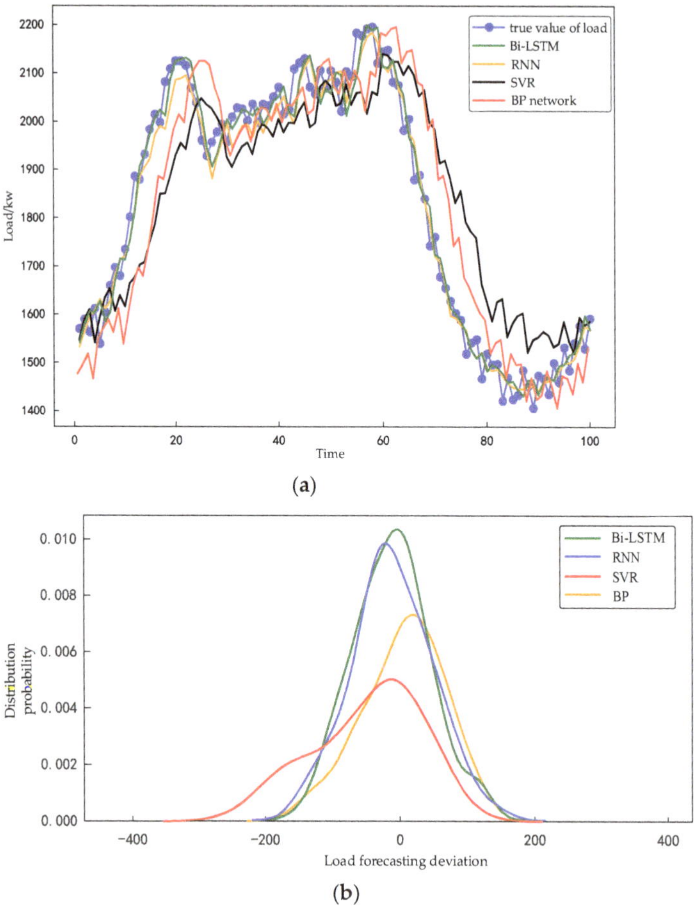

Figure 12. Comparison of prediction results of source domain: (**a**) Comparison of prediction results of different models; (**b**) Prediction error distribution curve of different models.

As can be seen from Figure 12a, compared with the other two models, the cyclic neural network can better fit the real value, but in terms of the details of the fitting curve and error curve, compared with RNN, the prediction result from the Bi-LSTM network is closer to the real value. Figure 12b shows the prediction error distribution curves of different models. It can be seen from the figure that the error of Bi-LSTM network is concentrated around 0, and the error fluctuation is smaller than that of other models.

Table 8 shows the comparison of source domain prediction errors of different models. It can be seen from the table that the prediction accuracy of Bi-LSTM model is better than that of other models in terms of RMSE and MAE. Compared with RNN, after setting the same parameters, RMSE and MAE decreased by 3.67% and 7.43% respectively; Compared with BP network and SVR, RMSE decreased by 46.52% and 22.21% respectively, and MAE decreased by 47.82% and 18.87% respectively.

Table 8. Source domain prediction error comparison.

Model	RMSE	MAE
Bi-LSTM	48.87	38.33
RNN	52.71	41.41
SVR	91.33	73.46
BP network	62.78	47.24

5.5. Comparative Analysis of Prediction Results of Target Domain Model

When the load data set changes, in order to improve the prediction accuracy of the model, it is necessary to extract the load information of the latest data set. However, in order to extract the information of the latest data, it is often necessary to retrain the model on the new data set, which is not only time-consuming, but also with the increase in the amount of data, the gradient of the model will disappear, underfitting and other problems will appear. In contrast, migration learning can quickly update the network weight and realize the efficient incremental training of prediction model.

Load the source domain model saved in Section 5.4, update the target domain model parameters with the target domain data, and compare the prediction results of the four models of unfixed network layer parameters (BiLSTM-TL0) and fixed network layer 1 to 3 parameters (BiLSTM-TL1~BiLSTM-TL3).

Figure 13a shows the prediction results of test sets obtained by fixing different network layers. It can be seen from the figure that all migration learning models can better fit the load curve, but in the details of the curve, the prediction results of BiLSTM-TL1 model obtained by fixing the network parameters of the first layer are closer to the true value. At this time, the load information extracted by the model is the most abundant. The data distribution of the target domain is similar to that of the source domain, and the amount of data in the target domain is small, so the BiLSTM-TL0 model is prone to over fitting. BiLSTM-TL3 and BiLSTM-TL2 models fix the deep network parameters and cannot effectively extract the overall characteristics of the target domain data.

Figure 13b shows the prediction error distribution curves of different transfer learning models. It can be seen from the figure that the errors of model BiLSTM-TL1 are concentrated around 0, and the distribution frequency with deviation of 0 is higher, which has less error fluctuation compared with other models. Therefore, the prediction accuracy of BiLSTM-TL1 model is better than other transfer learning models.

Table 9 shows the comparison of the prediction error results for the test sets of different transfer learning models. It can be seen from the table that the prediction results of BiLSTM-TL1 are the best in terms of RMSE and MAE. Compared with other transfer learning models, RMSE and MAE are reduced by 2.4% and 4.8%, respectively.

Table 9. Comparison of prediction errors of transfer learning model.

Model	RMSE	MAE
BiLSTM-TL0	47.06	36.82
BiLSTM-TL1	46.42	35.93
BiLSTM-TL2	48.80	37.47
BiLSTM-TL3	50.02	38.23

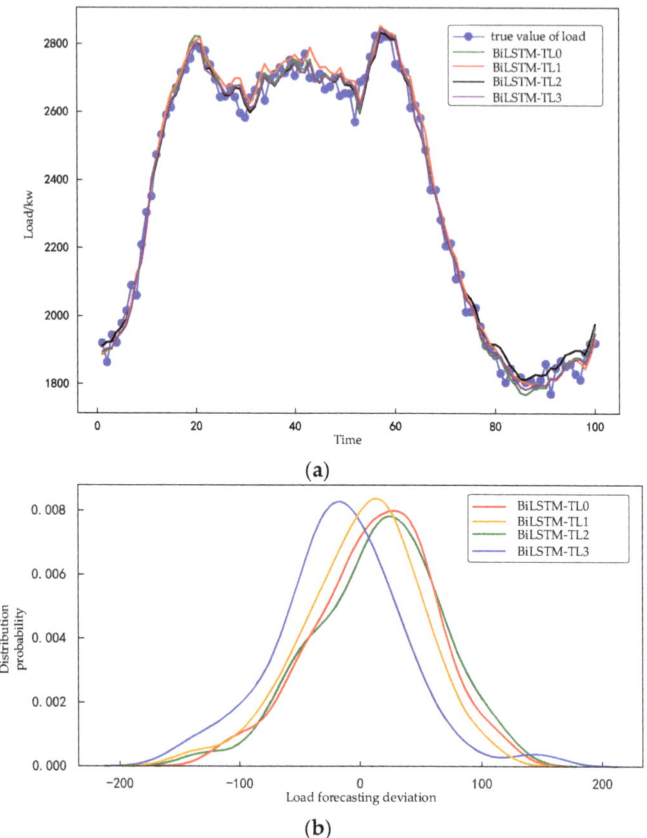

Figure 13. Comparison of prediction results of transfer learning model: (**a**) Comparison of prediction results of transfer learning model; (**b**) Prediction error distribution curve of transfer learning model.

Compare the BiLSTM-TL1 model with the Bi-LSTM network (take the data of source domain and target domain as the network training set), BiLSTM-S model (the source domain model saved in Section 5.4), BP network and SVR. The parameter settings of Bi-LSTM network, BP network and SVR are consistent with Section 5.4.

Figure 14a shows the prediction results for the test sets of different prediction models. It can be seen from the figure that the four prediction models can better fit the load change trend. However, from the detailed enlarged figure, it can be found that the predicted value of Bi-LSTM network is closer to the real value. Compared with BiLSTM-S model, the prediction effect of BiLSTM-TL1 model is improved, which is similar to Bi-LSTM network, indicating that there is no negative migration of BiLSTM-TL1 model. Figure 14b shows the prediction error distribution curves of different models in the target domain. It can be seen from the figure that the prediction error distribution curves of BiLSTM-TL1 model and Bi-LSTM network are very similar, which are concentrated near 0, and the error fluctuation is small compared with other models.

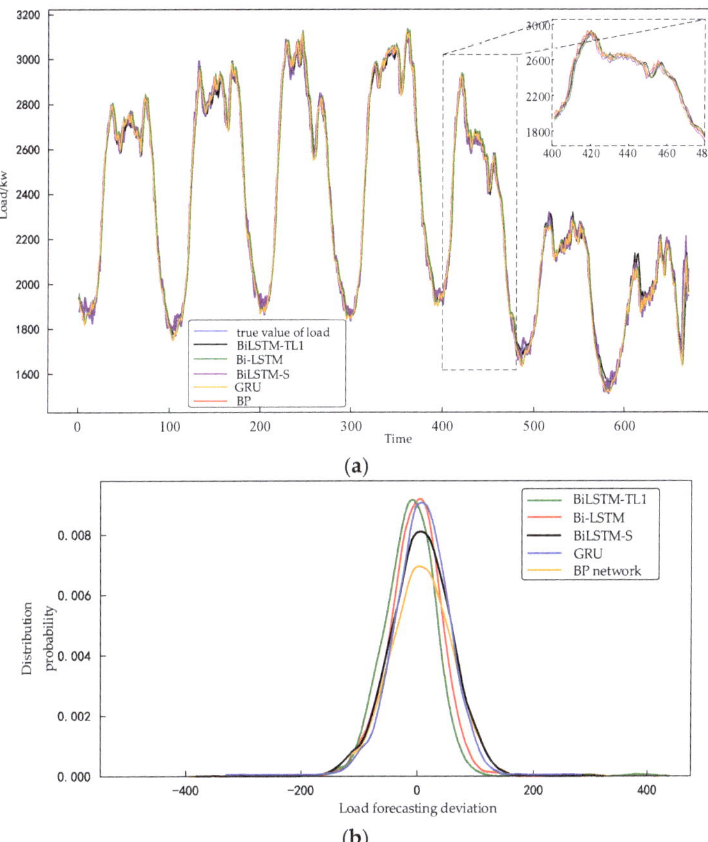

(**a**)

(**b**)

Figure 14. Comparison of prediction results of test set: (**a**) Comparison of prediction results of different model test sets; (**b**) Prediction error distribution curves of different model test sets.

From the prediction results of different model test sets in Table 10, it can be seen that the comprehensive prediction effect of BiLSTM-TL1 model is the best in the three evaluation indexes of RMSE, MAE, and training time. Compared with a Bi-LSTM network without transfer learning, on the premise of reducing the prediction accuracy by 3.9%, the training time of BiLSTM-TL1 model is shortened by 91.1%, and the training efficiency of the model is improved. Since the BiLSTM-TL1 model uses the latest sample data to update the network weight, compared with the BiLSTM-S model, RMSE and MAE are reduced by 8.8% and 7.3%, respectively. The load prediction accuracy of BiLSTM-TL1 is similar to that of GRU network, but BiLSTM-TL1 saves 87.6% training time. Compared with BP network, the training time of BiLSTM-TL1 model is similar, and the RMSE and MAE are reduced by 16.7% and 15.5%, respectively. Based on the above analysis, the BiLSTM-TL1 model, which introduces migration learning and fixes the first layer network parameters, can realize network incremental training, ensure the prediction accuracy and save a lot of model training time.

Table 10. Comparison of prediction errors in target domain.

Model	RMSE	MAE	Time/s
BiLSTM-TL1	46.42	35.93	64.8
Bi-LSTM	44.60	34.83	724.6
BiLSTM-S	50.92	38.78	0
GRU	45.93	35.12	523.8
BP network	55.76	42.67	62.7

6. Conclusions

This paper proposes an ultra-short-term load dynamic forecasting method based on model incremental training and considering abnormal data reconstruction. The main conclusions are:

(1) An abnormal data processing method based on IF-CGAN is proposed. This method can accurately eliminate the abnormal data, accurately fill in the abnormal point data, complete the load sequence, and effectively reduce the impact of abnormal data on the prediction results.

(2) An ultra-short-term load forecasting model based on a Bi-LSTM network is proposed that realizes the accurate prediction of high-dimensional long-term ultra-short-term load.

(3) A model incremental training method based on Bi-LSTM is proposed. By introducing transfer learning, the model weight can be adjusted quickly when the load data changes. Compared with the Bi-LSTM network without transfer learning, the proposed method can increase the prediction accuracy by 3.9%, shorten the model training time by 91.1%, and effectively improve the model training efficiency.

This paper mainly studies the accuracy and efficiency of ultra-short-term load forecasting under abnormal data. However, with the increasing scale of data collection and the massive and high-dimensional data, how to improve the accuracy of load abnormal data reconstruction and realize load online forecasting need to be further studied.

Author Contributions: Conceptualization, G.C. and Y.Z. (Yangfei Zhang).; methodology, G.C. and Y.W.; software, G.C. and Y.W; validation, G.C., Y.W., and L.Y.; formal analysis, G.C. and Y.W.; investigation, G.C. and Y.W; resources, G.C., G.L. and K.X.; data curation, Y.W., L.Y. and K.X.; writing—original draft preparation, Y.W.; writing—review and editing, G.C. and Y.W.; visualization, Y.W. and G.L.; supervision, G.C. and Y.Z. (Yuzhuo Zhang); project administration, G.C.; funding acquisition, G.C., L.Y. and K.X. All authors have read and agreed to the published version of the manuscript.

Funding: This research was funded by National Natural Science Foundation of China, grant number 52107098.

Data Availability Statement: Not applicable.

Conflicts of Interest: The authors declare no conflict of interest.

References

1. Li, B.; Zhang, J.; He, Y.; Wang, Y. Short-Term Load-Forecasting Method Based on Wavelet Decomposition With Second-Order Gray Neural Network Model Combined With ADF Test. *IEEE Access* **2017**, *5*, 16324–16331. [CrossRef]
2. Azeem, A.; Ismail, I.; Jameel, S.M.; Harindran, V.R. Electrical Load Forecasting Models for Different Generation Modalities: A Review. *IEEE Access* **2021**, *9*, 142239–142263. [CrossRef]
3. Shao, N.; Chen, Y. Abnormal Data Detection and Identification Method of Distribution Internet of Things Monitoring Terminal Based on Spatiotemporal Correlation. *Energies* **2022**, *15*, 2151. [CrossRef]
4. Zhu, X.; Shen, M. Based on the ARIMA model with grey theory for short term load forecasting model. In Proceedings of the 2012 International Conference on Systems and Informatics, Yantai, China, 19–20 May 2012. [CrossRef]
5. Ji, P.; Xiong, D.; Wang, P.; Chen, J. A study on exponential smoothing model for load forecasting. In Proceedings of the 2012 Asia-Pacific Power and Energy Engineering Conference, Shanghai, China, 27–29 March 2012. [CrossRef]
6. Jawad, M.; Nadeem, M.S.A.; Shim, S.O.; Khan, I.R.; Shaheen, A.; Habib, N.; Hussain, L.; Aziz, W. Machine Learning Based Cost Effective Electricity Load Forecasting Model Using Correlated Meteorological Parameters. *IEEE Access* **2020**, *8*, 146847–146864. [CrossRef]

7. Chen, K.; Chen, K.; Wang, Q.; He, Z.; Hu, J.; He, J. Short-term load forecasting with deep residual networks. *IEEE Trans. Smart Grid* **2019**, *10*, 3943–3952. [CrossRef]
8. Alquthami, T.; Zulfiqar, M.; Kamran, M.; Milyani, A.H.; Rasheed, M.B. A Performance Comparison of Machine Learning Algorithms for Load Forecasting in Smart Grid. *IEEE Access* **2022**, *10*, 48419–48433. [CrossRef]
9. Khan, A.H.; Li, S.; Cao, X.W. Tracking control of redundant manipulator under active remote center-of-motion constraints: An RNN-based metaheuristic approach. *Sci. China* **2021**, *64*, 149–166. [CrossRef]
10. Alhussein, M.; Aurangzeb, K.; Haider, S.I. Hybrid CNN-LSTM Model for Short-Term Individual Household Load Forecasting. *IEEE Access* **2020**, *8*, 180544–180557. [CrossRef]
11. Gao, X.Y.; Wang, Y.; Gao, Y.; Sun, C.Z.; Xiang, W.; Yue, Y.M. Short-term Load Forecasting Model of GRU Network Based on Deep Learning Framework. In Proceedings of the 2018 2nd IEEE Conference on Energy Internet and Energy System Integration, Beijing, China, 20–22 October 2018. [CrossRef]
12. Rafi, S.H.; Masood, N.A.; Deeba, S.R.; Hossain, E. A Short-Term Load Forecasting Method Using Integrated CNN and LSTM Network. *IEEE Access* **2021**, *9*, 32436–32448. [CrossRef]
13. Shohan, M.J.A.; Faruque, M.O.; Foo, S.Y. Forecasting of Electric Load Using a Hybrid LSTM-Neural Prophet Model. *Energies* **2022**, *15*, 2158. [CrossRef]
14. Mohammadi, A.; Saraee, M.H. Dealing with missing values in microarray data. In Proceedings of the 2008 4th International Conference on Emerging Technologies, Rawalpindi, Pakistan, 18–19 October 2008. [CrossRef]
15. Silva, L.O.; ZÁRATE, L.E. A brief review of the main approaches for treatment of missing data. *Intel. Data Anal.* **2014**, *18*, 1177–1198. [CrossRef]
16. Pessanha, J.; Melo, A.; Caldas, R.; Falcão, D. An Approach for Data Treatment of Solar Photovoltaic Generation. *IEEE Lat. Am. Trans.* **2020**, *18*, 1563–1571. [CrossRef]
17. Seaman, S.R.; White, I.R. Review of inverse probability weighting for dealing with missing data. *Stat. Methods Med. Res.* **2013**, *22*, 278–295. [CrossRef] [PubMed]
18. Batista, G.E.A.P.A.; Monard, M.C. An analysis of four missing data treatment methods for supervised learning. *Appl. Artif. Intell.* **2003**, *17*, 519–533. [CrossRef]
19. Hastie, T.; Mazumder, R.; Lee, J.; Zadeh, R. Matrix completion and low-rank SVD via fast alternating least squares. *J. Mach. Learn. Res.* **2014**, *16*, 3367–3402. [CrossRef]
20. Deng, W.; Guo, Y.; Liu, J.; Li, Y.; Liu, D.; Zhu, L. A missing power data filling method based on improved random forest algorithm. *Chin. J. Electr. Eng.* **2019**, *5*, 33–39. [CrossRef]
21. Goodfellow, I.; Pouget-Abadie, J.; Mirza, M.; Xu, B.; Warde-Farley, D.; Ozair, S.; Courville, A.; Bengio, Y. Generative Adversarial Nets. In Proceedings of the 27th International Conference on Neural Information Processing Systems, Montreal, QC, Canada, 8 December 2014. [CrossRef]
22. Oh, E.; Kim, T.; Ji, Y.; Khyalia, S. STING: Self-attention based Time-series Imputation Networks using GAN. In Proceedings of the 2021 IEEE International Conference on Data Mining, Auckland, New Zealand, 7–10 December 2021. [CrossRef]
23. Li, J.; He, H.; Li, L. CGAN-MBL for Reliability Assessment with Imbalanced Transmission Gear Data. *IEEE Trans. Instrum. Meas.* **2019**, *68*, 3173–3183. [CrossRef]
24. Hariri, S.; Kind, M.C.; Brunner, R.J. Extended Isolation Forest. *IEEE Trans. Knowl. Data Eng.* **2021**, *33*, 1479–1489. [CrossRef]
25. Gulrajani, I.; Ahmed, F.; Arjovsky, M.; Dumoulin, V.; Courville, A. Improved training of wasserstein GANs. In Proceedings of the Neural Information Processing Systems, Los Angeles, CA, USA, 4–9 December 2017.
26. Wei, Y.; Zhu, C.; Yang, Y.; Liu, Y. A Discrete Cosine Model of Light Field Sampling for Improving Rendering Quality of Views. In Proceedings of the 2020 IEEE International Conference on Visual Communications and Image Processing, Macau, China, 1–4 December 2020. [CrossRef]
27. Shao, L.; Zhu, F.; Li, X. Transfer Learning for Visual Categorization: A Survey. *IEEE Trans. Neural Netw. Learn. Syst.* **2015**, *26*, 1019–1034. [CrossRef] [PubMed]
28. Dang, W.; Li, H.J.; Ding, Z.M.; Nie, F.P.; Chen, J.Y.; Dong, X.; Wang, Z.H. Rethinking Maximum Mean Discrepancy for Visual Domain Adaptation. *IEEE Trans. Neural Netw. Learn. Syst.* **2015**, 1–14. [CrossRef]

Article

Resilience Maximization in Electrical Power Systems through Switching of Power Transmission Lines

Jaime Pilatásig [1,*], Diego Carrión [2] and Manuel Jaramillo [2]

[1] Master's Program in Electricity, Salesian Polytechnic University, Quito EC170702, Ecuador
[2] Smart Grid Research Group—GIREI (Spanish Acronym), Quito EC170702, Ecuador
* Correspondence: jpilatasigl@est.ups.edu.ec

Abstract: This research aims to maximize the resilience of an electrical power system after an $N - 1$ contingency, and this objective is achieved by switching the transmission lines connection using a heuristic that integrates optimal dc power flows (DCOPF), optimal transmission switching (OTS) and contingencies analysis. This paper's methodology proposes to identify the order of re-entry of the elements that go out of the operation of an electrical power system after a contingency, for which DCOPF is used to determine the operating conditions accompanied by OTS that seeks to identify the maximum number of lines that can be disconnected seeking the most negligible impact on the contingency index J. The model allows each possible line-switching scenario to be analyzed and the one with the lowest value of J is chosen as the option to reconnect, this process is repeated until the entire power system is fully operational. As study cases, the IEEE 14, 30 and 39 bus bars were selected, in which the proposed methodology was applied and when the OTS was executed, the systems improved after the contingency; furthermore, when an adequate connection order of the disconnected lines is determined, the systems are significantly improved, therefore, the resilience of power systems is maximized, guaranteeing stable, reliable and safe behavior within operating parameters.

Keywords: DC optimal power flow; transmission optimal commutation; contingency analysis; electrical power system; electrical power system restoration; resilience

Citation: Pilatásig, J.; Carrión, D.; Jaramillo, M. Resilience Maximization in Electrical Power Systems through Switching of Power Transmission Lines. *Energies* **2022**, *15*, 8138. https://doi.org/10.3390/en15218138

Academic Editors: Yuan Liao and Ke Xu

Received: 16 September 2022
Accepted: 26 October 2022
Published: 1 November 2022

1. Introduction

Extreme weather events endanger basic infrastructure, such as transportation, communications, water supply, gas and electricity, and these affectations can be so severe that infrastructure might collapse and is observable in the number and duration of energy interruptions caused [1], therefore, due to these unforeseen events, the reliability and operation of this infrastructure are compromised [2–5]. Among the different kinds of infrastructure that are compromised due to unforeseen events, recent research focuses on electricity supply infrastructure that is also vulnerable to natural events [6]. Extreme weather events or natural phenomena directly affect the electricity supply infrastructure, which has a direct impact on society and its economy [7], challenging aspects that require the development of models, methodologies and strategies to minimize the effect on electricity supply.

Technological development applied to electricity infrastructure is evidenced in the integration of renewable energy, distributed generation and smart grids in electrical systems, making planning, expansion and operation activities more challenging. The challenge of these new technologies lies in the dispatch of reactive power and how this affects the reconfiguration of power flows, for which in recent years several authors have focused on solving this problem [8].

The authors of [9] focus on solving the non-linearity problem that occurs in reactive power dispatch models through a backtracking search algorithm. The control of reactive power is not only a problem that occurs at the transmission and generation level, but in the distribution stage as well, and this also includes distributed generation. To solve this problem, the authors in [10] propose a manta ray foraging optimization algorithm, which is a bio-inspired technique for reactive control optimization. Other authors focus on controlling active power to improve voltage profiles in highly radialized distribution networks [11].

For [12], seven indicators are key to determining the resilience index of an electricity system, being: (a) the degree of dependence on fuel in generation, (b) measurement of efficiency of generation, (c) measurement of efficiency in electricity distribution, (d) measurement of carbon dioxide emissions in electricity generation, (e) the probability that electricity will be generated from different types of fuel, (f) redundant power for the available capacity in the event of high probability and low impact events, and (g) the degree of dependence on electricity imports.

A resilient power system must be able to execute preventive actions, mitigate the impact of extreme events, respond optimally through automated control procedures, reduce the time to restore electricity service and lead to significant economic savings, according to [13].

The traditional method to evaluate reliability is based on Monte Carlo simulation (MCS), and in [14], a method for contingency classification is presented that is incorporated to evaluate the reliability of the line switching operation. From these methodologies, a classification list of recommended lines, safe lines and critical lines resulting from the analysis of annual rates and events is obtained, determining that the switching of the recommended lines improves the system reliability [15,16].

The mathematical modeling of climate behavior represents various aspects, including the impact on electrical systems affected by extreme weather events, increased demand for electricity, decreased capacity and efficiency in generation and transmission, reduction of potential resources of water, wind and sun for electricity generation [17]. In [18], and research focused on the risks arising from extreme climate change and its impact on energy infrastructure is presented.

The productive environment is centralized in urban areas with consumption between 60% and 80% of the electricity generated, likewise, the economy and comfort of society depend on the continuity of the electricity supply, which is affected by climate change, cyber-attacks, terrorism, technical deficiencies, and unstable markets, among others, according to [19].

Resilience is defined as the ability of a material, mechanism or system to recover its initial state when the disturbance to which it had been subjected has ceased. The ability of a system to plan and prepare to absorb, recover and adapt to any circumstance that may affect it, such as natural phenomena, earthquakes, fires, floods, volcanic eruptions, hurricanes, and typhoons, among other unexpected phenomena, gives the system a resilient quality [20].

Given these adverse circumstances in [21], it is proposed that electrical systems must have sufficient capacity to withstand these disturbances and guarantee the continuity of the electricity supply; therefore, resilience is the ability to quickly recover from interruptions resulting from an extreme natural event or phenomenon.

Various investigations have emerged, and as part of their results, they agree on a definition of resilience. In [22], the researchers consider resilience as the capacity of a system to mitigate the consequences of rare events, predictable or not, but with a high impact.

Resilience also considers a critical aspect such as the cyber-security of electrical systems, and in [23], researchers establish the concept as the ability to maintain the continuity of the electricity supply within a load priority scheme.

No metrics have been standardized for resilience; however, key parameters can be identified from studies of small signal stability, transient stability, power flows and contingency analyses [24].

In [25], through the statistical analysis of event data, frequency and duration of interruptions, and system restoration times, among others, the resilience of the US electrical network is evaluated using a regression analysis through the Lewis–Robinson test. Simple methodologies are generally used to evaluate resilience and its interrelation in the network infrastructure, the fragility and performance of the infrastructure are considered as a measure of resilience, according to [26].

In [27], there is a compendium on resilience engineering that identifies tools for risk management based on the fragility and performance of the infrastructure, the modeling tends to determine the best strategies to minimize the impact on infrastructure energy systems.

A predominant aspect is the duration of the disturbance, which puts the electrical system to the test to absorb and resist the extreme event, a model that integrates contingency analysis, non-conventional renewable energies, smart grids, management and mitigation of risks inherent to climate events is treated in [28], which seeks to minimize the error to quantify the robustness of the electrical system.

In [29], researchers use probability density functions (PDF), the analysis is based on Markov chains, states and probabilistic scenarios for decision-making and operation strategies of the electrical system, the system operators evaluate and adjust these strategies ensuring a minimum cost. In [30], the authors highlight a fault chain-based model to determine a set of sensitive transmission lines, thus identifying them and improving the ability of a resilient system to prevent and resist extreme disturbances, through statistical analysis. From historical data of failures, load-ability and damage of the transmission system, the failure chains are predicted to quantify and evaluate their level of risk.

In [31], the author deals with the impact on the resilience of electrical systems due to excessive restoration times as a result of analysis through probability distribution of historical data with a period of at least ten years; it is identified that the restoration process depends on many factors, for example, the climate, location, type of failure and the availability of maintenance crews, with which the average restoration time tends to be highly variable.

New technologies of smart grids, micro-grids and wide area monitoring applications would allow rapid system restoration; also, ref. [32] determines the impacts of natural disasters on power systems, concluding that the improvement of resilience of the network involves interdisciplinary techniques, among them, statistics, meteorology, power systems engineering, optimization, communications, control, policies and regulations, stresses the importance of research in methods and tools for the forecast of disturbances related to natural disasters.

Finding a global optimal solution to the transmission line optimization problem is a complex task, since there are many local optimal solutions; in [33], the author compares modeling and optimization for transmission lines from classical techniques, heuristic techniques, meta-heuristic techniques and other promising techniques, the analysis considers from the design to the operation and maintenance of the transmission lines, which is complemented by the comparison of several methods of economic analysis, and better results can be obtained by combining two or more optimization techniques.

The EPS requires specialized analysis to minimize its operating costs, which is why the problems of economic dispatch based on optimal power flows become relevant. Authors in [34] propose a new methodology for the evaluation of the problem of optimal power flows using the gorilla troops optimization technique (GTOT).

Similarly, bioinspired heuristic techniques can be used for the problem of optimal dispatch of generation plants, for example in [35], a methodology for the electro-thermal dispatch of power plants is proposed employing a jellyfish search algorithm. Additionally, the authors in [36] propose a multi-objective solution for the electro-thermal dispatch of generation plants.

In [37], the authors consider reliability maximization as a multi-objective optimization problem that aims to increase transfer capacity, improve voltage profiles, and improve power system reliability. Transmission Expansion Planning (TEP) is a large-scale, mixed-integer, non-linear, non-convex problem; Ref. [38] addresses the transmission expansion planning problem through mixed-integer linear programming based on the linear power flow model for losses and costs of the generator from planning and operation, construction costs of lines, also consider security restrictions $N-1$, the balance of power in each node, state of the line, if it exists or should be built, charge-ability of the line, angular difference, generation dispatch limit, the linear model of generation losses and costs, and its objective function aims to minimize the investment cost.

In [39], transmission switching in expansion planning is treated from the approach of a master problem that uses the set of lines and generation units that are candidates for investment, with two sub-problems to alleviate any violation in transmission and optimal dispatch of generation units. In [40], the probabilistic planning of transmission expansion focuses on random and non-random uncertainties especially related to wind load and generation; likewise, the security $N - 1$ is evaluated to find the structure of optimal transmission to meet the peak load demand with minimal investment and losses due to energy not supplied, however, the restrictions increase the cost of transmission investment that can be mitigated by incorporating limits on load shedding.

In [41], the authors use genetic algorithms to separately solve two models that consider the uncertainty of the total demand and the demand in each load bar, the methodology provides an optimal plan for expansion of the transmission network, including more accessible plans. From the economic aspect, they satisfy the uncertainty in the demand; likewise, it allows the analysis of the possible location of demand in each bus and thus determines if it is possible to supply a new load in a specific bus.

The purpose of this research is to identify the order of re-connection of transmission lines in a power system after being affected by a contingency $N - 1$, for this, a heuristic is proposed with optimal flows of dc power, optimal switching of transmission and contingency analysis; therefore, as a result, the resilience of the power system is maximized by switching transmission lines.

This article is organized as follows: Section 2 defines some aspects of optimal dc power flows and optimal transmission switching; Section 3 formulates the problem, while Section 4 analyzes and compares results in the IEEE test systems. Finally, Section 6 presents the conclusions of this research.

2. Electrical Power System Operation

2.1. Optimal Power Flow

Optimal power flow is a technique that has been widely used in recent years in the analysis and optimization of electrical power systems.
The authors in [42] developed a method for the optimal flow of fully distributed DC power, sub-problems are generated at the level of neighboring bus bars, the study is developed in the RTS96 and IEEE118 systems, obtaining the optimal solution even under a cold start scenario; however, the initial conditions must be close to the optimum.

The authors in [43] considered interconnected weak power systems that tend to disconnect in the face of extreme events due to their low inertia, as well as the incidence of low and high frequency; for this, based on the optimal power flow limited by separation events, modeled by linear programming of mixed integers, in the test system the improvement of the resilience of the system results from conditions of high and low frequency and levels of inertia in each area.

In [44], the authors present an optimization through optimal stochastic DC power flows with uncertain load and renewable generation capacity, where the reserve of said generators is used within its limits, the results are evidenced in the test systems of 6 and 118 bars in generation costs lower than expected.

In [45], the researchers propose a model that improves the accuracy of the calculation through optimal flows of dc power focused on practical marginal local price markets, the model uses the curve improvement fitting technique that evaluates the losses in the parameters with total lines and lightly loaded.

2.2. Optimal Transmission Switching

In [46], through mixed integer linear programming, the problem of optimal power flows and optimal transmission switching (OTS) is addressed. To modify the transmission topology and optimal generation dispatch binary variables are used to represent the states that describe the physical system. The objective function maximizes the total cost of the system with a number j of lines enabled to switch in the transmission system:

$$TC_j = Max \sum_k c_{nk} P_{nk} \tag{1a}$$

$$\theta_n^{min} \leq \theta_n \leq \theta_n^{max} \tag{1b}$$

$$P_{ng}^{min} \leq P_{ng} \leq P_{ng}^{max} \tag{1c}$$

$$P_{nk}^{min} z_k \leq P_{nk} \leq P_{nk}^{max} z_k \tag{1d}$$

$$-\sum_k P_{nk} - \sum_g P_{ng} - \sum_d P_{nd} = 0 \tag{1e}$$

$$B_k(\theta_n - \theta_m) - P_{nk} + (1 - z_k)M \geq 0 \tag{1f}$$

$$B_k(\theta_n - \theta_m) - P_{nk} + (1 - z_k)M \leq 0 \tag{1g}$$

$$\sum_k (1 - z_k) \leq j \tag{1h}$$

where, k is the transmission lines, n, m are the nodes, g is the generators, d is the loads, θ_n is the voltage angle at the node n, P_{nk}, P_{ng}, P_{nd} are the active power flow to or from line k, generator g, or loads d to node n, and z_k is a binary variable indicating if the line k has been removed from the system. The restrictions consider voltage angle limits, power in generators, power flow limit through the transmission line, the power balance in each node and limit in the number of disconnected transmission lines, as a result, in the test case of 118 bars, a saving of up to 25% is obtained in the economic dispatch.

In [47], the objective function in the approximate OTS model considers minimizing the total costs of electricity generation:

$$Min, P_3 = \sum_n C_n g_n + C' \sum_k \frac{\varepsilon_k}{M_k} \tag{2}$$

where, C_n is the operating cost of the generator n, and g_n corresponds to the power generated by the generator n, ε_k is a decision variable, C is a constant number, and M_k is a very large value. The restrictions contemplate: that the power that enters and leaves a node is equal, the voltage angles in bars, the power flow in lines, the thermal limit of the lines, limits of the generators, and limits of angular difference in bars. The approximate model of OTS in test systems of 14, 39, 57, IEEE 118 and 2383 bus bars provides similar results, but with fewer switched transmission lines. Through the mixed integer programming model with binary variables, ref. [47] addresses the optimal transmission switching under contingency $N - 1$, the formulation of the optimal DC power flow under contingency $N - 1$ seeks to minimize the costs under the physical constraints of the system:

$$Min, TC = \sum_g c_{ng0} P_{ng0} \tag{3}$$

Being: for $c = 0$ there is no contingency, for $c > 0$ there is a contingency $N - 1$; c_{ngc} is the production cost from generator g in state c; P_{ngc} is the power supply from generator g at node n for state c; z_k is the binary variable for the transmit element, 0 open and 1 closed.

The objective function is subject to the angle of the voltage at node n for state c, power balance at the nodes, and limits of transmission lines and generators for each state, from which a reduction of approximately 15% in the cost under contingency $N - 1$, switching transmission lines in the generation dispatch.

For [48], in the optimal switching of transmission lines, the computational complexity can be reduced by considering in the algorithm implemented through mixed integer programming a limit of lines selected to be disconnected.

Through mixed integer programming, ref. [49] covers the problem of optimal switching of transmission lines, based on two heuristics: linear programming is solved first and then mixed integer programming, and the procedure removes one line at a time. At the same time, both heuristics depend on the problem of optimal DC power flows to determine a classification of transmission lines for switching.

For [50], transmission congestion management is possible through optimal transmission switching formulated as a mixed-integer nonlinear programming problem, thereby relieving congestion while maintaining voltage safety, identifying the order of the switching operations aside from optimal transmission switching strategies for handling transmission congestion.

The optimal transmission line switching problem treated in [49] is improved through heuristics based on optimal AC power flows [51], and despite being an exact method, it increases the calculation time that is impractical in real situations.

Transmission switching alters the topology of the transmission network, and this is an effective way to reduce operating costs; however, this is achieved by determining a set of suitable transmission lines to be switched, and this process requires in the formulations the inclusion of restrictions and decision variables. Ref. [52] proposes among the restrictions to avoid electrical islands as a result of a set of optimal lines to be switched; however, in an OTS, the computational requirement increases, with the relaxation of the problem in addition that the reliability is not reduced, the computational time improves.

In [53], the author states mixed-integer second-order programming via the AC power flow model; however, the AC transmission switching problem is relaxed by including inequalities in the optimal power flow problem of AC power. However, applying an ACOPF would become a mathematically challenging problem due to the restrictions of the power flow and the variables for commutation that would give rise to two non-convexities.

The optimal transmission switching in, ref. [54], is analyzed from the safety aspect in cases of transmission line overloads and voltage stability, the safety level decreases if transmission lines are disconnected in different locations, and the expected cost per outage tends to increase, the economic benefit is expected to be positive by identifying the topology of the network by incorporating probabilistic security constraints in the optimal transmission switching problem.

In [55], the mixed-integer nonlinear programming model through optimal AC power flow addresses optimal transmission switching and security evaluation considering the $N - 1$ contingency criterion, the analysis is sequential for contingencies $N - 1$ and energy not supplied, this allows estimating a security level for each solution.

From real-time contingency analysis developed in [56], the researchers provide the corrective AC transmission switching, and the methodology also runs time-domain simulations to verify the dynamic stability of the corrective transmission switching solution; that is, through the contingency analysis, the critical contingency that causes potential violations to the system is identified, and the corrective algorithm finds the effective corrective actions to mitigate these violations, therefore, resulting in the reduction of operating costs, improvement of system reliability and the incorporation of renewable resources.

In [57], the authors study the optimal stochastic transmission switching with optimal DC and AC power flows, and compared with the optimal deterministic transmission switching, the stochastic formulations are faster, and among them, using the optimal AC power flow obtains the lowest operational cost and with less calculation time.

In [58], the research covers the optimal transmission switching problem, poses several mathematical models that consider avoiding the switching of unnecessary lines, avoiding the formation of electrical islands in the system, and justifies the occurrence of the Braess paradox in the OTS problem, the results being the reduction of disconnected lines and the non-formation of electrical islands.

Ref. [59] considers the optimal transmission commutation avoiding electric islands and the real load-ability conditions of the lines based on real climate data, the proposed procedure allows the maximization of the reliability and the minimization of the total cost of generation. The study results on the IEEE RTS-96 system to determine the improvement in the system performance and decrease in generation cost.

3. Problem Formulation

In electrical systems, from the user's point of view, the systems must satisfy the demand for active and reactive power at any moment and for the required time, thus the system operators are responsible for the operational planning of the generation, transmission and distribution resources.

Electrical networks are susceptible to low-probability risks, but whose consequences are of great impact if they materialize. As an example, there can be earthquakes, atmospheric discharges, loss of one or several elements of the system, and cyber-attacks, among others. From the technical–operational point of view, electrical systems can have partial disconnections of generation, transmission and distribution links, and large load centers, until finally the collapse of the entire system, a condition that is not desired by any system operator.

Resilience in electrical systems is referred to any internal or external disturbance, is established as any available capacity of the system that can contribute to avoiding total collapse, going from a relatively normal state to a new disturbed state, but depending on its rapid response allows a new quasi-normal state to be reached, and from the technical aspect, the slightest improvement in the systems and subsystems contributes to the electrical system improving its resilience.

This research aims to maximize the resilience of electrical systems, focusing on the heuristics that incorporate the optimal dc power flow (DCOPF), the optimal transmission switching (OTS), the contingency index, and the switching of the lines disconnected looking for the lowest rate of contingencies concerning the variable δ to establish the order of connection of the disconnected lines, improving the response of the system.

In this work development, the following premises are established: transformers and generators are not considered switching elements, no formation of isolated systems, linear cost of generators, resistance and shunt capacitance of lines is zero, and losses and reactive power are ignored. As a stability criterion, a range is established for the voltage angle δ in bars of $\pm\pi/3$ that allows a dynamic behavior to satisfy the recovery of the system by connecting disconnected lines.

In an electrical system, the contingencies are presented from the output of lines, transformers, generators and even loads to order the severity or impact that occurs in the electrical system, and contingency indices are established with which a ranking of contingencies can be determined based on their severity. From the technique proposed by [60] for ranking and selection of contingencies, a variation for the J index of contingencies presented in Equation (4) is considered.

$$J = \sum_{i=1}^{l} \frac{W_i}{m} \left(\frac{\delta_i}{\delta_{i_{max}}}\right)^m \tag{4}$$

where δ_i represents voltage angle of the voltage at the node i, $\delta_{i_{max}}$ is the maximum angular deviation that is allowed for the voltage at the nodes, which has been established to be 0.6 [rad]. W_i is a factor that represents the importance of the element in the EPS that, for the proposed methodology, it is considered that all the elements are equally important, and for this reason, it is assigned a value of 1, and m is an integer value greater than 1 that allows the elimination of negativity conditions, and for the present study, it was assigned the value of 2.

4. Methodology

The proposed heuristic seeks to maximize the resilience of electrical power systems by switching electrical transmission lines, through the optimal flow of dc power, the initial conditions that will allow comparing results are determined, followed by a contingency of the type $N-1$ in transmission lines and DCOPF plus OTS is executed, and depending on the variable δ, the contingency index is determined for $N-1$ and the SW lines disconnected by the OTS.

The procedure has the following steps: to evaluate the connection options of the disconnected SW lines, to determine the first connection option, one line is restored at a

Table 3. 30 bus bar system generators data.

Node	Pmax [MW]	Pmin [MW]	Qmax [Mvar]	Qmin [Mvar]	Cost [MVA]
1	80	0	150	−20	2
2	80	0	60	−20	1.75
13	40	0	44.7	−15	3
22	50	0	62.5	−15	1
23	30	0	40	−10	3
27	55	0	48.7	−15	3.25

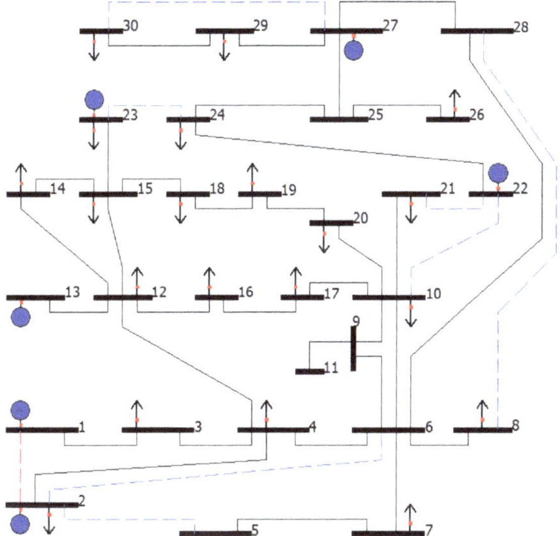

Figure 2. 30 bus bar system, contingency $N-1$ and lines SW disconnected.

To establish the re-connection order of the SW lines, the connection of each line is evaluated when executing the DCOPF, from which the connection with the lowest contingency index J is identified, and in Table 4, the order of re-connection is detailed.

Table 4. Line reconnect options for the 30 bus bar system.

Re-Connection Option	Line	J	SW
1ra	L_{23-24}	0.3745	7 disconnected lines
2da	L_{10-22}	0.3185	6 disconnected lines
3ra	L_{2-5}	0.3053	5 disconnected lines
4ta	L_{1-2}	0.2105	4 disconnected lines
5ta	L_{2-6}	0.1453	3 disconnected lines
6ta	L_{27-30}	0.1379	2 disconnected lines
7ma	L_{8-28}	0.1363	1 disconnected lines
8va	L_{21-22}	0.3186	0 disconnected lines

Figure 3 represents the different combination possibilities, likewise the line with the lowest contingency index J that determines the connection sequence is identified.

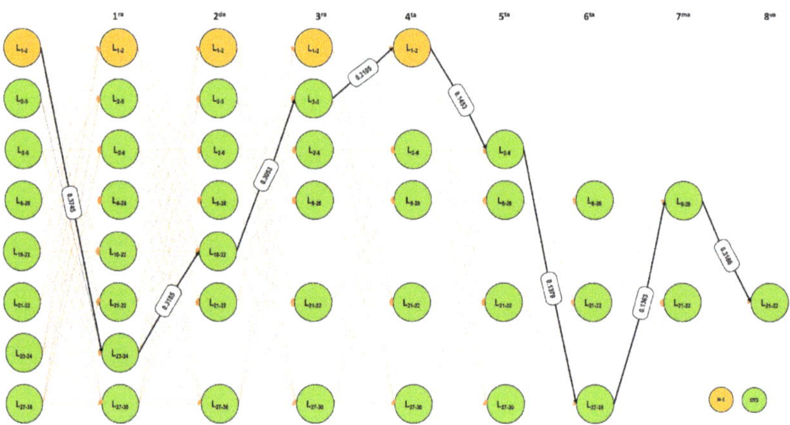

Figure 3. Re-connection lines for the 30 bus bar system.

The variation in the angle of voltage δ in bus bars is presented in Figure 4 for the 30-bus system in a normal state. Additionally, a new contingency state $N - 1$, the change of state with the result of the OTS, and finally the connection options of the disconnected lines SW for system restoration are shown as well.

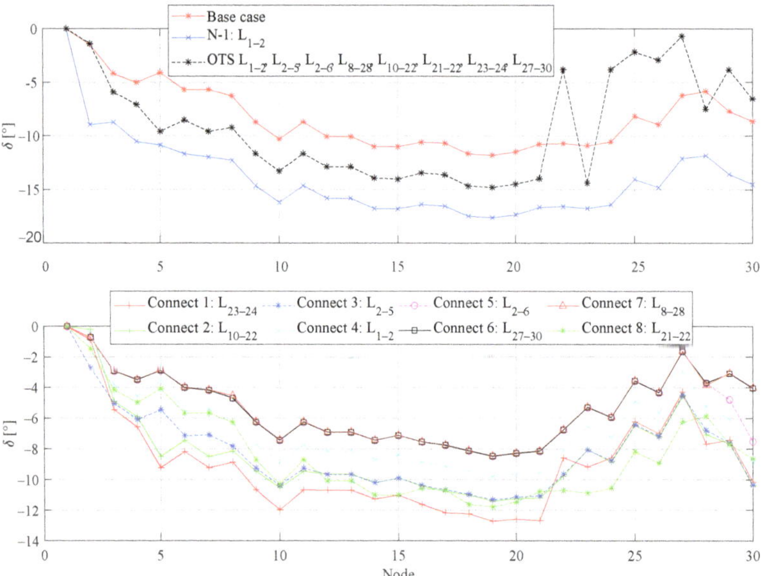

Figure 4. Behavior of the voltage angle δ in the 30 bus bar system.

Regarding power flows through the transmission lines in a normal state of the system, the total demand is satisfied, when changing the topology of the system when the contingency of the L_{1-2} occurs, it is evident that power flows are redistributed without affecting demand when the OTS is executed, and lines are disconnected, which improves the system response, power flows are redistributed, and load capacity is reduced on some of the lines, to name a few, L_{1-3}, 32%, L_{3-4}, 33%, see Figure 5; likewise, the load-ability in other lines of the system is increased. By seeking a connection order for the disconnected lines, the condition of the system is improved and the power flows are redistributed in a better way, an aspect that maximizes the resilience of the system without affecting the demand in the various cases.

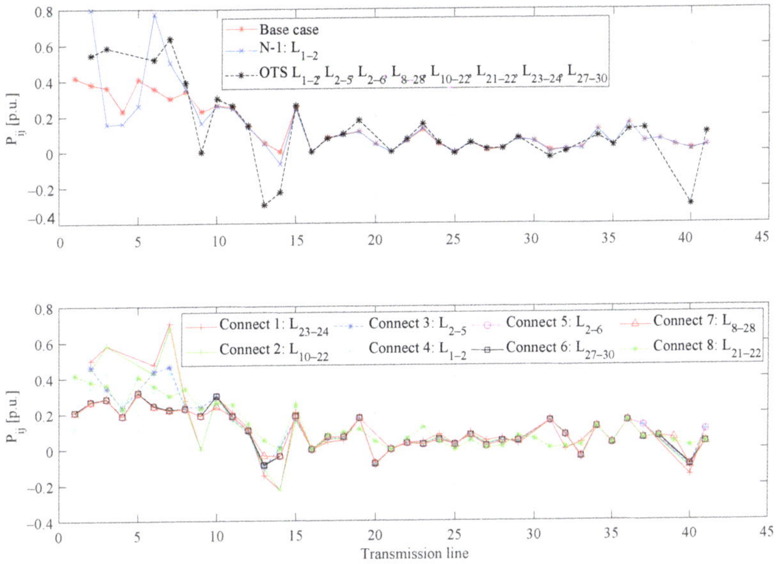

Figure 5. Power flows behavior in the 30-bar system.

The generators of the 30-bar system are dispatched based on their costs; however, in all cases, they satisfy the entire demand of the system.

Through the analysis of the results obtained from the IEEE test systems, the application of the heuristic that links the DCOPF, OTS and index J of contingencies allows us to maximize the resilience of power systems through the switching of transmission lines after a $N-1$ contingency, respecting the operating restrictions of the system.

In the IEEE 14-bar test system, the contingency $N-1$ involving the L_{2-3} presents the highest contingency index J of 0.5243, see Table 5, the OTS determines commutable to the L_{12-13}, the new contingency index J is set at 0.5239, the generation power remains at 2.59 in p.u., for the re-connection order the L_{2-3} is the first option in reason that the contingency index J as a function of the variable δ is 0.3634.

For the IEEE 39 bus test system, the contingency $N-1$ with the L_{2-25} produces a contingency index J of 0.3737, the OTS determines the lines SW disconnected L_{3-18}, L_{16-24} and L_{26-29}; therefore, the system presents a new index J of 0.6832, the power generated is 61.5 in p.u., the order of re-connection to maximize resilience corresponds to: (1) L_{2-25} with a contingency index J of 0.3544, (2) L_{16-24} with a contingency index J of 0.401, (3) L_{26-29} with a contingency index J of 0.315, and (4) L_{3-18} with a contingency index J of 0.3836.

Table 5. Test systems to maximize resiliency.

System	$N-1$	J_{N-1}	OTS	J_{OTS}
IEEE 14 bus bars	L_{2-3}	0.5243	1 line disconnected	0.5239
IEEE 30 bus bars	L_{1-2}	0.8612	7 lines disconnected	0.4403
IEEE 39 bus bars	L_{2-25}	0.3737	3 lines disconnected	0.6832

In the analysis of the results, lines that presented a higher contingency index J were discarded because they included the disconnection of transformers, which is not considered in this investigation. Although the result of the OTS improves the system after a $N-1$ contingency, by determining the line re-connection order that allows the electrical power system to be strengthened, by identifying the lines that improve the system, and above all the order of connection allows us to maximize the resilience.

Concerning the generation in the test systems, the costs of the generators condition the dispatch and participation in each of the analyzed scenarios. This is because the DCOPF seeks to minimize costs, and in this sense, the order of priority depends on the declared costs for each generator.

6. Conclusions

The optimal switching of transmission lines is currently a technical aspect inherent to the operation of power systems, and several investigations have supported its benefits; however, in relatively small power systems it is not even considered an applicable option, probably due to the technical–operational procedures in force for decades, of the heuristics applied in the test systems with the execution of the OTS the systems change to a new relatively stable state, the restrictions and assumptions are respected, followed by the maximization of the resilience of the voltage angle behavior when connecting each of the disconnected lines.

From the operational point of view in quasi-real-time, instantaneous simulations are required, hence several investigations propose the application of optimal DC power flows, the results being quite successful compared to optimal AC power flows; likewise, the DCOPF requires less computational time.

Although electrical power systems can withstand the output of certain defined elements, they must have the capacity and resources to recover, this being the purpose of resilient systems. The system's dynamic behavior has been prioritized by limiting the angle of the bus voltage in a range of $\pm\pi/3$ with which stability problems are avoided; likewise, by applying contingency $N-1$ and evaluating its impact on the elements of the system, determining its severity through the ranking of contingencies, guarantees the system reliability. On the other hand, further degrading the system is avoided by not forming isolated areas or zones that allow for service security.

The development of robust power systems does not always guarantee economy and reliability; in other words, their resilience. This aspect encompasses several methods or techniques that favor the recovery of the system under certain disturbances; among many, switching has been analyzed for transmission lines, identifying the order of re-connection of the disconnected elements, favoring dynamic, reliable and safe behavior.

It is important to consider future studies that include transformer switching, interconnection lines between large systems, reactive compensation equipment, using optimal ac power flows, and real-time analysis.

Author Contributions: J.P., D.C.: conceptualization, methodology, validation, writing—review and editing. J.P.: methodology, software, writing—original draft. D.C.: data curation, formal analysis. D.C.: supervision. M.J.: writing—review and editing. All authors have read and agreed to the published version of the manuscript.

Funding: This work was supported by Universidad Politécnica Salesiana and GIREI-Smart Grid Research Group under the project Optimal operation of electrical power systems considering new technologies and energy sustainability criteria.

Data Availability Statement: Not applicable.

Conflicts of Interest: The authors declare no conflict of interest.

References

1. Liévanos, R.S.; Horne, C. Unequal resilience: The duration of electricity outages. *Energy Policy* **2017**, *108*, 201–211. [CrossRef]
2. Panteli, M.; Mancarella, P. Influence of extreme weather and climate change on the resilience of power systems: Impacts and possible mitigation strategies. *Electr. Power Syst. Res.* **2015**, *127*, 259–270. [CrossRef]
3. Panteli, M.; Mancarella, P.; Trakas, D.N.; Kyriakides, E.; Hatziargyriou, N.D. Metrics and Quantification of Operational and Infrastructure Resilience in Power Systems. *IEEE Trans. Power Syst.* **2017**, *32*, 4732–4742. [CrossRef]

4. Panteli, M.; Pickering, C.; Wilkinson, S.; Dawson, R.; Mancarella, P. Power System Resilience to Extreme Weather: Fragility Modeling, Probabilistic Impact Assessment, and Adaptation Measures. *IEEE Trans. Power Syst.* **2017**, *32*, 3747–3757. [CrossRef]
5. Panteli, M.; Trakas, D.N.; Mancarella, P.; Hatziargyriou, N.D. Power Systems Resilience Assessment: Hardening and Smart Operational Enhancement Strategies. *Proc. IEEE* **2017**, *105*, 1202–1213. [CrossRef]
6. Huang, G.; Wang, J.; Chen, C.; Qi, J.; Guo, C. Integration of Preventive and Emergency Responses for Power Grid Resilience Enhancement. *IEEE Trans. Power Syst.* **2017**, *32*, 4451–4463. [CrossRef]
7. Dehghanian, P.; Aslan, S.; Dehghanian, P. Maintaining Electric System Safety Through An Enhanced Network Resilience. *IEEE Trans. Ind. Appl.* **2018**, *54*, 4927–4937. [CrossRef]
8. Carrión, D.; García, E.; Jaramillo, M.; González, J.W. A Novel Methodology for Optimal SVC Location Considering N-1 Contingencies and Reactive Power Flows Reconfiguration. *Energies* **2021**, *14*, 6652. [CrossRef]
9. Shaheen, A.M.; El-Sehiemy, R.A.; Farrag, S.M. Optimal reactive power dispatch using backtracking search algorithm. *Aust. J. Electr. Electron. Eng.* **2016**, *13*, 200–210. [CrossRef]
10. Shaheen, A.M.; Elsayed, A.M.; El-Sehiemy, R.A.; Ginidi, A.R.; Elattar, E. Optimal management of static volt-ampere-reactive devices and distributed generations with reconfiguration capability in active distribution networks. *Int. Trans. Electr. Energy Syst.* **2021**, *31*, e13126. [CrossRef]
11. Lemus, A.; Carrión, D.; Aguire, E.; González, J.W. Location of distributed resources in rural-urban marginal power grids considering the voltage collapse prediction index. *Ingenius* **2022**, *28*, 25–33. [CrossRef]
12. Molyneaux, L.; Wagner, L.; Froome, C.; Foster, J. Resilience and electricity systems: A comparative analysis. *Energy Policy* **2012**, *47*, 188–201. [CrossRef]
13. Wang, J.; Gharavi, H. Power Grid Resilience [Scanning the Issue]. *Proc. IEEE* **2017**, *105*, 1199–1201. [CrossRef]
14. Zhao, S.; Singh, C. A hybrid method for reliability evaluation of line switching operations. *Electr. Power Syst. Res.* **2018**, *163*, 365–374. [CrossRef]
15. Pinzón, S.; Carrión, D.; Inga, E. Optimal Transmission Switching Considering N-1 Contingencies on Power Transmission Lines. *IEEE Lat. Am. Trans.* **2021**, *19*, 534–541. [CrossRef]
16. Quinteros, F.; Carrión, D.; Jaramillo, M. Optimal Power Systems Restoration Based on Energy Quality and Stability Criteria. *Energies* **2022**, *15*, 2062. [CrossRef]
17. Craig, M.T.; Cohen, S.; Macknick, J.; Draxl, C.; Guerra, O.J.; Sengupta, M.; Haupt, S.E.; Hodge, B.M.; Brancucci, C. A review of the potential impacts of climate change on bulk power system planning and operations in the United States. *Renew. Sustain. Energy Rev.* **2018**, *98*, 255–267. [CrossRef]
18. Varianou Mikellidou, C.; Shakou, L.M.; Boustras, G.; Dimopoulos, C. Energy critical infrastructures at risk from climate change: A state of the art review. *Safety Sci.* **2017**, *110*, 110–120. [CrossRef]
19. Sharifi, A.; Yamagata, Y. Principles and criteria for assessing urban energy resilience: A literature review. *Renew. Sustain. Energy Rev.* **2016**, *60*, 1654–1677. [CrossRef]
20. Masache, P.; Carrión, D.; Cárdenas, J. Optimal Transmission Line Switching to Improve the Reliability of the Power System Considering AC Power Flows. *Energies* **2021**, *14*, 3281. [CrossRef]
21. Bie, Z.; Lin, Y.; Li, G.; Li, F. Battling the Extreme: A Study on the Power System Resilience. *Proc. IEEE* **2017**, *105*, 1253–1266. [CrossRef]
22. Gholami, A.; Shekari, T.; Amirioun, M.H.; Aminifar, F.; Amini, M.H.; Sargolzaei, A. Toward a consensus on the definition and taxonomy of power system resilience. *IEEE Access* **2018**, *6*, 32035–32053. [CrossRef]
23. Arghandeh, R.; Von Meier, A.; Mehrmanesh, L.; Mili, L. On the definition of cyber-physical resilience in power systems. *Renew. Sustain. Energy Rev.* **2016**, *58*, 1060–1069. [CrossRef]
24. Eshghi, K.; Johnson, B.K.; Rieger, C.G. Metrics required for power system resilient operations and protection. In *Proceedings of the 2016 Resilience Week, RWS 2016, Chicago, IL, USA, 16–18 August 2016*; pp. 200–203. [CrossRef]
25. Shen, L.; Cassottana, B.; Tang, L.C. Statistical trend tests for resilience of power systems. *Reliab. Eng. Syst. Saf.* **2018**, *177*, 138–147. [CrossRef]
26. Reed, D.A.; Kapur, K.C.; Christie, R.D. Methodology for assessing the resilience of networked infrastructure. *IEEE Syst. J.* **2009**, *3*, 174–180. [CrossRef]
27. Patriarca, R.; Bergström, J.; Di Gravio, G.; Costantino, F. Resilience engineering: Current status of the research and future challenges. *Saf. Sci.* **2018**, *102*, 79–100. [CrossRef]
28. Panteli, M.; Mancarella, P. Modeling and evaluating the resilience of critical electrical power infrastructure to extreme weather events. *IEEE Syst. J.* **2017**, *11*, 1733–1742. [CrossRef]
29. Wang, C.; Hou, Y.; Qiu, F.; Lei, S.; Liu, K. Resilience Enhancement with Sequentially Proactive Operation Strategies. *IEEE Trans. Power Syst.* **2017**, *32*, 2847–2857. [CrossRef]
30. Yang, J.; Jiang, K. The sensitive line identification in resilient power system based on fault chain model. *Int. J. Electr. Power Energy Syst.* **2017**, *92*, 212–220. [CrossRef]
31. Kancherla, S.; Dobson, I. Heavy-tailed transmission line restoration times observed in utility data. *IEEE Trans. Power Syst.* **2018**, *33*, 1145–1147. [CrossRef]
32. Wang, Y.; Chen, C.; Wang, J.; Baldick, R. Research on Resilience of Power Systems under Natural Disasters—A Review. *IEEE Trans. Power Syst.* **2016**, *31*, 1604–1613. [CrossRef]

33. Kishore, T.S.; Singal, S.K. Optimal economic planning of power transmission lines: A review. *Renew. Sustain. Energy Rev.* **2014**, *39*, 949–974. [CrossRef]
34. Shaheen, A.; Ginidi, A.; El-Sehiemy, R.; Elsayed, A.; Elattar, E.; Dorrah, H.T. Developed Gorilla Troops Technique for Optimal Power Flow Problem in Electrical Power Systems. *Mathematics* **2022**, *10*, 1636. [CrossRef]
35. Ginidi, A.; Elsayed, A.; Shaheen, A.; Elattar, E.; El-Sehiemy, R. An innovative hybrid heap-based and jellyfish search algorithm for combined heat and power economic dispatch in electrical grids. *Mathematics* **2021**, *9*, 2053. [CrossRef]
36. Sarhan, S.; Shaheen, A.; El-Sehiemy, R.; Gafar, M. A Multi-Objective Teaching–Learning Studying-Based Algorithm for Large-Scale Dispatching of Combined Electrical Power and Heat Energies. *Mathematics* **2022**, *10*, 2278. [CrossRef]
37. Salkuti, S.R. Congestion management using optimal transmission switching. *IEEE Syst. J.* **2018**, *12*, 3555–3564. [CrossRef]
38. Zhang, H.; Vittal, V.; Heydt, G.T.; Quintero, J. A mixed-integer linear programming approach for multi-stage security-constrained transmission expansion planning. *IEEE Trans. Power Syst.* **2012**, *27*, 1125–1133. [CrossRef]
39. Khodaei, A.; Shahidehpour, M.; Kamalinia, S. Transmission switching in expansion planning. *IEEE Trans. Power Syst.* **2010**, *25*, 1722–1733. [CrossRef]
40. Orfanos, G.A.; Georgilakis, P.S.; Hatziargyriou, N.D. Transmission expansion planning of systems with increasing wind power integration. *IEEE Trans. Power Syst.* **2013**, *28*, 1355–1362. [CrossRef]
41. Silva, I.d.J.; Rider, M.J.; Romero, R.; Murari, C.A.F. Transmission network expansion planning considering uncertainty in demand. *IEEE Trans. Power Syst.* **2006**, *21*, 1565–1573. [CrossRef]
42. Zhang, Q.; Sahraei-Ardakani, M. Distributed DCOPF with flexible transmission. *Electr. Power Syst. Res.* **2018**, *154*, 37–47. [CrossRef]
43. Püschel-Løvengreen, S.; Ghazavi Dozein, M.; Low, S.; Mancarella, P. Separation event-constrained optimal power flow to enhance resilience in low-inertia power systems. *Electr. Power Syst. Res.* **2020**, *189*, 106678. [CrossRef]
44. Kannan, R.; Luedtke, J.R.; Roald, L.A. Stochastic DC optimal power flow with reserve saturation. *Electr. Power Syst. Res.* **2020**, *189*, 106566. [CrossRef]
45. Vaishya, S.; Sarkar, V. Accurate loss modelling in the DCOPF calculation for power markets via static piecewise linear loss approximation based upon line loading classification. *Electr. Power Syst. Res.* **2019**, *170*, 150–157. [CrossRef]
46. Fisher, E.B.; O'Neill, R.P.; Ferris, M.C. Optimal Transmission Switching. *IEEE Trans. Power Syst.* **2008**, *23*, 1346–1355. [CrossRef]
47. Hedman, K.W.; O'Neill, R.P.; Fisher, E.B.; Oren, S.S. Optimal Transmission Switching With Contingency Analysis. *IEEE Trans. Power Syst.* **2009**, *24*, 1577–1586. [CrossRef]
48. Barrows, C.; Blumsack, S. Transmission Switching in the RTS-96 Test System. *IEEE Trans. Power Syst.* **2011**, *27*, 1134–1135. [CrossRef]
49. Fuller, J.D.; Ramasra, R.; Cha, A. Fast heuristics for transmission-line switching. *IEEE Trans. Power Syst.* **2012**, *27*, 1377–1386. [CrossRef]
50. Khanabadi, M.; Ghasemi, H.; Doostizadeh, M. Optimal transmission switching considering voltage security and N-1 contingency analysis. *IEEE Trans. Power Syst.* **2013**, *28*, 542–550. [CrossRef]
51. Soroush, M.; Fuller, J.D. Accuracies of optimal transmission switching heuristics based on DCOPF and ACOPF. *IEEE Trans. Power Syst.* **2014**, *29*, 924–932. [CrossRef]
52. Ostrowski, J.; Wang, J.; Liu, C. Transmission switching with connectivity-ensuring constraints. *IEEE Trans. Power Syst.* **2014**, *29*, 2621–2627. [CrossRef]
53. Kocuk, B.; Dey, S.S.; Sun, X.A. New Formulation and Strong MISOCP Relaxations for AC Optimal Transmission Switching Problem. *IEEE Trans. Power Syst.* **2017**, *32*, 4161–4170. [CrossRef]
54. Henneaux, P.; Kirschen, D.S. Probabilistic security analysis of optimal transmission switching. *IEEE Trans. Power Syst.* **2016**, *31*, 508–517. [CrossRef]
55. Tarafdar Hagh, M.; Zamani Gargari, M.; Vahid Pakdel, M.J. Sequential analysis of optimal transmission switching with contingency assessment. *IET Gener. Transm. Distrib.* **2017**, *12*, 1390–1396. [CrossRef]
56. Li, X.; Balasubramanian, P.; Sahraei-Ardakani, M.; Abdi-Khorsand, M.; Hedman, K.W.; Podmore, R. Real-Time Contingency Analysis with Corrective Transmission Switching. *IEEE Trans. Power Syst.* **2017**, *32*, 2604–2617. [CrossRef]
57. Lan, T.; Zhou, Z.; Huang, G.M. Modeling and Numerical Analysis of Stochastic Optimal Transmission Switching with DCOPF and ACOPF. *IFAC-PapersOnLine* **2018**, *51*, 126–131. [CrossRef]
58. Flores, M.; Macedo, L.H.; Romero, R. Alternative Mathematical Models for the Optimal Transmission Switching Problem. *IEEE Syst. J.* **2021**, *15*, 1245–1255. [CrossRef]
59. EL-Azab, M.; Omran, W.; Mekhamer, S.; Talaat, H. Congestion management of power systems by optimizing grid topology and using dynamic thermal rating. *Electr. Power Syst. Res.* **2021**, *199*, 107433. [CrossRef]
60. Ejebe, G.C.; Wollenberg, B.F. Automatic Contingency Selection. *IEEE Trans. Power Appar. Syst.* **1979**, *PAS-98*, 97–109. [CrossRef]
61. Zimmerman, R.D.; Murillo-Sánchez, C.E.; Thomas, R.J. MATPOWER: Steady-State Operations, Planning, and Analysis Tools for Power Systems Research and Education. *IEEE Trans. Power Syst.* **2011**, *26*, 12–19. [CrossRef]

Article

Single and Multi-Objective Optimal Power Flow Based on Hunger Games Search with Pareto Concept Optimization

Murtadha Al-Kaabi *, Virgil Dumbrava and Mircea Eremia

Department of Power Systems, Faculty of Energy, University Politehnica of Bucharest, 060029 Bucharest, Romania
* Correspondence: mmsk.1986s@gmail.com; Tel.: +40-964-7500580065

Abstract: In this study, a new meta-heuristic optimization method inspired by the behavioral choices of animals and hunger-driven activities, called hunger games search (HGS), is suggested to solve and formulate the single- and multi-objective optimal power flow problem in power systems. The main aim of this study is to optimize the objective functions, which are total fuel cost of generator, active power losses in transmission lines, total emission issued by fossil-fueled thermal units, voltage deviation at PQ bus, and voltage stability index. The proposed HGS approach is optimal and easy, avoids stagnation in local optima, and can solve multi-constrained objectives. Various single-and multi-objective (conflicting) functions were proposed simultaneously to solve OPF problems. The proposed algorithm (HGS) was developed to solve the multi-objective function, called the multi-objective hunger game search (MOHGS), by incorporating the proposed optimization (HGS) with Pareto optimization. The fuzzy membership theory is the function responsible to extract the best compromise solution from non-dominated solutions. The crowding distance is the strategies carried out to determine and ordering the Pareto non-dominated set. Two standard tests (IEEE 30 bus and IEEE 57 bus systems) are the power systems that were applied to investigate the performance of the proposed approaches (HGS and MOHGS) for solving single and multiple objective functions with 25 studied cases using MATLAB software. The numerical results obtained by the proposed approaches (HGS and MOHGS) were compared to other optimization algorithms in the literature. The numerical results confirmed the efficiency and superiority of the proposed approaches by achieving an optimal solution and giving the faster convergence characteristics in single objective functions and extracting the best compromise solution and well-distributed Pareto front solutions in multi-objective functions.

Keywords: multi-objective optimal power flow (MOOPF); hunger games search (HGS); multi-objective hunger games search (MOHGS); Pareto concept; fuzzy set theory; fuel cost; active power losses; emission; voltage deviation; voltage stability index

Citation: Al-Kaabi, M.; Dumbrava, V.; Eremia, M. Single and Multi-Objective Optimal Power Flow Based on Hunger Games Search with Pareto Concept Optimization. *Energies* **2022**, *15*, 8328. https://doi.org/10.3390/en15228328

Academic Editors: Yuan Liao and Ke Xu

Received: 17 October 2022
Accepted: 2 November 2022
Published: 8 November 2022

Publisher's Note: MDPI stays neutral with regard to jurisdictional claims in published maps and institutional affiliations.

1. Introduction

Power flow is one of the most important tools used to analyze and operate the power system by finding out active power losses in transmission lines and reactive power injection on lines and calculating the voltage at different buses. Optimal power flow (OPF) is a non-convex, nonlinear, and large-scale problem. The main aim of using OPF in power systems is to obtain the optimal objective functions by setting the control variables. These control variables involve real power output of generation units except the slack bus, the voltages magnitude at PV bus, regulating tap setting at transformers, and the source VAR compensator connected to transmission lines with satisfying the equality and inequality constraints. The objective functions that will be optimized are the total fuel cost of generation units, active power losses in transmission lines, total emission, voltage deviation, and the voltage stability index of system. The first presentation of the OPF problem formulation was presented by Dommel and Tinney 1968 [1].

Traditional and metaheuristic optimization methods are the types used to solve OPF problems in the power system. Linear programming, Newton methods, gradient based method, quadratic programming, and dynamic programming are the traditional methods that were proposed to solve OPF problems [2]. To overcome these weaknesses, many metaheuristic optimization methods were developed to solve OPF problems, such as differential evolution (DE) [3], genetic algorithm (GA) [4], modified artificial bee colony (MABC) [5], improved differential evolution (IDE) [6,7], adaptive constraint differential evolution (ACDE) [8], enhanced adaptive differential evolution with self-adaptive (EJADE-SP) [9], crisscross search based grey wolf optimizer (GS-GWO) [10], Harris hawks optimization [11], and fruit fly optimization (FFO) [12]. The previous metaheuristic optimization methods were applied to single-objective OPF problems.

Recently, multi-objective OPF problems (MOOPF) are solved by several optimization methods. These problems are described as nonlinear and large-scale. For example, improved differential evolution algorithm (IDEA) have been developed to solve multi-objective functions, called the multi-objective improved differential evolution algorithm (MOIDEA) [13]. The multi-objective slime mould algorithm (MOSMA) was proposed as a new optimization method to solve multi-objective optimal power flow (MOOPF) problems [14]. The multi-objective backtracking search algorithm (MOBSA) was suggested as a new multi-objective approach to solve and formulate MOOPF problems on power systems [15]. The multi-objective evolutionary algorithm (MOEA) was proposed as novel hybrid decomposition and local dominance to solve MOOPF problems with conflicting objectives [16]. The author in Ref. [17] suggested a new multi-objective method, called the multi-objective search group algorithm (MOSGA), which efficiently solves MOOPF problems in power systems.

In this work, the hunger games search algorithm was proposed to solve single objective OPF problems. In addition, the proposed algorithm HGS was developed to solve multi-objective optimal power flow (MOOPF) problems, namely, the multi-objective hunger games search (MOHGS). The proposed approach integrates the Pareto concept with the fuzzy membership approach to determine the nondominated solutions and extract the best compromise solution, respectively. The main aim of this approach is to optimize the objective function by satisfying the equality and inequality constraints. The objective functions that are minimized are total fuel cost, active power losses on transmission lines, total emission, voltage deviation at PQ busses, and the voltage stability index of whole system. In a multi-objective function. these objective functions that will be optimized may be conflicting simultaneously, which means when the first objective function is minimized, the second objective function will be maximized and vice versa. Therefore, the main goal in using the multi-objective optimal power flow (MOOPF) is to find out the compromise solution among multiple objective functions.

The hunger games search (HGS) is a stochastic optimization algorithm inspired by the behavior of social animal cooperation, which is proportional to their level of hunger, by Y. Yang et al. in 2021 [18]. The HGS is characterized by high convergence speed, powerful local exploitation, and global exploration. The main contributions of this paper can be summarized as follows:

- The hunger games search (HGS) and the developed approach, multi-objective hunger game search (MOOPF), are proposed to solve for single-and multi-objective optimal power flow (MOOPF) in power systems. Pareto concept (PC), fuzzy membership function (FMF), and crowding distance (CD) are the approaches used to find out non-dominated Pareto fronts (NDPF), extract the best compromise solution (BCS), and rank and reduce the Pareto repository, respectively.
- The IEEE 30-bus and IEEE 57-bus tests are the power systems applied to 25 studied cases by using single and multiple (Bi, Tri, Quad, and Quinta) objective functions. The numerical results obtained by the proposed approaches (HGS and MOHGS) will be compared with known optimization methods in the literature.

The rest of the paper will be arranged as follows: Section 2 introduces the formulation of OPF problems. Section 3 is focused on single- and multi-objective OPF frameworks. The simulation results of proposed approaches (HGS and MOHGS) are presented in Section 4. Finally, this paper will be finished with the conclusion.

2. The Formulation of the OPF Problem

The main aim of applied single- and multi-objective OPF in power systems is to optimize the objective functions of single and multiple objectives (Bi, Tri, Quad, and Quinta) through setting optimal control variables (active power output of generators except for active power output of slack bus, voltage magnitude of PV bus, Source VAR compensator, and tap setting regulating of transformers) with satisfying equality and inequality constraints, simultaneously. The mathematical model can be formulated as follows:

$$
\begin{aligned}
\text{Optimize} \quad & f(x,u) = f_1(x,u), f_2(x,u), \ldots, f_{N_{obj}}(x,u) \\
\text{subjected to} \quad & g(x,u) = 0 \\
& h(x,u) \leq 0
\end{aligned}
\tag{1}
$$

where $f(x, u)$, $g(x, u)$, and $h(x, u)$ represent the objective function to be optimized, equality, and inequality constraints, respectively. x and u are the state and control variables, respectively.

2.1. The State and Variables

The mathematical expression of the state variable x is:

$$
x = \left[P_{G_1}, \left| V_{L_1} \right|, \cdots, \left| V_{L_{N_{PQ}}} \right|, Q_{G_1}, \cdots, Q_{G_{N_{PV}}}, S_{l_1}, \cdots, S_{l_{N_l}} \right]
\tag{2}
$$

where P_{G1} is the active power at slack bus; V_L is the voltage magnitude at load bus (PQ buses); Q_G is the reactive power at PV bus; and S_l is the transmission line loading; N_{PV}, N_{PQ}, and N_l denote numbers of generators, load buses, transmission lines, respectively. The mathematical expression of the state variable u is:

$$
u = \left[P_{G_2}, \cdots, P_{G_{N_G}}, \left| V_{G_1} \right|, \cdots, \left| V_{G_{N_G}} \right|, T_1, \cdots, T_{N_T}, Q_{C_1}, \cdots, Q_{C_{N_C}} \right]
\tag{3}
$$

where P_G is the active power at the PV bus except the slack bus. V_G is the voltage magnitude at PV buses. T denotes the tap settings regulating transformers. Q_c is the source VAR compensator. N_G, N_T, and N_C are the numbers of generators, source VAR (shunt) compensators, and regulating transformers, respectively.

2.2. Objective Constraints

The constraints of OPF can be divided into equality and inequality constraints. The equality constraints represent the physical structure (the power balance equations) of the whole system.

$$
\begin{aligned}
P_{Gi} - P_{Di} - |V_i| \sum_{j=1}^{N_B} |V_j| \left(G_{ij} \cos(\theta_{ij}) + B_{ij} \sin(\theta_{ij}) \right) = 0 \quad & \forall i \in N \\
Q_{Gi} + Q_{Ci} - Q_{Di} - |V_i| \sum_{j=1}^{N_B} |V_j| \left(G_{ij} \sin(\theta_{ij}) + B_{ij} \cos(\theta_{ij}) \right) = 0 \quad & \forall i \in N
\end{aligned}
\tag{4}
$$

where N_B denotes the number of buses. P_G, P_D are the active power output of generator and load demand, resepectively. Q_G, Q_C, and Q_D are the reactive power output of generator, shunt VAR compensator, and load demand, respectively. G_{ij} and B_{ij} are the conductance and susceptance between bus i and bus j, respectively. θ_{ij} is the voltage angle difference between bus i and bus j.

The inequality constraints represent the operation limit of the equipment (generators, transformers, shunt compensators, and security). It can be presented as:

- Generator constraints:

$$P_{Gi}{}^{min} \leq P_{Gi} \leq P_{Gi}{}^{max} \qquad i = 1, 2, \ldots, N_G \tag{5}$$

$$Q_{Gi}{}^{min} \leq Q_{Gi} \leq Q_{Gi}{}^{max} \qquad i = 1, 2, \ldots, N_G \tag{6}$$

$$V_{Gi}{}^{min} \leq V_{Gi} \leq V_{Gi}{}^{max} \qquad i = 1, 2, \ldots, N_G \tag{7}$$

where P_G^{max}, Q_G^{max}, and V_G^{max} denote the maximum limit of active power, reactive power, and voltage magnitude of generators, respectively. P_G^{min}, Q_G^{min}, and V_G^{min} are the minimum limit of active power, reactive power, and voltage magnitude of generators, respectively.

- Transformer constraints:

$$T_j{}^{min} \leq T_j \leq T_j{}^{max} \qquad j = 1, 2, \ldots, N_T \tag{8}$$

where T^{max} and T^{min} represent the maximum and minimum limit of tap regulating at transformers, respectively. N_T is the number of tap changers.

- Shunt compensator constraints:

$$Q_{C_k}{}^{min} \leq Q_{C_k} \leq Q_{C_k}{}^{max} \qquad k = 1, 2, \ldots, N_C \tag{9}$$

where Q^{max} and Q^{min} denote the maximum and minimum limit of the shunt compensator. N_C is the number of shunt capacitor sources.

- Security constraints:

$$V_{Li}{}^{min} \leq V_{Li} \leq V_{Li}{}^{max} \qquad i = 1, 2, \ldots, N_L \tag{10}$$

$$S_{L_m} \leq S_{L_m}{}^{max} \qquad m = 1, 2, \ldots, N_{nl} \tag{11}$$

where S_L^{max} is the maximum limit of MVA in the transmission line. N_{nl} is the number of transmission lines.

2.3. Objective Functions

In this paper, the five most common objective functions were optimized to solve OPF problems, which are fuel cost, losses, emission, voltage deviation, and the voltage stability index.

- Total Fuel Cost [\$/h]

The mathematical formulation for the total fuel cost in the power system can be expressed as follows [19]:

$$F_C = \sum_{i=1}^{N_G} f_i(P_{G_i}) = \sum_{i=1}^{N_G} \left(a_i P_{G_i}^2 + b_i P_{G_i} + c_i \right) \quad [\$/h] \tag{12}$$

a_i, b_i, and c_i denote fuel cost coefficients for generators.

- Active power losses [MW]

The second objective function in this study is active power losses in transmission lines. It can be expressed as [19]:

$$F_{loss} = \sum_{k=1}^{N_{nl}} G_i \left(V_i^2 + V_j^2 - 2V_i V_j \cos \delta_{i,j} \right) \quad [MW] \tag{13}$$

where F_{loss} is the total real losses in transmission lines. G_i is the transfer conductance.

- Total emission [ton/h]

This objective function aims to minimize the emission level in the atmosphere by reducing pollution gases, such as NO_x and SO_2. The mathematical description of this objective function can be expressed as [7]:

$$F_{em} = \sum_{i=1}^{N_G} 10^{-2}\left(\alpha_i + \beta_i P_{G_i} + \gamma_i P_{G_i}^2\right) + \zeta_i \exp\left(\lambda_i P_{G_i}\right) \quad [\text{ton/h}] \tag{14}$$

α_i, β_i, γ_i, ζ_i, and λ_i denote fuel emission coefficients in generators.

- Voltage deviation [p.u.]

The goal of this case is to optimize the voltage quality in the system by minimizing the voltage deviations at the PQ bus from 1.0 [p.u.] The mathematical expression of this objective is expressed as [20]:

$$F_{VD} = \sum_{i=1}^{N_{PQ}} |V_i - 1.0| \quad [\text{p.u.}] \tag{15}$$

- Voltage stability index

The last objective function is to enhance the voltage stability index in the whole system by minimizing the maximum value of the voltage stability indicator (L-index). The mathematical formulation of the L-index is defined as [20]:

$$L_j = \left|1 - \sum_{i=1}^{N_G}\left(F_{ji} \times \frac{V_i}{V_j}\right)\right| \quad j = 1, 2, \ldots, Nl \tag{16}$$

$$F_{ji} = -[Y_1]^{-1} \times [Y_2] \tag{17}$$

where Y_1 and Y_2 denote the sub-matrices of the admittance matrix (Y_{bus}). It can describe this objective function as:

$$F_{VSI} = Max\left(L_j\right) \quad j = 1, 2, \ldots, Nl \tag{18}$$

The above objective functions are related as a single objective function, and they can be obtained by the proposed HGS approach. In addition to the single objective function, the multi-objective function was proposed to solve the multi-objective optimal power flow (MOOPF). In this article, many objective functions were optimized (they may be conflicting) simultaneously while satisfying equality and inequality constraints. Generally, a multi-objective function can be expressed by the following equation:

$$f_m(x, u) = \sum_{m=1}^{N} \omega_m f_m(x, u) = \omega_1 f_1 + \omega_2 f_2 + \ldots + \omega_m f_m \quad m = 1, 2, \ldots, N \tag{19}$$

where f denotes the MOF with No. of functions, f_m is individual OF, and ω is the weight coefficient at the ranges [0,1], N denotes the No. of OF. The sum of weight coefficients is equal to 1. In this article, the weight coefficient of each OF is inverse the number of OF. If number of OF equal 2 (Bi objective function), the weight coefficient equal to 0.5, and so on. In other words, there is no preference for any OF over the others. Therefore, we do not have any prime significance for any OF.

3. Single- and Multi-Objective OPF Frameworks

3.1. Conventional Hunger Games Search

Hunger games search (HGS) is a new optimization technique inspired by the behavior of social animal cooperatives, which is proportional to their level of hunger. The processes that characterized this algorithm into two stages can be summarized as follows:

1. Approach food

The mathematical model of this process is described as follows:

$$Y(t+1) = \begin{cases} Y(t) \cdot (1 + rand(1)), r_1 < l \\ V_1 \cdot Y_b + S \cdot V_2 \cdot |Y_b - Y(t)|, r_1 > l, r_2 > e \\ V_1 \cdot Y_b - S \cdot V_2 \cdot |Y_b - Y(t)|, r_1 > l, r_2 < e \end{cases} \tag{20}$$

where $Y(t+1)$ is the position of an individual in the next iteration; $Y(t)$ is the position of an individual in the current iteration; Y_b denotes the position of the individual chosen randomly in optimal individuals; $rand$ (1), r_1, and r_2 are the random numbers within $[0, 1]$; V_1 and V_2 represent the hunger weights; S is the number within $[-a, a]$; l is the parameter setting experiment value; The formula of e is:

$$e = Sech(|f(i) - bf|) \tag{21}$$

where $f(i)$ denotes the fitness value; and bf is the best fitness. *Sech* is a hyperbolic function and can be described as:

$$\left(Sech(x) = \frac{2}{e^x + e^{-x}} \right) \tag{22}$$

S can be formulated as:

$$S = 2 \times A \times r - A$$
$$A = 2 \times \left(1 - \frac{t}{Max\_iter} \right) \tag{23}$$

where r is a random number within $[0, 1]$; and Max_iter is the maximum iteration.

2. Hunger role

The characteristics of starvation can be expressed as a mathematical formulation as follows:

$$V_1(l) = \begin{cases} hun(i) \cdot \frac{N}{SHun} \times r_4 & r_3 < l \\ 1 & r_3 > l \end{cases} \tag{24}$$

$$V_2(l) = (1 - \exp(-|hun(i) - SHun|)) \times r_5 \times 2 \tag{25}$$

where N is the number of populations, and $SHun$ is the sum of populations that feel hunger. r_3, r_4, and r_5 are random numbers within $[0, 1]$. hun denotes the population's hunger and can be formulated as:

$$hun = \begin{cases} 0 & Allfitness(i) == bf \\ hun(i) + h & Allfitness(i)! = bf \end{cases}$$

$$h = \begin{cases} Lh \times (1 + r), & Th < Lh \\ Th & Th \geq Lh \end{cases} \tag{26}$$

$$Th = \frac{f(i) - bf}{wf - bf} \times r_6 \times 2 \times (ub - lb)$$

where $Allfitness(i)$ and $f(i)$ are the fitness of each population; h denotes the hunger sensation; r_6 is a random number within $[0, 1]$; bf and wf are the best and the worst fitness; ub and lb refer to the upper and lower bounds; and Lh is the lower bound.

To describe the process of the proposed approach, HGS, to solve optimal power flow (OPF), a pseudo-code and flowchart are presented in Algorithm 1 and Figure 1.

Algorithm 1: Pseudo-code of HGS.

1.	Input all parameters N, $Max\_iter, l, D, SHun$
2.	Determine the location of population X_i ($i = 1, 2, \cdots, N$)
3.	While ($T \leq Max\_iter$)
4.	Calculate $f(i)$
5.	Update bf, wf, Y_b, bl
6.	Calculate the population's hunger hun by Equation (26)
7.	Calculate the hunger weight (1) V_1 by Equation (24)
8.	Calculate the hunger weight (2) V_2 by Equation (25)
9.	For each population
10.	Calculate e by Equation (21)
11.	Update S by Equation (22)
12.	Update location by Equation (20)
13.	End (For)
14.	$T = T + 1$
15.	End (While)
16.	Return bf, Y_b

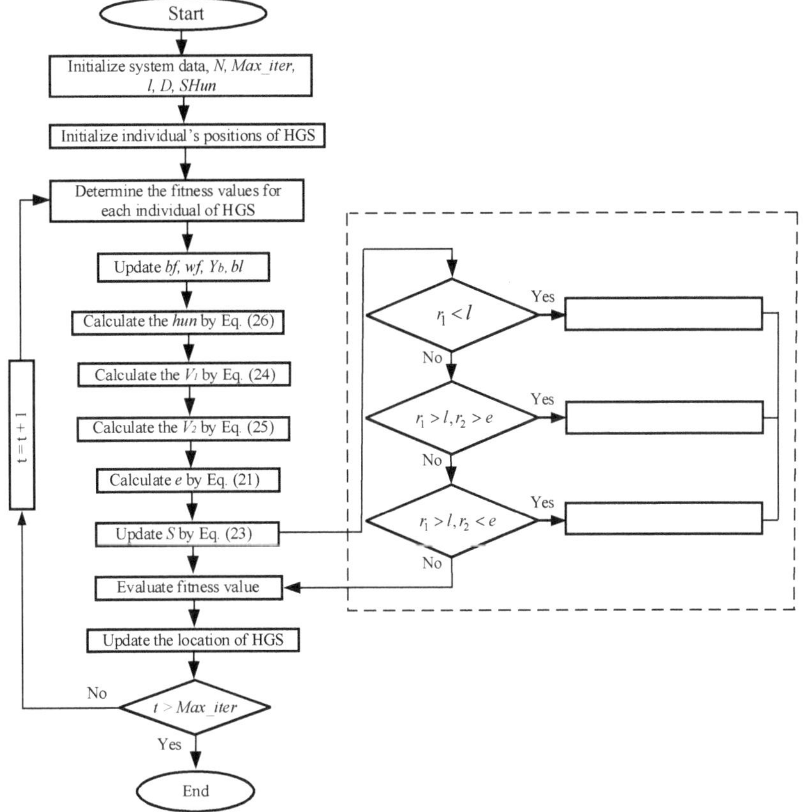

Figure 1. Flowchart of the HGS.

3.2. Multi-Objective Hunger Games Search (MOHGS)

A new approach was proposed to solve MOOPF problems (two or more objective functions) and optimize simultaneously, namely, the multi-objective hunger games search (MOHGS). Based on the number of objective functions, the Pareto concept (PC) is the proposed approach to find out the dominant and non-dominated solutions. The decision maker is responsible to determine the best compromise solution (BCS) from non-dominated solutions (NDS). Many approaches were suggested to determine the BCS, such as the fuzzy set theory, entropy criterions, and centroid concept [21]. The most popular approach used to find out the BCS is the fuzzy decision maker [21–24]. In this study, the fuzzy membership function (FMF) is the equation used to extract the BCS from NDS. Finally, the specific strategy that is employed to reduce and arrange the NDPF is the crowding distance (CD).

1.   Pareto concept (PC)

The Pareto concept (PC) is the method used to find the NDS. In general, a solution x_1 dominates solution x_2 when:

$$\begin{aligned}
\forall i \in \{1,2,\ldots,n\} : f_i(x_1) \le f_i(x_2) \\
\forall j \in \{1,2,\ldots,n\} : f_j(x_1) < f_j(x_2)
\end{aligned} \tag{27}$$

2.   The best compromise solution (BCS)

The equation below is used to determine the BCS based on fuzzy membership function (FMF) (Figure 2).

$$u_i^k = \begin{cases} 1 & F_i \le F_i^{min} \\ \dfrac{F_i^{max}-F_i}{F_i^{max}-F_i^{min}} & F_i^{min} < F_i < F_i^{max} \\ 0 & F_i \ge F_i^{max} \end{cases} \tag{28}$$

$$u^k = \frac{\sum\limits_{i=1}^{N_{obj}} u_i^k}{\sum\limits_{k=1}^{M}\sum\limits_{i=1}^{N_{obj}} u_i^k} \tag{29}$$

where F_i^{min} and F_i^{max} represent the minimum and maximum values of NDS. u_i^k denotes the normalized membership function of each NDS. u^k is the BCS when having the maximum value [25].

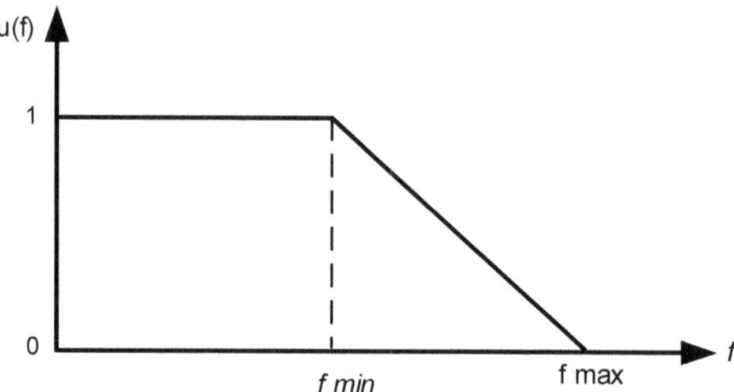

Figure 2. The membership functions.

3.   Crowding distance strategy

This strategy is employed to rank and reduce NDS in NDPF. The average of two neighboring NDS represents the value of crowding distance (CD). First, the fitness of for each population should be sorted in a descending order. Thereafter, the boundary solutions are determined as an infinite value for each objective function. For the remainder solutions, the corresponding diagonal length of the intermediate solutions is assigned. Figure 3 represents the CD of an individual solution as the diagonal length of the cuboid. It is expressed as follows:

$$cd_i = \prod_{n=1}^{m} \frac{|f_n(x_i + 1) - f_n(x_i - 1)|}{f_{n,min}}, \qquad i = 1, 2, \cdots, n_b \tag{30}$$

where cd_i represents the crowding distance; $F_{n,\,min}$ denotes the minimum value of the nth objective function; N_b is the No. of candidate solutions.

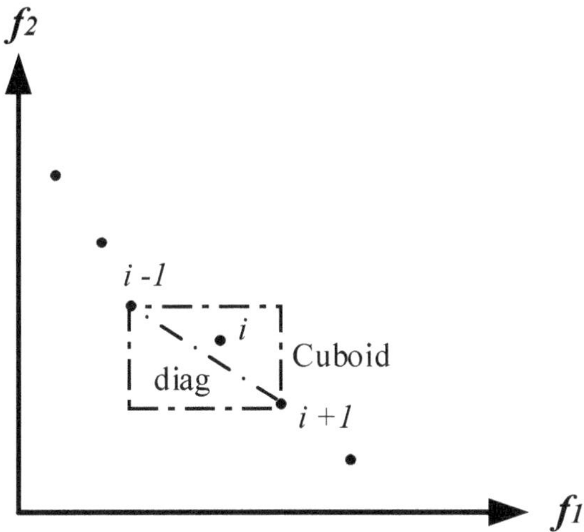

Figure 3. The crowding distance estimation.

4. Stages of multi-objective HGS

The phases of MOHGS can be summarized as follows:

Step 1: Initialize system data such as max iteration, No. of population, the sum of populations that feel hunger, No. of NDS, No. of control variables, the parameter setting experiment, etc.

Step 2: Initialize individual's positions in the HGS.

Step 3: Calculate fitness values for each individual in the HGS.

Step 4: Sort NDS according to the fitness value and store it in the initial repository.

Step 5: Calculate the best and the worst fitness, the upper and lower bounds, and the position individual is chosen randomly.

Step 6: Calculate the *hun* by (25).

Step 7: Calculate e, S, and the hunger weights (V_1 *and* V_2) by (20), (22), (23), and (24).

Step 8: Update the position of the individual in the next iteration (Y(t+1)), based on (19).

Step 9: Calculate the fitness value of each HGS.

Step 10: Update the position of HGS.

Step 11: Sort the NDS of HGS and store it in the HGS repository.

Step 12: Combine the NDS in the initial and HGS repository with the NDS of the HGS, updated to find new NDS.

Step 13: The stopping criteria, if the number of non-dominated solutions is equal to 500, end. Otherwise, store the non-dominated solutions in the initial repository, return to step 3.

The processes used to solve MOOPF problems with the developed approach MOHGS can be expressed in Figure 4.

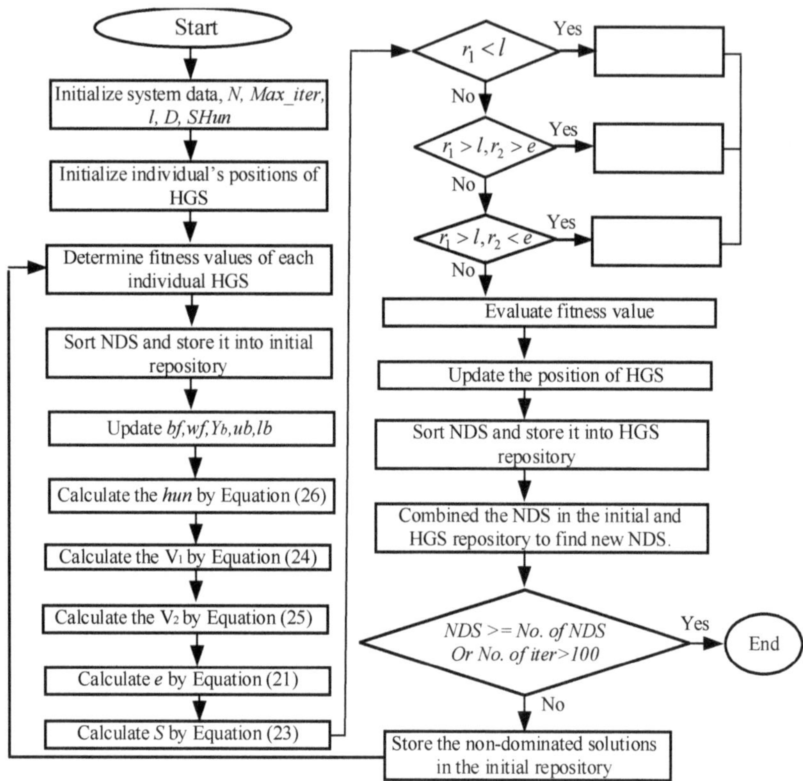

Figure 4. Flowchart of MOHGS.

4. Simulation Results and Discussion

In this study, two bus power systems were tested, IEEE 30 and IEEE 57 bus, and twenty-five case studies were investigated to prove the viability and efficiency of the proposed approaches (HGS and MOHGS). Table 1 presents the characteristics of these systems. The cases studied are summarized in Table 2. The simulation results were carried out on Intel Core (TM) i5-10500H 2.5GHz (12 CPUs) and 16384 (64 bit) MB RAM. The active power output of the generator without the slack bus, the voltage of the generator bus, tap setting regulating of transformers, and shunt capacitors VAR, which are connected on transmission lines, are the control variables that were set to obtain the optimal objective function.

4.1. IEEE 30-Bus Power System

In this article, the IEEE 30-bus system represents the first test system that was applied to validate the effectiveness and superiority of the proposed approaches (HGS and MO-HGS), as shown in Figure 5. The control variable for the OPF problem for the IEEE 30-bus system is 24. The cost and emission coefficients of this system are illustrated in Table 3. The maximum and minimum limits of voltage magnitude for generators and load bus are in the range [0.9–1.1] and [0.95–1.05], respectively.

4.1.1. Single-Objective OPF

Five objective functions were applied to demonstrate the efficiency and superiority of the proposed algorithm HGS. In a single OF, the objective functions that are optimized are total fuel cost [\$/h], total emission [ton/h], active power losses [MW], voltage deviation [p.u.], and voltage stability index. The parameters chosen to solve a single OF are 1000 iterations and 250 population size.

Table 1. The main characteristics of the systems.

System Characteristics	IEEE 30-Bus	IEEE 57-Bus
Buses	30	57
Branches	41	80
Generators	9 (Buses:1, 2, 5, 8, 11, and 13)	7 (Buses: 1, 2, 3, 6, 8, 9, 12)
Generator voltage limits	0.90–1.1 [p.u.]	0.90–1.1 [p.u.]
Load voltage limits	0.95–1.05 [p.u.]	0.94–1.06 [p.u.]
Limit of tap changer setting	0.90–1.1 [p.u.]	0.90–1.1 [p.u.]
Limit of VAR	0–5 [p.u.]	0–20 [p.u.]
Shunts	9 (Buses: 10, 12, 15, 17, 20, 21, 23, 24, and 29)	3 (Buses: 18, 25, 53)
Transformers	4 (Buses: 11, 12, 15, and 36)	17 (Buses: 19, 20, 31, 35, 36, 37, 41, 46, 54, 58, 59, 65, 66, 71, 73, 76, 80)
MW demand	283.4 [MW]	1250.8 [MW]
Control variables	24	33

Table 2. The studied cases with different types.

System	Type of OF(s)	Case #	FC	Em	APL	VD	VSI
IEEE 30-bus	Single OF (s)	#1	√				
		#2			√		
		#3		√			
		#4				√	
		#5					√
	Bi-OF(s)	#6	√	√			
		#7	√		√		
		#8	√			√	
		#9	√				√
		#10		√		√	
		#11			√	√	
		#12				√	√
	Triple-OF(s)	#13	√	√	√		
		#14	√	√		√	
		#15	√		√	√	
		#16	√	√		√	
		#17	√		√		√
		#18	√	√			√
		#19	√			√	√
	Quad-OF(s)	#20	√	√	√	√	
		#21	√	√	√		√
	Quinta-OF(s)	#22	√	√	√	√	√
IEEE 57-bus	Single OF (s)	#23	√				
		#24		√			
		#25			√		

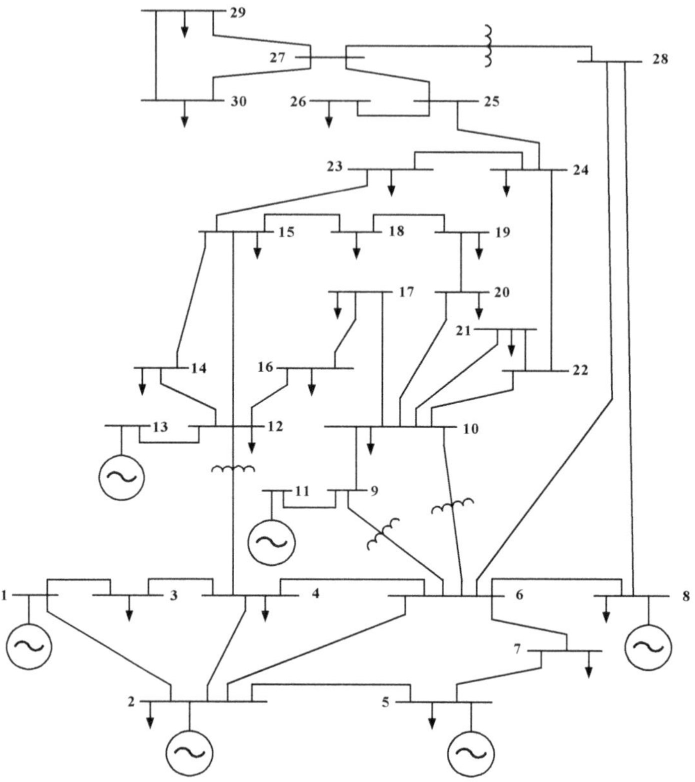

Figure 5. Single-line diagram of the IEEE 30 bus system.

Table 3. The cost and emission coefficients of generators for the IEEE 30 bus system.

	Coefficient Generating Unit					
	G1	G2	G5	G8	G11	G13
	Fuel cost coefficient					
a	0	0	00	0	0	0
b	2	1.75	1	3.25	3	3
c	0.00375	0.0175	0.0625	0.00834	0.025	0.025
	Emission coefficient					
α	4.091	2.543	4.258	5.326	4.258	6.131
β	-5.554	-6.047	-5.094	-3.55	-5.094	-5.555
γ	6.49	5.638	4.586	3.38	4.586	5.151
ζ	2.00×10^{-4}	5.00×10^{-4}	1.00×10^{-6}	2.00×10^{-3}	1.00×10^{-6}	1.00×10^{-5}
λ	2.857	3.33	8	2	8	6.67

Case #1: In this case, the rate of fuel cost minimization [$/h] is the objective function that was optimized by the HGS algorithm. The convergence plot of fuel cost [$/h] for the IEEE 30 bus system is depicted in Figure 6a. The total fuel costs are reduced from 901.6391 [$/h] (initial case) to 799.2202 [$/h] (best case) with a reduction loss of 11.36%, as listed in Table 4.

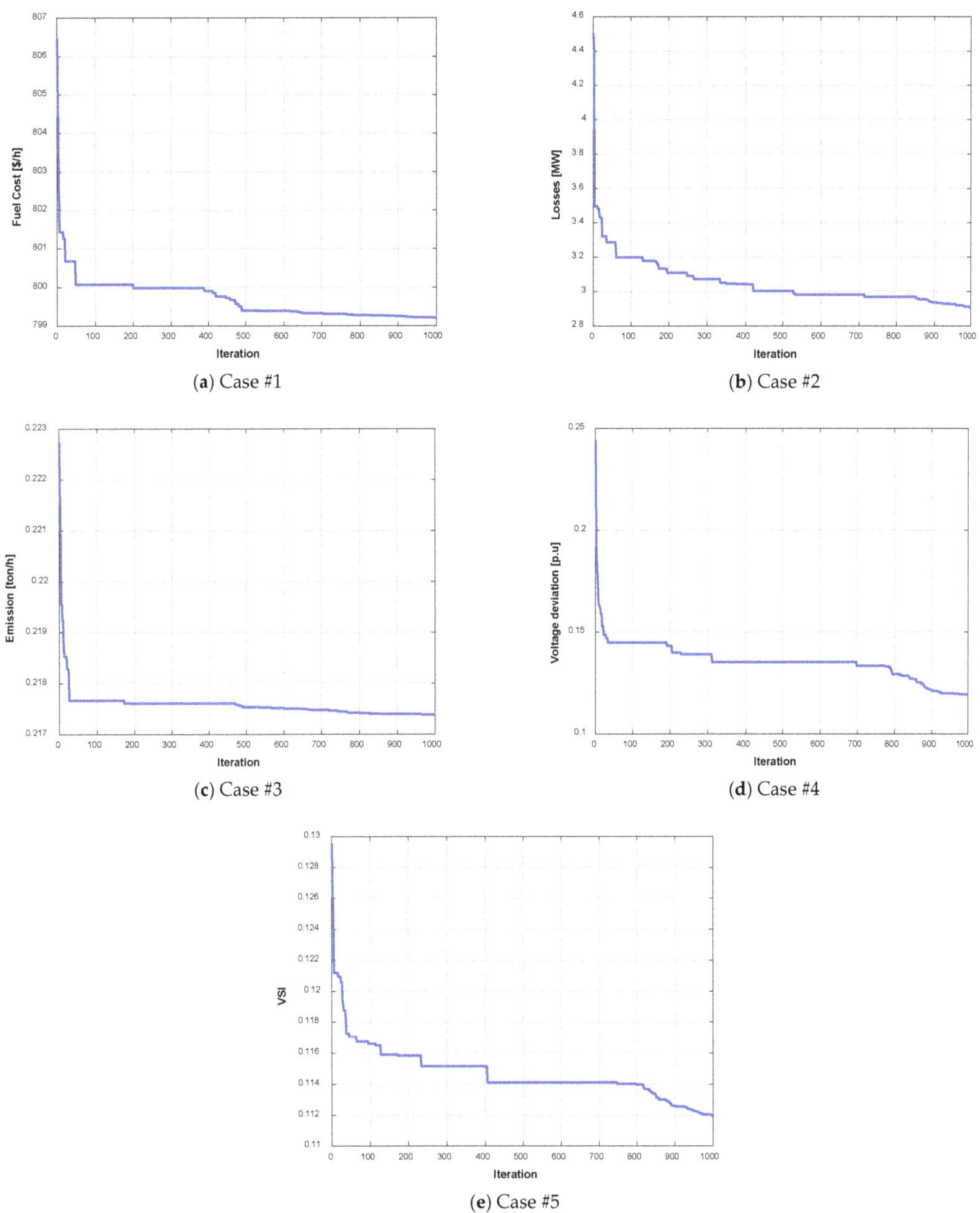

Figure 6. Convergence plots obtained by HGS for cases #(1–5).

Table 4. The best control variables obtained by HGS for cases #(1–5).

Item		Limit		Initial	Case				
		Max	Min	[26]	#1	#2	#3	#4	#5
[MW]	P_1	50	200	99.223	176.73	51.518	67.9326	99.3206	75.4018
	P_2	20	80	80	48.6285	79.9472	71.0846	56.2324	72.3195
	P_5	15	50	50	21.3556	49.9963	49.999	48.5423	49.6165
	P_8	10	35	20	20.9611	34.8738	34.999	34.7508	32.8099
	P_{11}	10	30	20	12.3146	29.9886	29.9995	19.9674	28.8962
	P_{13}	12	40	20	12.047	39.9871	32.7369	29.7117	27.9364
[p.u.]	V_1	0.95	1.1	1.05	1.10	1.09957	1.07979	1.00733	1.09893
	V_2	0.95	1.1	1.04	1.0879	1.09575	1.02963	0.99464	1.09953
	V_5	0.95	1.1	1.01	1.08115	1.07676	1.06095	1.06787	1.09397
	V_8	0.95	1.1	1.01	1.08895	1.09501	1.09796	1.05785	1.09956
	V_{11}	0.95	1.1	1.05	1.09935	1.09985	1.09540	1.05771	1.09950
	V_{13}	0.95	1.1	1.05	1.09998	1.09975	1.06955	0.99197	1.09772
[MVAr]	Q_{c10}	0	5	0	4.40056	4.21049	4.05070	3.89253	4.96503
	Q_{C12}	0	5	0	4.24071	4.19982	3.13443	1.68208	2.09009
	Q_{c15}	0	5	0	4.47918	4.26252	3.70849	2.33112	4.61987
	Q_{17}	0	5	0	4.98546	4.31378	3.62262	3.14785	3.53632
	Q_{c20}	0	5	0	4.62076	4.26326	4.13636	4.99605	4.95659
	Q_{21}	0	5	0	4.97743	4.27526	1.42028	4.06093	4.94313
	Q_{c23}	0	5	0	4.41930	4.16337	2.22773	4.97509	4.66820
	Q_{24}	0	5	0	4.98744	4.23189	4.69879	4.99004	4.99429
	Q_{29}	0	5	0	3.78063	4.03580	3.42316	2.55515	4.99515
Tap Position	T_{11}	0.9	1.1	1.078	1.02609	1.01962	1.03789	0.97665	1.04588
	T_{12}	0.9	1.1	1.069	0.99766	1.04051	1.02467	0.98726	1.03403
	T_{15}	0.9	1.1	1.032	0.99149	0.95436	1.02072	0.99930	0.95605
	T_{36}	0.9	1.1	1.068	0.98371	0.99075	1.01205	0.96666	0.95025
Fuel cost [$/h]				901.639	**799.220**	966.837	933.703	889.278	920.183
Active power loss [MW]				5.6891	8.6424	**2.9109**	3.3515	5.1253	3.5803
Emission [ton/h]				0.2253	0.3673	0.2216	**0.2174**	0.2343	0.2204
Voltage deviation [p.u.]				1.1747	1.5671	1.7078	0.7971	**0.1195**	1.9699
Voltage stability index				0.1727	0.1199	0.1191	0.1314	0.1375	**0.1119**
Reduction rate				-	11.36%	48.83%	3.51%	89.83%	35.20%
Average resolution time [s/iter]				-	1.4834	1.631	1.613	1.575	1.609

The bold represents the best values of OF.

Case #2: The main aim of this case was to reduce the active power loss [MW] in the transmission lines using the HGS algorithm. The reduction rate in losses is 48.83% (reducing from 5.6891 [MW] to 2.9109 [MW]). The convergence speed of this case is illustrated in Figure 6b.

Case #3: Due to increasing environmental pollution, the new studies drew growing attention in emission. In this paper, this objective function deals with emission minimization. The variations of emission over iteration are described in Figure 6c. The best value of emission that can be achieved by the HGS algorithm is 0.2174 [ton/h]. The total emissions are reduced from 0.2253 [ton/h] to 0.2174 [ton/h] with a reduction rate equal to 3.51%, as illustrated in Table 4.

Case #4: To improve the voltage at PQ bus, the voltage deviation is the objective function that is considered. The sum voltage deviation reduced from 1.1747 to 0.1195 (reduction rate is 89.83%), as seen in Table 4. The convergence speed for this case is shown in Figure 6d.

Case #5: The fifth objective function is to enhance voltage stability of the whole system by minimizing the maximum value of the voltage stability indicator (L-max) using the HGS algorithm. The convergence characteristics of VSI using the proposed approach of HGS is depicted in Figure 6e. The reduction rate of this case is equal to 35.20% (compared between the initial case, which is 0.1727, and the optimal case using HGS, which is 0.1119) as given in Table 4.

The best control variables for cases (1–5) using the proposed algorithm HGS on the IEEE 30 bus system are depicted in Table 4. To prove the superiority and effectiveness of the performance of the HGS algorithm, the best results obtained by the HGS algorithm for fuel cost [\$/h], active power losses [MW], emission [ton/h], voltage deviation [p.u.], and voltage stability index are compared with other recent optimization methods results reported in the literature, as seen in Tables 5 and 6. The bold line represents the best values of OF obtained by HGS.

Table 5. Comparison of the best results obtained by HGS with other algorithms for cases (1,2).

Case #1		Case #2	
Method	**FC [\$/h]**	**Method**	**APL [MW]**
Initial	901.6391	Initial	5.830
SCA [27]	800.1018	SSO [28]	3.8239
DSA [29]	800.3887	EM [30]	3.1775
JAYA [31]	800.479	GPU-PSO [32]	3.2601
MSA [33]	800.5099	EGA-DQLF [34]	3.2008
SP-DE [35]	800.4131	ASO [36]	3.1600
MGOA [37]	800.4744	EGA-EA [38]	3.2601
AMTPG-Jaya [39]	800.1946	GWO [10]	4.2905
TLBO [39]	800.4604	PSO [40]	5.1957
ABC [41]	800.6850	HPSO-DE [40]	5.1476
IABC [41]	800.4215	FAHSPSO-DE [40]	4.9989
SSO [28]	802.2580	IPSO [42]	5.0732
GPU-PSO [32]	800.53	HHO [43]	3.49
EGA [44]	802.06	SSA [43]	3.50
IGA [45]	800.805	WOA [43]	3.50
AGAPOP [46]	799.8441	MF [43]	3.50
ABC [47]	800.66	GWO [43]	3.51
PSOGSA [48]	800.49859	SMA [14]	2.9934
GWO [10]	802.7924	HGS	**2.9109**
ISSA [49]	800.4752		
MFO [49]	800.7134		
IHS [49]	800.5202		
GA [49]	800.5272		
SOS [50]	801.5733		
SMA [14].	799.2557		
HGS	**799.2202**		

The bold line represents the best values of OF obtained by HGS.

Table 6. Comparison of the best results obtained by HGS with other algorithms for cases (3–5).

Case #3		Case #4		Case #5	
Method	**Em [ton/h]**	**Method**	**VD [p.u]**	**Method**	**VSI**
Initial	0.3661	Initial	1.1747	Initial	0.1727
BSA [51]	0.2425	HFPSO [52]	0.1467	Jaya [31]	0.1243
SSO [28]	0.2315	EJADE-SP [9]	0.3752	AMTPG-Jaya [39]	0.1240
HHO [43]	0.2850	MABC [53]	0.1292	SSO [28]	0.1267
SSA [43]	0.2950	HGS	**0.1195**	NISSO [28]	0.12547
WOA [43]	0.2950			TLBO [39]	0.12444
MF [43]	0.2950			ARCBBO [54]	0.1369
GWO [43]	0.2960			ECHT-DE [55]	0.13632
SMA [14]	0.2175			SPEA [56]	0.1247
HGS	**0.2174**			DE [57]	0.1246
				SMA [14]	0.1136
				HGS	**0.1119**

The bold represents the best values of objective OF obtained by the HGS.

4.1.2. Bi-Objective OPF

In this subsection, two objective functions were optimized simultaneously to achieve the best compromise solution (BCS) of non-dominated solutions (NDS). The function used to find the BCS was the fuzzy membership function (FMF). The crowding distance (CD) was the method employed to rank and reduce NDS of the NDPF. The developed approach MOHGS was applied in power systems to demonstrate its performance to solve MOOPF problems. In this subsection, seven case studies were suggested to prove the efficiency and superiority of the MOHGS. The population size was 500 pollinators; the program was stopped when the number of NDS was equal to the number of NDS suggested, which was 500, or a number of iterations equal to 100 iterations. These cases can be summarized as follows:

Case #6: In the first case of Bi-OF, the total fuel cost and total emission were optimized simultaneously by using the proposed approach MOHGS. The BCS of fuel cost and emission were 827.735 [$/h] and 0.2587 [ton/h], respectively.

Case #7: The second case of this type, the total fuel cost and active power losses were considered as objective functions and optimized simultaneously. The BCS of fuel cost and losses were 826.842 [$/h] and 5.5946 [MW].

Case #8: The third case of bi-objectives functions that were optimized in this work were fuel cost and voltage deviation. The BCS of fuel cost and voltage deviation are 803.094 [$/h] and 0.1813 [p.u.]

Case #9: The fuel cost and voltage stability index are the objective functions that were minimized simultaneously in this case. The BCS was 801.491 [$/h] and 0.1213 of fuel cost and voltage stability index, respectively.

Case #10: The fifth case of Bi-OF was minimizing the total emission and voltage deviation simultaneously. The BCS of emission and voltage deviation were 0.2245 [ton/h] and 0.1508 [p.u.].

Case #11: The active power losses and voltage deviation represented the objective functions that were optimized simultaneously in this case. The BCS of losses and voltage deviation were 3.8636 [MW] and 0.2048 [p.u.].

Case #12: The voltage deviation and voltage stability index represented the last case of Bi objective OPF. The BCS of voltage deviation and voltage stability index were 0.3690 [p.u.] and 0.1280.

Figure 7a–g shows Pareto front solutions of the bi-objective functions obtained by the MOHGS on the IEEE 30-bus system. The best compromise solution of Bi OF for cases 6–12 is indicated by red diamonds, as shown in Figure 7.

4.1.3. Triple Objective OPF

In this type, three objective functions were considered simultaneously to obtain the BCS from NDS in the NDPF. Seven case studies are proposed; the population size is 500 pollinators. The stopping criteria of simulation running when the number of iterations equal to 100 iterations, or the number of non-dominated solutions equal to 500 pollinators. These cases can be summarized as follows:

Case #13: In the first case in this type, the fuel cost, emission, and losses combined were optimized simultaneously by using the proposed approach MOHGS to achieve the Pareto set solutions. The BCS of fuel cost, emission, and losses were 845.2988 [$/h], 0.2416 [ton/h], and 5.0562 [MW], respectively.

Case #14: In this case, the fuel cost, emission, and voltage deviation were considered objective functions and optimized simultaneously. The BCS of fuel cost, emission, and voltage deviation were 825.1766 [$/h], 0.2667 [ton/h], and 0.1953 [p.u.], respectively.

Case #15: The third case of triple-objective functions were minimization of fuel cost, losses, and voltage deviation. The BCS of fuel cost, losses, and voltage deviation were 817.1199 [$/h], 7.3484 [MW], and 0.1767 [p.u.], respectively.

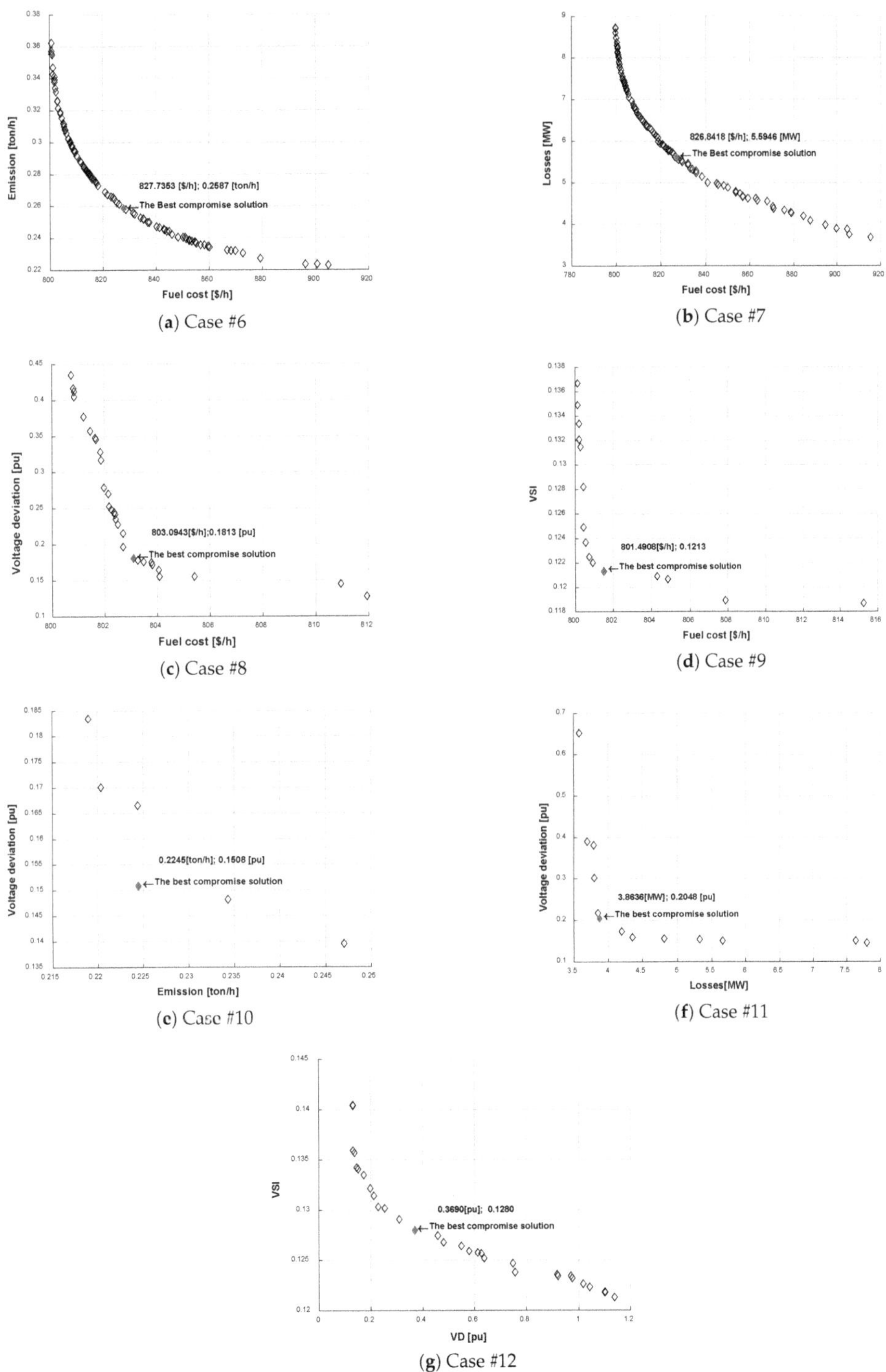

Figure 7. The Pareto fronts obtained by MOSMA for cases #(6–12).

Case #16: The active power losses, emission, and voltage deviation were the objective functions that were optimized simultaneously. The BCS was 3.8246 [MW], 0.2196 [ton/h], 0.2122 [p.u.] for losses, emission, and voltage deviation, respectively.

Case #17: The seventeenth case in this paper was minimization in fuel cost, losses, and voltage stability index, simultaneously. The BCS of fuel cost, losses, and voltage stability index were 834.4639 [$/h], 5.4740 [MW], 0.1198, respectively.

Case #18: In this case, the objective functions that were optimized included the fuel cost, emission, and voltage stability index. The BCS of fuel cost, losses, and voltage stability index were 835.0571 [$/h], 0.2504 [ton/h], 0.1192, respectively.

Case #19: Fuel cost, voltage deviation, and voltage stability index were the objective functions that were minimized simultaneously. The BCS of fuel cost, voltage deviation, and voltage stability index were 802.5842 [$/h], 0.4813 [p.u.], 0.1279, respectively.

The three-dimensional Pareto fronts set that were obtained by the developed approach MOHGS are illustrated in Figure 8a–g. The best compromise solutions of Tri OF for cases 13–19 are indicated by red diamonds, as shown in Figure 8.

4.1.4. Quad and Quinta Objective OPF

The last type of objective functions applied on the IEEE 30-bus system represents the Quad and Quinta objective functions. Two case studies on Quad objective functions and one case study on Quinta objective function are the suggested cases to solve MOOPF in this type. The stopping criteria are achieved when the No. of NDS is equal to 500 pollinators, or the No. of iterations reaches 500. It can be summarized as follows:

Case #20: Total fuel cost, losses, emission, and voltage deviation are the objective functions that were optimized in this case. The BCS obtained by MOHGS were 845.4721 [$/h], 5.7458 [MW], 0.2507 [ton/h], and 0.1400 [p.u.] for fuel cost, losses, emission, and voltage deviation., respectively.

Case #21: The objective functions that were optimized simultaneously were fuel cost, losses, emission, and voltage stability index. The BCS were 819.4061 [$/h], 7.4137 [MW], 0.2898 [ton/h], and 0.1389 for fuel cost, losses, emission, and voltage stability index, respectively.

Case #22: In this case, five objective functions were optimized simultaneously, which were fuel cost, losses, emission, voltage deviation, and voltage stability index. The best results of objective functions obtained by MOHGS were 818.7575 [$/h], 7.4471 [MW], 0.2912 [ton/h], 0.3272 [p.u.], and 0.1399 for fuel cost, losses, emission, voltage deviation, and voltage stability index, respectively.

Tables 7 and 8 represent the better control variables obtained by MOHGS for Bi, Tri, Quad, and Quinta objective functions that are illustrated by cases (20–22).

The voltage profiles at each load bus are charted in Figure 9a. From Figure 9a, it can be concluded that the voltage magnitude in the load bus for cases (1, 2, 3, and 5) exceeds the maximum limit (1.05 [p.u.]) for some buses, while the voltage magnitude of PQ bus for case 4 (when the voltage deviation is considered as an objective function) have not exceeded the minimum and maximum limit (0.95–1.05) [p.u.]). In Bi objective functions, the voltage profiles of this type are illustrated in Figure 9b. The numerical results obtained by MOHGS illustrate that the voltage magnitude of PQ bus exceeds the maximum limit in cases when the voltage deviation is not considered an objective function (cases (1,2,4)). However, the voltages magnitude has not exceeded the minimum and maximum limit for the cases in which the voltage deviation is considered an objective function (cases 3,5,6,7)). Figure 9c illustrates that the voltage magnitude values of some buses in cases 8, 12, 13 (the cases where the voltage deviation is not considered an objective function) exceed the maximum [1.05 p.u.], while the voltage magnitude values in cases (9, 10, 11, 14) have not exceeded the minimum and maximum limit [0.95–1.05 p.u.]. The voltage profiles for types Quad and Quinta are illustrated in Figure 9d. Figure 9d illustrates that the voltage magnitude of the load bus for these cases has not violated the boundary limit [0.95–1.05] [p.u.].

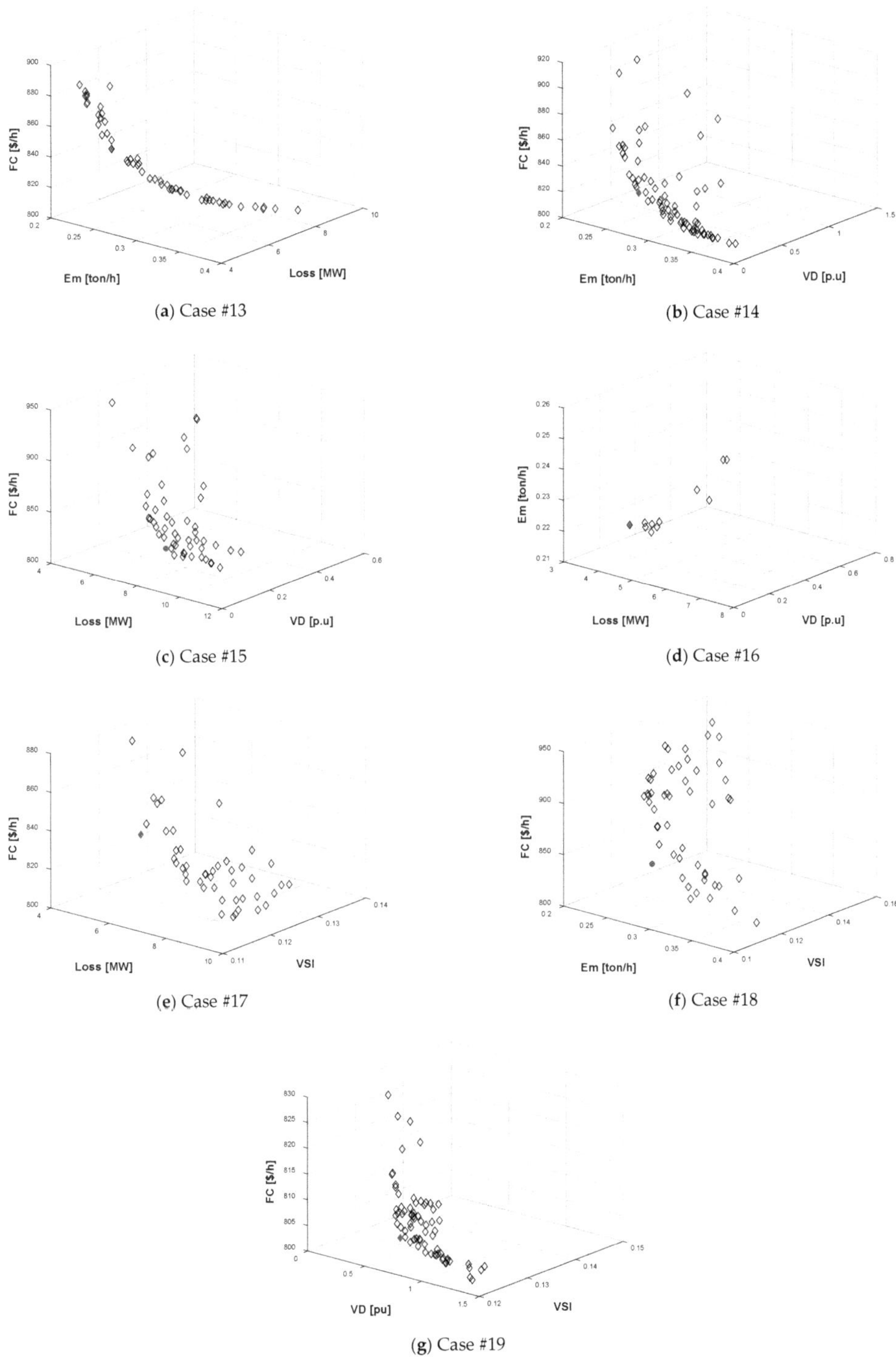

Figure 8. The Pareto fronts obtained by MOSMA for cases #(13–19).

Table 7. Better control variables obtained by HGS for cases (6−15).

Item		#6	#7	#8	#9	#10	#11	#12	#13	#14	#15
						Case					
[MW]	P_1	120.99	125.59	177.232	176.954	87.646	62.266	145.554	107.558	127.609	141.273
	P_2	60.834	48.558	46.335	44.891	64.306	75.233	78.062	61.963	59.373	52.885
	P_5	28.719	27.875	21.539	18.819	47.005	48.636	24.003	32.594	26.056	29.708
	P_8	34.476	34.619	21.650	21.417	34.672	32.980	18.529	33.168	34.366	30.809
	P_{11}	21.773	26.675	13.122	14.813	27.983	29.414	10.806	29.236	27.365	22.131
	P_{13}	22.520	25.678	13.130	15.373	26.889	38.735	15.923	23.937	15.486	13.942
[p.u.]	V_1	1.092	1.097	1.051	1.095	1.003	1.031	1.053	1.097	1.072	1.068
	V_2	1.053	1.099	1.015	1.099	1.026	1.060	1.043	1.086	1.000	1.047
	V_5	1.093	1.063	1.053	1.099	1.076	1.002	0.972	1.082	1.068	1.040
	V_8	1.055	1.077	1.061	1.084	1.063	1.054	1.097	1.068	1.037	1.004
	V_{11}	1.095	1.099	1.037	1.085	0.986	1.040	1.083	1.099	1.055	1.037
	V_{13}	1.090	1.076	1.085	1.082	1.028	1.010	1.052	1.073	1.015	0.979
[MVAr]	Q_{c10}	1.354	2.316	3.047	2.628	1.773	0.075	3.266	0.699	0.658	0.688
	Q_{C12}	0.804	2.958	0.160	2.868	1.243	0.000	0.278	3.942	0.730	4.276
	Q_{c15}	3.354	0.888	4.338	0.836	0.512	0.240	4.486	0.400	1.395	0.405
	Q_{17}	2.167	2.201	0.342	3.719	2.748	0.188	3.587	2.653	0.360	0.627
	Q_{c20}	1.133	1.648	3.588	3.687	1.528	0.782	3.673	1.737	1.704	4.744
	Q_{21}	0.261	1.228	4.004	3.411	2.983	4.688	3.797	2.421	0.715	1.176
	Q_{c23}	0.503	2.613	3.848	1.022	3.508	4.964	1.118	4.808	1.550	0.123
	Q_{24}	0.908	3.930	2.127	4.944	4.474	1.196	4.449	1.649	4.883	4.128
	Q_{29}	1.623	1.035	0.545	4.305	0.898	0.893	4.948	2.707	4.010	1.663
Tap Position	T_{11}	0.977	1.038	1.057	0.965	0.972	0.961	1.043	0.993	0.970	0.952
	T_{12}	0.962	0.983	0.971	0.974	0.950	0.963	1.098	0.972	0.978	0.971
	T_{15}	1.060	1.054	0.958	1.009	0.968	0.984	0.989	1.088	0.959	0.950
	T_{36}	1.000	0.997	0.955	0.953	0.954	0.951	0.951	0.972	0.973	0.954
GFC [\$/h]		**827.74**	**826.84**	**803.094**	**801.491**	899.074	946.099	826.338	**845.299**	**825.177**	**817.12**
RPL [MW]		5.9178	**5.5946**	9.6092	8.8720	5.1007	**3.8636**	9.4757	**0.2416**	6.8612	**7.3484**
Em [ton/h]		**0.2587**	0.2619	0.3678	0.3668	**0.2245**	0.2208	0.3077	**5.0562**	**0.2667**	0.2868
VD [p.u.]		0.8944	0.9559	**0.1813**	1.3536	**0.1508**	**0.2048**	**0.3690**	1.0547	**0.1953**	**0.1767**
VSI		0.1373	0.1333	0.1410	**0.1213**	0.1413	0.1445	**0.1280**	0.1282	0.1392	0.1412
Average resolution time [s/iter]		2.723	4.7027	2.7262	2.6784	2.70	2.68	2.703	2.79	2.734	2.718

The bold represents the BCS of OF.

Table 8. The better control variables obtained by HGS for cases (16–22).

Item		#16	#17	#18	#19	#20	#21	#22
					Case			
[MW]	P_1	67.936	121.170	113.261	179.444	119.182	142.724	143.244
	P_2	68.222	55.959	65.029	45.132	48.973	52.684	57.063
	P_5	49.162	34.921	29.736	19.931	40.045	33.242	32.352
	P_8	34.370	33.035	34.654	23.632	24.910	29.195	24.328
	P_{11}	29.153	27.184	22.918	11.624	18.664	16.913	20.928
	P_{13}	38.381	16.605	23.210	13.170	39.105	16.055	12.932
[p.u.]	V_1	1.015	1.083	1.099	1.086	1.006	1.052	1.074
	V_2	0.987	1.076	1.090	1.087	1.086	0.998	1.004
	V_5	1.032	1.099	1.079	1.026	1.088	1.005	1.026
	V_8	1.048	1.087	1.093	1.099	1.097	1.093	1.076
	V_{11}	1.094	1.083	1.087	0.980	0.996	1.096	1.013
	V_{13}	1.071	1.096	1.091	0.990	1.084	1.085	0.981

Table 8. *Cont.*

Item		Case						
		#16	**#17**	**#18**	**#19**	**#20**	**#21**	**#22**
[MVAr]	Q_{c10}	2.414	3.784	4.171	3.803	0.229	2.182	0.888
	Q_{C12}	2.485	3.543	1.272	4.502	1.402	1.709	1.529
	Q_{c15}	3.754	4.974	1.142	0.108	2.792	4.237	3.093
	Q_{17}	0.880	1.844	0.188	0.204	1.085	0.831	1.063
	Q_{c20}	1.670	4.075	2.411	0.782	0.612	4.614	0.894
	Q_{21}	0.349	4.448	4.984	4.005	3.912	2.728	3.500
	Q_{c23}	1.894	2.264	4.077	2.077	4.720	3.717	2.649
	Q_{24}	4.431	3.851	3.015	4.868	3.889	0.812	2.482
	Q_{29}	0.494	3.948	4.108	4.796	2.432	3.919	3.497
Tap Position	T_{11}	1.010	0.950	0.959	0.998	1.044	1.012	0.963
	T_{12}	0.981	1.026	0.980	0.979	0.976	1.075	1.024
	T_{15}	1.001	0.962	0.974	1.003	0.984	0.957	0.993
	T_{36}	0.963	0.954	0.953	0.952	1.028	1.013	0.976
GFC [\$/h]		936.4713	**834.4639**	**835.0571**	**802.5842**	**845.4721**	**819.4061**	818.7575
RPL [MW]		**3.8246**	**5.4740**	5.4107	9.5336	**5.7458**	7.4137	7.4471
Em [ton/h]		**0.2196**	0.2559	**0.2504**	0.3746	**0.2507**	0.2898	0.2912
VD [p.u.]		**0.2122**	1.5161	1.6370	**0.4813**	**0.1400**	0.3514	0.3272
VSI		0.1444	**0.1198**	**0.1192**	**0.1279**	0.1456	**0.1389**	**0.1399**
Average resolution time [s/iter]		2.7456	2.718	2.7652	2.7479	2.7334	2.7668	2.8734

The bold represents the BCS of OF.

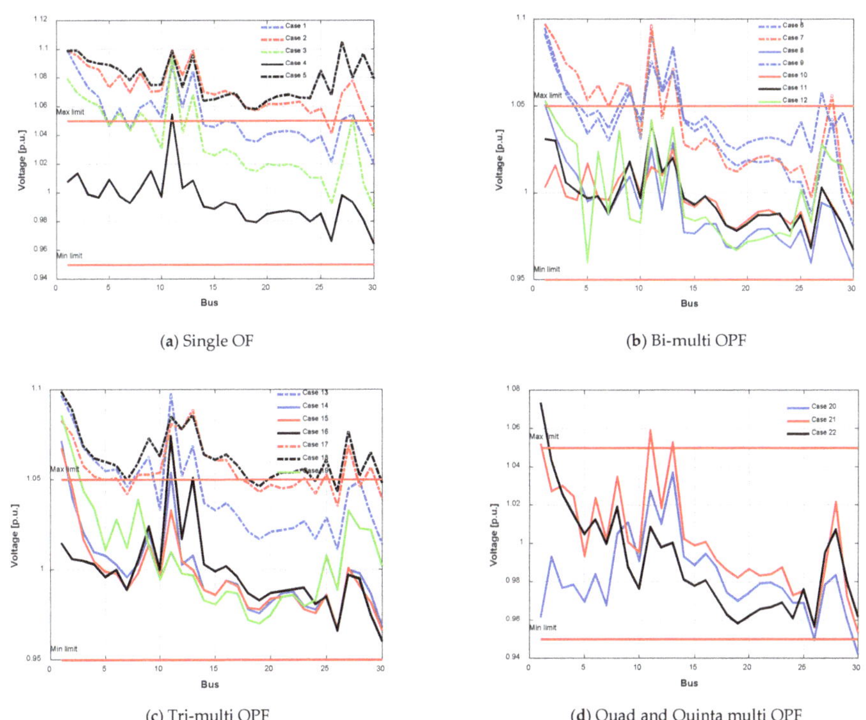

(**a**) Single OF

(**b**) Bi-multi OPF

(**c**) Tri-multi OPF

(**d**) Quad and Quinta multi OPF

Figure 9. The voltage profiles for cases 1 to case 22.

Figure 10a–f shows the values of the reactive power output from the generators for all cases (1–22). It can be observed that the MVAr of some generators (1, 5, and 8) exceeds the minimum and maximum limit in cases (1, 2, 5, 10, 12, 19, and 20). While the remaining generators have not violated the constraints in all cases.

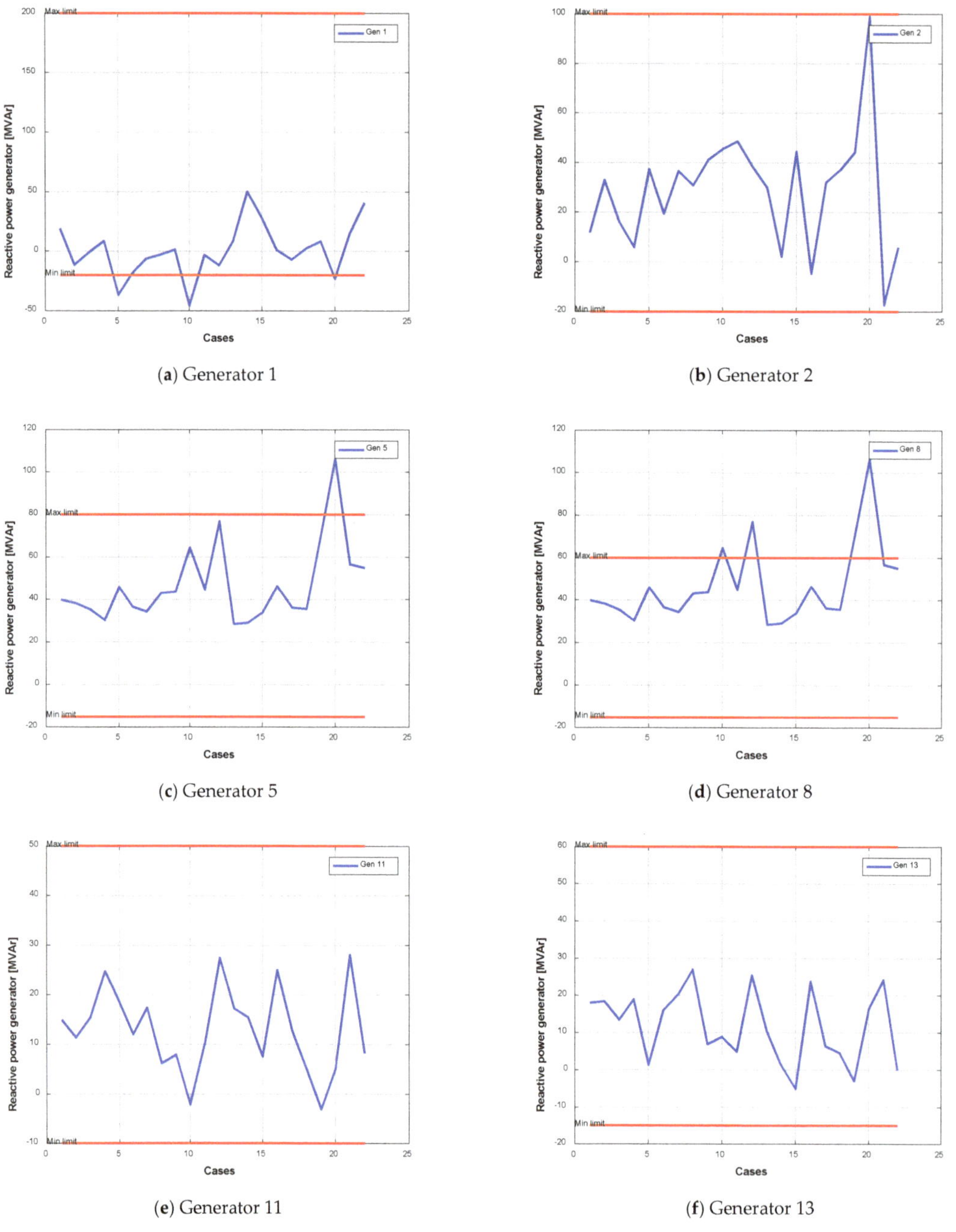

(**a**) Generator 1

(**b**) Generator 2

(**c**) Generator 5

(**d**) Generator 8

(**e**) Generator 11

(**f**) Generator 13

Figure 10. Reactive power output from generators for cases (1−22).

4.2. IEEE 57-Bus Power System

The second test system applied in the paper to validate the effectiveness and superiority of the proposed algorithm HGS is the IEEE 57 bus power system, as shown in Figure 11. The control variables in the OPF problem for the IEEE 57-bus system are 33. The cost and emission coefficients of IEEE 57 bus power system are illustrated in Table 9. The optimal value control variables obtained by the proposed algorithm HGS are reported in Table 10.

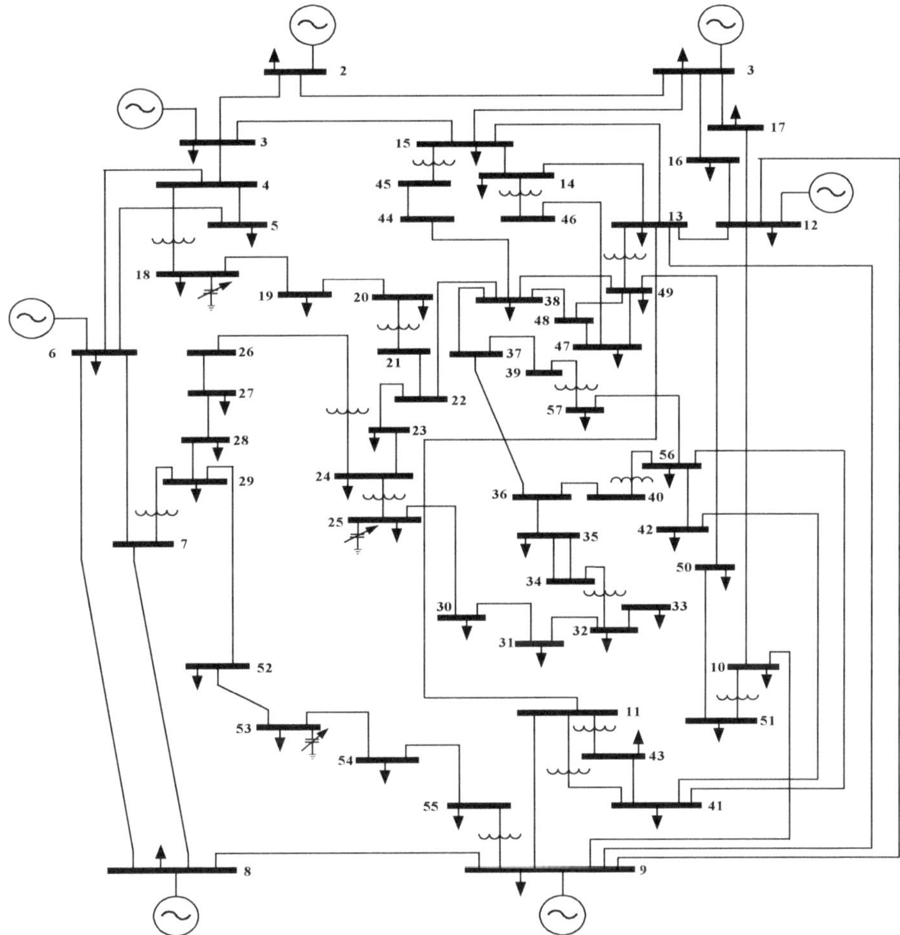

Figure 11. Single line diagram of the IEEE 57-bus system.

Single Objective OPF

The objective functions that are minimized are fuel cost [\$/h], active power losses [MW], and emission [ton/h]. The parameters chosen in the proposed algorithm HGS to solve OPF problem are 100 iterations and 100 population sizes.

Case #23: Total fuel cost was minimized from 51,353 [\$/h] (initial) to 41,679.4 [\$/h] (the better), and the reduction rate was 18.84%, as illustrated in Table 10. The convergence plot of fuel cost [\$/h] is depicted in Figure 12a.

Table 9. The cost and emission coefficients for generators for the IEEE 57-bus test system.

	\multicolumn{7}{c}{Coefficient Generating Unit}						
	G1	G2	G3	G6	G8	G9	G12
	\multicolumn{7}{c}{Fuel cost coefficient}						
a	0	0	0	0	0	0	0
b	2	1.75	3	2	1	1.75	3.25
c	0.00375	0.0175	0.025	0.00375	0.0625	0.0195	0.00834
	\multicolumn{7}{c}{Emission coefficient}						
α	4.091	2.543	6.131	3.491	4.258	2.754	5.326
β	-5.554	-6.047	-5.555	-5.754	-5.094	-5.847	-3.555
γ	6.49	5.638	5.151	6.39	4.586	5.238	3.38
ζ	2.0×10^{-4}	5.0×10^{-4}	1.0×10^{-5}	3.0×10^{-4}	1.0^{-6}	4.0×10^{-4}	2.0×10^{-3}
λ	2.857×10^{-1}	3.33×10^{-1}	6.67×10^{-1}	2.66×10^{-1}	8.0×10^{-1}	2.88×10^{-1}	2.0×10^{-1}

Table 10. Better control variables obtained by HGS for cases (23−25).

	Item	Limit		Initial	Case		
		Max	Min		#23	#24	#25
[MW]	P_1	0.0	576	478	147.65	180.33	191.87
	P_2	30.0	100	0	92.19	28.74	99.34
	P_3	40.0	140	40	46.14	124.68	138.93
	P_6	30.0	100	0	68.59	33.74	99.88
	P_8	100.0	550	450	463.81	390.93	282.28
	P_9	30.0	100	0	84.31	94.41	99.84
	P_{12}	100.0	410	310	363.11	409.80	353.99
[p.u.]	V_1	0.95	1.10	1.040	1.07	1.07	1.02
	V_2	0.95	1.10	1.010	1.07	1.08	1.02
	V_3	0.95	1.10	0.985	1.06	1.07	1.02
	V_6	0.95	1.10	0.980	1.06	1.06	1.02
	V_8	0.95	1.10	1.005	1.09	1.08	1.04
	V_9	0.95	1.10	0.980	1.08	1.06	1.04
	V_{12}	0.95	1.10	1.015	1.06	1.05	1.01
Tap Position	T_{4-18}	0.90	1.10	0.97	1.02	1.02	1.01
	T_{4-18}	0.90	1.10	0.978	1.02	1.05	1.04
	T_{21-20}	0.90	1.10	1.043	0.98	0.99	1.01
	T_{24-25}	0.90	1.10	1	1.03	1.04	0.99
	T_{24-25}	0.90	1.10	1	1.04	1.03	0.99
	T_{24-26}	0.90	1.10	1.043	1.06	1.00	1.01
	T_{7-29}	0.90	1.10	0.967	1.05	1.05	0.99
	T_{34-32}	0.90	1.10	0.975	0.98	0.97	0.98
	T_{11-41}	0.90	1.10	0.955	1.02	1.09	1.03
	T_{15-45}	0.90	1.10	0.955	1.00	1.05	1.01
	T_{14-46}	0.90	1.10	0.9	1.01	1.05	1.00
	T_{10-51}	0.90	1.10	0.93	1.03	1.09	0.97
	T_{13-46}	0.90	1.10	0.895	0.95	1.02	1.00
	T_{11-43}	0.90	1.10	0.958	1.03	1.02	1.08
	T_{40-56}	0.90	1.10	0.958	1.00	0.98	1.05
	T_{39-57}	0.90	1.10	0.98	0.99	1.00	1.02
	T_{9-55}	0.90	1.10	0.94	1.06	1.03	1.02
[MVAr]	Q_{c18}	0.00	20	10	12.58	26.08	11.25
	Q_{25}	0.00	20	5.9	22.16	21.56	10.78
	Q_{53}	0.00	20	6.3	12.78	10.89	18.02

Table 10. *Cont.*

Item	Limit		Initial	Case		
	Max	**Min**		**#23**	**#24**	**#25**
GFC [$/h]			51353	**41679.4**	43740	45118
RPL [MW]			2.4129	15.24	**11.829**	15.6796
Em [ton/h]			27.868	1.3746	1.3048	**0.9616**
Reduction rate			-	18.84%	57.55%	60.15%
Average resolution time [s/iter]			-	12.257	12.225	12.24

The bold represents the better values of OF obtained by the proposed approach.

Case #24: Active power losses are minimized from 27.868 [MW] (initial) to 11.829 [MW] (the better); the reduction rate is 57.55%. The convergence speed of this case is illustrated in Figure 12b.

Case #25: Figure 12c illustrates the variations of emission over iteration. The better value of emission is 0.962 [ton/h]. The reduction rate is 60.15%.

The better control variables for the IEEE 57 bus system that are obtained by the HGS algorithm are depicted in Table 10. The bold represents the better values of OF obtained by the proposed approach.

Table 11 compares the numerical results obtained by HGS with other recent optimization methods for fuel cost [$/h], active power losses [MW], and emission [ton/h].

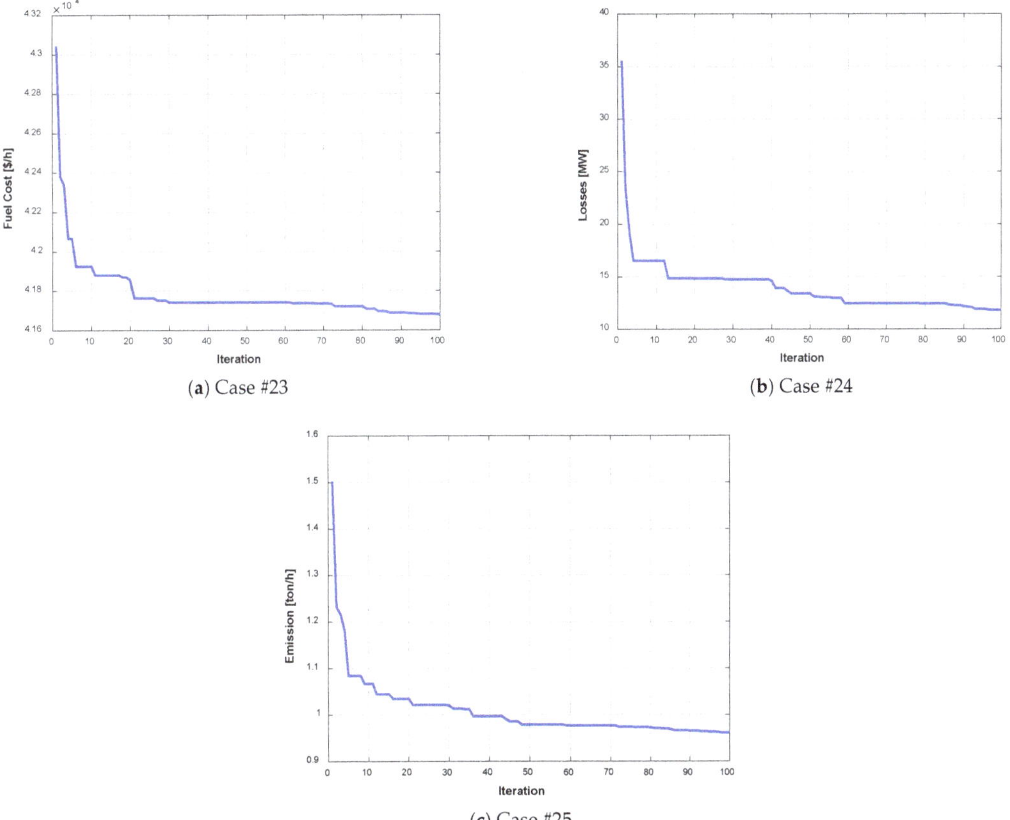

(**a**) Case #23

(**b**) Case #24

(**c**) Case #25

Figure 12. Convergence plot of HGS on the IEEE 57-bus system.

Table 11. Comparison of the numerical results obtained by HGS with other algorithms for cases (23–25).

Case #23		Case #24		Case #25	
Method	**FC [ton/h]**	**Method**	**RPL [p.u]**	**Method**	**Em [ton/h]**
GOA [37]	41680	ABC [41]	12.6260	ABC [41]	1.2048
GA [37]	41685	MICA [58]	11.8826	IABC [41]	1.0484
TLBO [37]	41683	GA [59]	13.3983	Jaya [59]	1.1111
MO-DEA [23]	41683	PSO [59]	13.6673	GA [59]	1.189
ABC [47]	41694	**HGS**	**11.83**	PSO [59]	1.19
GSA [60]	41695			MTLBO [61]	1.0772
NPSO [62]	41699.52			MICA [58]	1.2246
KHA [63]	41709.3			GBBICA [64]	1.1724
EADDE [65]	41713.6			SKHA [66]	1.08
fuzzy GA [67]	41716.3			SSO [28]	1.7024
HGS	**41679.4**			NISSO [28]	1.03927
				HGS	**0.9616**

The bold represents the better values of OF obtained by the proposed HGS approach.

The voltage profiles at each load bus for the IEEE 57 bus test system are charted in Figure 13. Voltage limits [0.94–1.06 p.u.] were exceeded on some buses. The exceeding of admissible voltage deviation at load buses may be attributed to several reasons. One of the most important reasons is the fact that we have a single objective function for these cases (23–25), and the voltage deviation was not considered an objective function for these cases. The reactive power of generators (2, 6, and 9) has violated the limits, while for the reactive power output for the generators (1, 3, 8, and 12), the constraints were not violated. The reactive power output for the generators in cases (23–25) is shown in Figure 14.

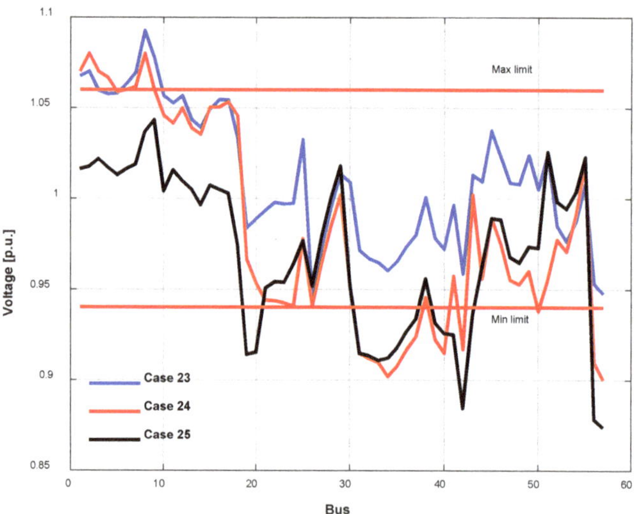

Figure 13. The voltage profiles for cases (23–25).

4.3. Discussions

To demonstrate the efficiency and the performance of the proposed HGS algorithm and of the developed MOHGS approach, the proposed approaches were applied in power systems to solve single and multiple OPF. For solving multi-objective optimal power flow (MOOPF) problems, the authors have developed the proposed HGS algorithm considering a new approach, namely, the multi-objective hunger game search (MOHGS), by merging with Pareto concept optimization to obtain a better distribution of non-dominated solutions in the Pareto front set. In this article, the numerical results listed in Table 4 confirmed the ability of the proposed HGS algorithm to find good convergence and the better solution compared with state-of-the-art computational algorithms. Furthermore, the simulation results drawn in Figures 7 and 8 confirmed the very good capability of the developed MOHGS approach to achieve the convergence characteristics and the well-distribution in the Pareto front set.

The single and multi-objective OPF problems faced many challenges, the most important of which were the accurate convergence to reach the global optimal solution and uniform distribution of the Pareto front set (high coverage). In other words, a balance should be found between coverage and convergence. Based on the free lunch theorem (NFL) [68], can one heuristic meta algorithm solve all optimization problems? This reason, in addition to convergence and coverage, represents the main reasons concerning the non-superiority of some algorithms over others. The proposed approaches provide more diversification in the search space and highly boost exploitation by founding very good quality solutions and a very good distribution as well. Furthermore, the proposed MOHGS approach can solve more complex problems with less computational effort. Briefly, the obtained results by HGS and MOHGS can provide most convenient solutions of Pareto front set with good coverage and convergence along one, two, and three objectives when applied to solve OPF problems.

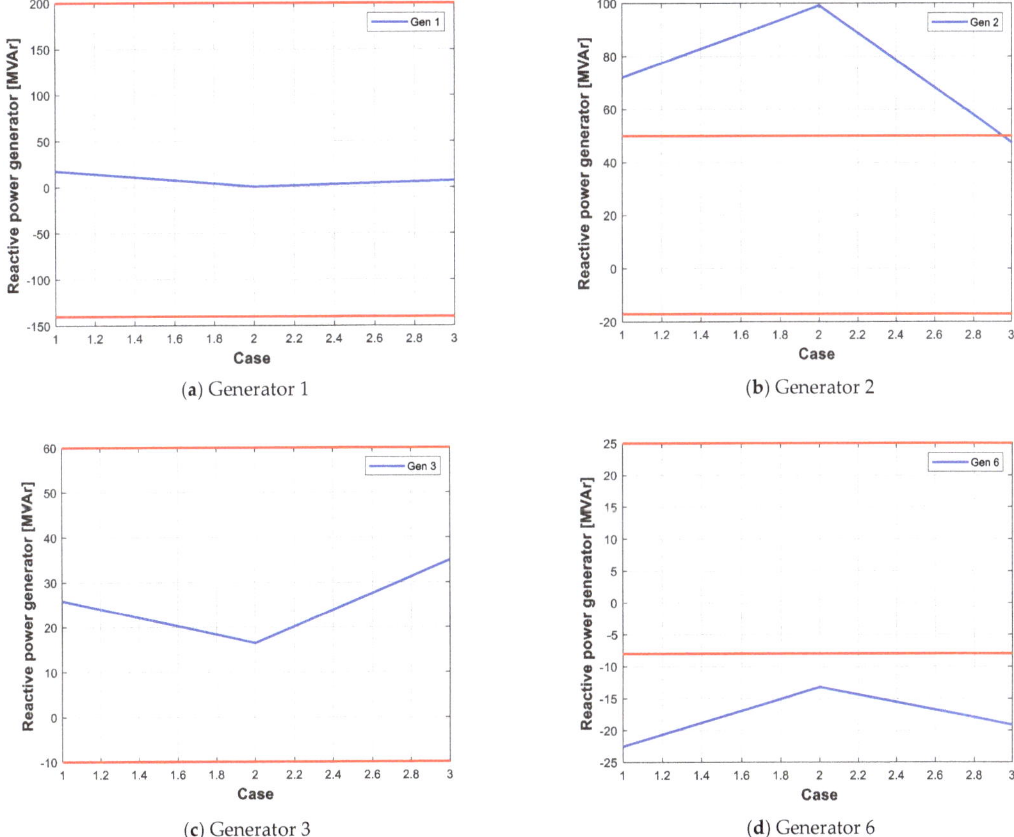

(**a**) Generator 1

(**b**) Generator 2

(**c**) Generator 3

(**d**) Generator 6

Figure 14. *Cont.*

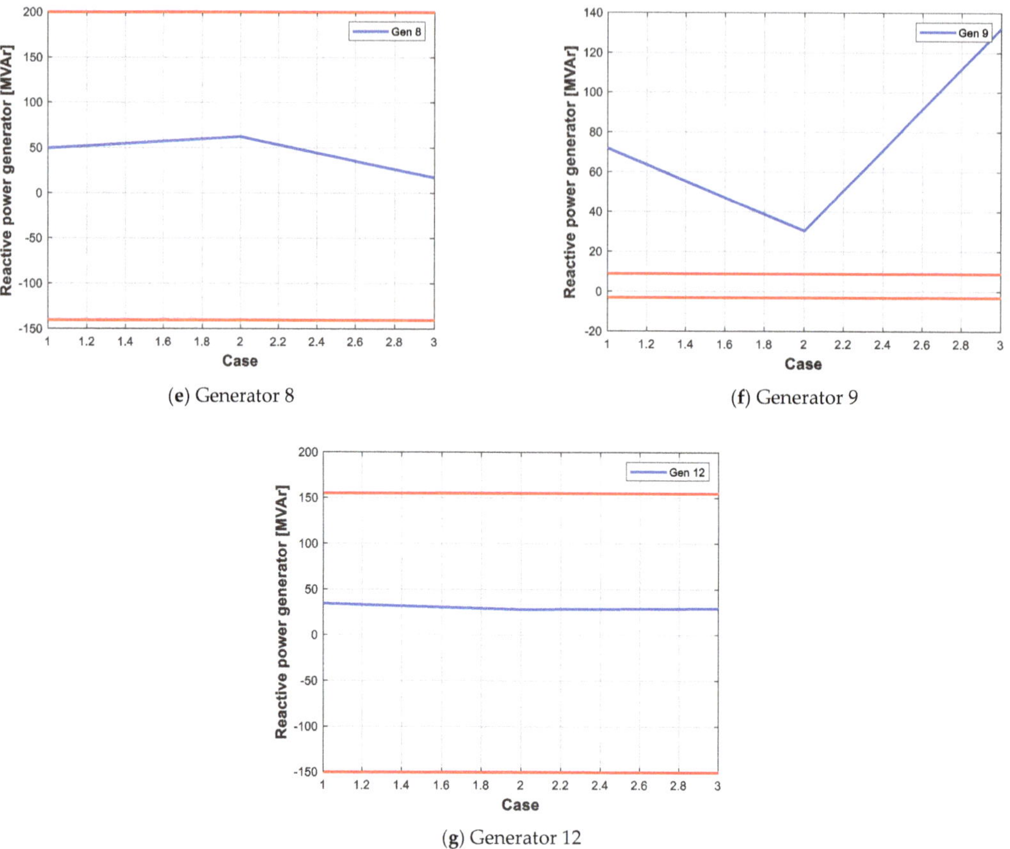

(**e**) Generator 8

(**f**) Generator 9

(**g**) Generator 12

Figure 14. Reactive power output for generators in cases (23−25).

5. Conclusions

In this study, a new optimization algorithm was tested, namely, the hunger games search (HGS), inspired by the behavior of social animal cooperation that is proportional to their level of hunger. This algorithm is proposed for solving the single objective optimal power flow problems in power systems to achieve the economical, technical, and environmental benefits. The HGS algorithm was developed to obtain better solutions for multiple conflicting objective functions simultaneously, namely, the multi-objective hunger games search (MOHGS). Various objective functions were considered, which are fuel cost [\$/h], active power losses [MW], emission [ton/h], voltage deviation [p.u.], and voltage stability index. The optimization method that was used to determine the non-dominated solutions in the Pareto front set was the Pareto concept. The strategies that were used to extract BCS and rerank the non-dominated solutions were fuzzy set theory and crowding distance. To demonstrate the efficiency and capability of the proposed HGS algorithm and developed MOHGS approach, two power systems were used (IEEE 30-bus and IEEE 57-bus) for various single and multiple objective functions. The simulation results, illustrating the performance of the proposed approaches, are characterized by good convergence speed, well-distribution, and high efficiency. The comparison results with other well-known optimization methods confirmed the quality and robustness of the proposed approaches. Finally, the MOHGS is a reliable and robust optimization method to solve a multi-objective optimal power flow problem.

Author Contributions: Conceptualization, M.A.-K. and V.D.; methodology, M.A.-K.; software, M.A.-K.; validation, M.A.-K., V.D. and M.E.; formal analysis, M.A.-K.; investigation, M.A.-K. and V.D.; data curation, M.A.-K.; writing—original draft preparation, M.A.-K.; writing—review and editing, V.D. and M.E.; visualization, M.A.-K. and V.D.; supervision, V.D. and M.E.; project administration, M.E.; funding acquisition, V.D. All authors have read and agreed to the published version of the manuscript.

Funding: This research was funded by University POLITEHNICA of Bucharest.

Data Availability Statement: The data supporting the reported results are available in the manuscript.

Acknowledgments: This project was supported by the University POLITEHNICA of Bucharest in Romania, funded by the University POLITEHNICA of Bucharest.

Conflicts of Interest: The authors declare no conflict of interest.

References

1. Carpentier, J. Contribution to the economic dispatch problem. *Bull. Soc. Fr. Electr.* **1962**, *3*, 431–447.
2. Frank, S.; Steponavice, I.; Rebennack, S. Optimal power flow: A bibliographic survey I. *Energy Syst.* **2012**, *3*, 221–258. [CrossRef]
3. Al-Bahrani, L.; Al-kaabi, M.; Al-saadi, M.; Dumbrava, V. Optimal power flow based on differential evolution optimization technique. *U.P.B. Sci. Bull. Ser. C* **2020**, *82*, 378–388.
4. Bouktir, T.; Slimani, L.; Belkacemi, M. A genetic algorithm for solving the optimal power flow problem. *Leonardo J. Sci.* **2004**, *4*, 44–58.
5. Al-kaabi, M.; Al-Bahrani, L. Modified Artificial Bee Colony Optimization Technique with Different Objective Function of Constraints Optimal Power Flow. *Int. J. Intell. Eng. Syst.* **2020**, *13*, 378–388. [CrossRef]
6. Al-Kaabi, M.; Al-Bahrani, L.; Dumbrava, V.; Eremia, M. Optimal Power Flow with Four Objective Functions using Improved Differential Evolution Algorithm: Case Study IEEE 57-bus power system. In Proceedings of the 2021 10th International Conference Energy and Environment, Bucharest, Romania, 14–15 October 2021; pp. 1–5. [CrossRef]
7. Al-Bahrani, L.; Al-Kaabi, M.; Al-Hasheme, J. Solving Optimal Power Flow Problem Using Improved Differential Evolution Algorithm. *Int. J. Electr. Electron. Eng. Telecommun.* **2022**, *11*, 146–155. [CrossRef]
8. Li, S.; Gong, W.; Hu, C.; Yan, X.; Wang, L.; Gu, Q. Adaptive constraint differential evolution for optimal power flow. *Energy* **2021**, *235*, 12136. [CrossRef]
9. Li, S.; Gong, W.; Wang, L.; Yan, X.; Hu, C. Optimal power flow by means of improved adaptive differential evolution. *Energy* **2020**, *198*, 117314. [CrossRef]
10. Meng, A.; Zeng, C.; Wang, P.; Chen, D.; Zhou, T.; Zheng, X.; Yin, H. A high-performance crisscross search based grey wolf optimizer for solving optimal power flow problem. *Energy* **2021**, *225*, 120211. [CrossRef]
11. Al-Kaabi, M.; Al Hasheme, J.; Dumbrava, V.; Eremia, M. Application of Harris Hawks Optimization (HHO) Based on Five Single Objective Optimal Power Flow. In Proceedings of the 2022 14th International Conference on Electronics, Computers and Artificial Intelligence (ECAI), Ploesti, Romania, 30 June–1 July 2022; pp. 1–8. [CrossRef]
12. Arsyad, H.; Suyuti, A.; Said, S.M.; Akil, Y.S. Multi-objective dynamic economic dispatch using Fruit Fly Optimization method. *Arch. Electr. Eng.* **2021**, *70*, 351–366. [CrossRef]
13. Al-Kaabi, M.; Al-Bahrani, L.; Al Hasheme, J. Improved Differential Evolution Algorithm to solve multi-objective of optimal power flow problem. *Arch. Electr. Eng.* **2022**, *71*, 641–657. [CrossRef]
14. Al-Kaabi, M.; Dumbrava, V.; Eremia, M. A Slime Mould Algorithm Programming for Solving Single and Multi-Objective Optimal Power Flow Problems with Pareto Front Approach: A Case Study of the Iraqi Super Grid High Voltage. *Energies* **2022**, *15*, 7473. [CrossRef]
15. Daqaq, F.; Ouassaid, M.; Ellaia, R. A new meta-heuristic programming for multi-objective optimal power flow. *Electr. Eng.* **2021**, *103*, 1217–1237. [CrossRef]
16. Avvari, R.K.; Vinod Kumar, D.M. Multi-Objective Optimal Power Flow with efficient Constraint Handling using Hybrid Decomposition and Local Dominance Method. *J. Inst. Eng. Ser. B* **2022**, *103*, 1643–1658. [CrossRef]
17. Huy, T.H.B.; Kim, D.; Vo, D.N. Multiobjective Optimal Power Flow Using Multiobjective Search Group Algorithm. *IEEE Access* **2022**, *10*, 77837–77856. [CrossRef]
18. Yang, Y.; Chen, H.; Heidari, A.A.; Gandomi, A.H. Hunger games search: Visions, conception, implementation, deep analysis, perspectives, and towards performance shifts. *Expert Syst. Appl.* **2021**, *177*, 114864. [CrossRef]
19. Khunkitti, S.; Siritaratiwat, A.; Premrudeepreechacharn, S. Multi-Objective optimal power flow problems based on Slime mould algorithm. *Sustainability* **2021**, *13*, 7448. [CrossRef]
20. Premkumar, M.; Jangir, P.; Sowmya, R.; Elavarasan, R.M. Many-Objective Gradient-Based Optimizer to Solve Optimal Power Flow Problems: Analysis and Validations. *Eng. Appl. Artif. Intell.* **2021**, *106*, 104479. [CrossRef]
21. Barocio, E.; Regalado, J.; Cuevas, E.; Uribe, F.; Zúñiga, P.; Torres, P.J.R. Modified bio-inspired optimisation algorithm with a centroid decision making approach for solving a multi-objective optimal power flow problem. *IET Gener. Transm. Distrib.* **2017**, *11*, 1012–1022. [CrossRef]

22. Niknam, T.; rasoul Narimani, M.; Jabbari, M.; Malekpour, A.R. A modified shuffle frog leaping algorithm for multi-objective optimal power flow. *Energy* **2011**, *36*, 6420–6432. [CrossRef]
23. Shaheen, A.M.; El-Sehiemy, R.A.; Farrag, S.M. Solving multi-objective optimal power flow problem via forced initialised differential evolution algorithm. *IET Gener. Transm. Distrib.* **2016**, *10*, 1634–1647. [CrossRef]
24. Sivasubramani, S.; Swarup, K.S. Multi-objective harmony search algorithm for optimal power flow problem. *Int. J. Electr. Power Energy Syst.* **2011**, *33*, 745–752. [CrossRef]
25. Varadarajan, M.; Swarup, K.S. Solving multi-objective optimal power flow using differential evolution. *IET Gener. Transm. Distrib.* **2008**, *2*, 720–730. [CrossRef]
26. Bouchekara, H. Optimal power flow using black-hole-based optimization approach. *Appl. Soft Comput.* **2014**, *24*, 879–888. [CrossRef]
27. Attia, A.; El Sehiemy, R.; Hasanien, H. Optimal power flow solution in power systems using a novel Sine-Cosine algorithm. *Int. J. Electr. Power Energy Syst.* **2018**, *99*, 331–343. [CrossRef]
28. Nguyen, T. A high performance social spider optimization algorithm for optimal power flow solution with single objective optimization. *Energy* **2019**, *171*, 218–240. [CrossRef]
29. Abaci, K.; Yamacli, V. Differential search algorithm for solving multi-objective optimal power flow problem. *Int. J. Electr. Power Energy Syst.* **2016**, *79*, 1–10. [CrossRef]
30. El-Hana Bouchekara, H.R.; Abido, M.A.; Chaib, A.E. Optimal power flow using an improved electromagnetism-like mechanism method. *Electr. Power Compon. Syst.* **2016**, *44*, 434–449. [CrossRef]
31. Warid, W.; Hizam, H.; Mariun, N.; Abdul-Wahab, N.I. Optimal power flow using the Jaya algorithm. *Energies* **2016**, *9*, 678. [CrossRef]
32. Roberge, V.; Tarbouchi, M.; Okou, F. Optimal power flow based on parallel metaheuristics for graphics processing units. *Electr. Power Syst. Res.* **2016**, *140*, 344–353. [CrossRef]
33. Mohamed, A.; Mohamed, Y.S.; El-Gaafary, A.A.M.; Hemeida, A.M. Optimal power flow using moth swarm algorithm. *Electr. Power Syst. Res.* **2017**, *142*, 190–206. [CrossRef]
34. Kumari, M.S.; Maheswarapu, S. Enhanced genetic algorithm based computation technique for multi-objective optimal power flow solution. *Int. J. Electr. Power Energy Syst.* **2010**, *32*, 736–742. [CrossRef]
35. Biswas, P.P.; Suganthan, P.N.; Mallipeddi, R.; Amaratunga, G.A.J. Optimal power flow solutions using differential evolution algorithm integrated with effective constraint handling techniques. *Eng. Appl. Artif. Intell.* **2018**, *68*, 81–100. [CrossRef]
36. Abdel-Rahim, A.M.; Shaaban, S.A.; Raglend, I.J. Optimal Power Flow Using Atom Search Optimization. In Proceedings of the 2019 Innovations in Power and Advanced Computing Technologies (i-PACT), Vellore, India, 22–23 March 2019; pp. 1–4. [CrossRef]
37. Taher, M.A.; Kamel, S.; Jurado, F.; Ebeed, M. Modified grasshopper optimization framework for optimal power flow solution. *Electr. Eng.* **2019**, *101*, 121–148. [CrossRef]
38. Surender Reddy, B.P. Efficiency improvements in meta-heuristic algorithms to solve the optimal power flow problem. *Int. J. Electr. Power Energy Syst.* **2016**, *82*, 288–302. [CrossRef]
39. Warid, W. Optimal power flow using the AMTPG-Jaya algorithm. *Appl. Soft Comput.* **2020**, *91*, 106252. [CrossRef]
40. Naderi, E.; Pourakbari-Kasmaei, M.; Cerna, F.V.; Lehtonen, M. A novel hybrid self-adaptive heuristic algorithm to handle single-and multi-objective optimal power flow problems. *Int. J. Electr. Power Energy Syst.* **2021**, *125*, 106492. [CrossRef]
41. He, X.; Wang, W.; Jiang, J.; Xu, L. An improved artificial bee colony algorithm and its application to multi-objective optimal power flow. *Energies* **2015**, *8*, 2412–2437. [CrossRef]
42. Niknam, T.; Narimani, M.R.; Aghaei, J.; Azizipanah-Abarghooee, R. Improved particle swarm optimisation for multi-objective optimal power flow considering the cost, loss, emission and voltage stability index. *IET Gener. Transm. Distrib.* **2012**, *6*, 515–527. [CrossRef]
43. Islam, M.Z.; Wahab, N.I.A.; Veerasamy, V.; Hizam, H.; Mailah, N.F.; Guerrero, J.M.; Mohd Nasir, M.N. A Harris Hawks Optimization Based Single-and Multi-Objective Optimal Power Flow Considering Environmental Emission. *Sustainability* **2020**, *12*, 5248. [CrossRef]
44. Bakirtzis, A.G.; Biskas, P.N.; Zoumas, C.E.; Petridis, V. Optimal power flow by enhanced genetic algorithm. *IEEE Trans. Power Syst.* **2002**, *17*, 229–236. [CrossRef]
45. Lai, L.L.; Ma, J.T.; Yokoyama, R.; Zhao, M. Improved genetic algorithms for optimal power flow under both normal and contingent operation states. *Int. J. Electr. Power Energy Syst.* **1997**, *19*, 287–292. [CrossRef]
46. Attia, A.; Al-Turki, Y.A.; Abusorrah, A.M. Optimal power flow using adapted genetic algorithm with adjusting population size. *Electr. Power Compon. Syst.* **2012**, *40*, 1285–1299. [CrossRef]
47. Adaryani, M.R.; Karami, A. Artificial bee colony algorithm for solving multi-objective optimal power flow problem. *Int. J. Electr. Power Energy Syst.* **2013**, *53*, 219–230. [CrossRef]
48. Radosavljević, J.; Klimenta, D.; Jevtić, M.; Arsić, N. Optimal Power Flow Using a Hybrid Optimization Algorithm of Particle Swarm Optimization and Gravitational Search Algorithm. *Electr. Power Compon. Syst.* **2015**, *43*, 1958–1970. [CrossRef]
49. Abd El-sattar, S.; Kamel, S.; Ebeed, M.; Jurado, F. An improved version of salp swarm algorithm for solving optimal power flow problem. *Soft Comput.* **2021**, *25*, 4027–4052. [CrossRef]

50. Duman, S. Symbiotic organisms search algorithm for optimal power flow problem based on valve-point effect and prohibited zones. *Neural Comput. Appl.* **2017**, *28*, 3571–3585. [CrossRef]
51. Chaib, A.E.; Bouchekara, H.; Mehasni, R.; Abido, M.A. Optimal power flow with emission and non-smooth cost functions using backtracking search optimization algorithm. *Int. J. Electr. Power Energy Syst.* **2016**, *81*, 64–77. [CrossRef]
52. Khan, A.; Hizam, H.; bin Abdul Wahab, N.I.; Lutfi Othman, M. Optimal power flow using hybrid firefly and particle swarm optimization algorithm. *PLoS ONE* **2020**, *15*, e0235668. [CrossRef]
53. Khorsandi, A.; Hosseinian, S.H.; Ghazanfari, A. Modified artificial bee colony algorithm based on fuzzy multi-objective technique for optimal power flow problem. *Electr. Power Syst. Res.* **2013**, *95*, 206–213. [CrossRef]
54. Kumar, A.R.; Premalatha, L. Optimal power flow for a deregulated power system using adaptive real coded biogeography-based optimization. *Int. J. Electr. Power Energy Syst.* **2015**, *73*, 393–399. [CrossRef]
55. Bouchekara, H.; Abido, M.A.; Boucherma, M. Optimal power flow using teaching-learning-based optimization technique. *Electr. Power Syst. Res.* **2014**, *114*, 49–59. [CrossRef]
56. Abido, M.A. Multiobjective optimal power flow using strength Pareto evolutionary algorithm. In Proceedings of the 39th International Universities Power Engineering Conference, Bristol, UK, 6–8 September 2004; pp. 457–461.
57. Abido, M.A.; Al-Ali, N.A. Multi-objective differential evolution for optimal power flow. In Proceedings of the 2009 International Conference on Power Engineering, Energy and Electrical Drives, Lisbon, Portugal, 18–20 March 2009; pp. 101–106. [CrossRef]
58. Ghasemi, M.; Ghavidel, S.; Ghanbarian, M.M.; Gharibzadeh, M.; Vahed, A.A. Multi-objective optimal power flow considering the cost, emission, voltage deviation and power losses using multi-objective modified imperialist competitive algorithm. *Energy* **2014**, *78*, 276–289. [CrossRef]
59. Abd El-Sattar, S.; Kamel, S.; El Sehiemy, R.A.; Jurado, F.; Yu, J. Single-and multi-objective optimal power flow frameworks using Jaya optimization technique. *Neural Comput. Appl.* **2019**, *31*, 8787–8806. [CrossRef]
60. Duman, S.; Güvenç, U.; Sönmez, Y.; Yörükeren, N. Optimal power flow using gravitational search algorithm. *Energy Convers. Manag.* **2012**, *59*, 86–95. [CrossRef]
61. Shabanpour-Haghighi, A.; Seifi, A.R.; Niknam, T. A modified teaching–learning based optimization for multi-objective optimal power flow problem. *Energy Convers. Manag.* **2014**, *77*, 597–607. [CrossRef]
62. Selvakumar, A.I.; Thanushkodi, K. A new particle swarm optimization solution to nonconvex economic dispatch problems. *IEEE Trans. Power Syst.* **2007**, *22*, 42–51. [CrossRef]
63. Roy, P.K.; Paul, C. Optimal power flow using krill herd algorithm. *Int. Trans. Electr. Energy Syst.* **2015**, *25*, 1397–1419. [CrossRef]
64. Ghasemi, M.; Ghavidel, S.; Ghanbarian, M.M.; Gitizadeh, M. Multi-objective optimal electric power planning in the power system using Gaussian bare-bones imperialist competitive algorithm. *Inf. Sci.* **2015**, *294*, 286–304. [CrossRef]
65. Vaisakh, K.; Srinivas, L.R. Evolving ant direction differential evolution for OPF with non-smooth cost functions. *Eng. Appl. Artif. Intell.* **2011**, *24*, 426–436. [CrossRef]
66. Pulluri, H.; Naresh, R.; Sharma, V. A solution network based on stud krill herd algorithm for optimal power flow problems. *Soft Comput.* **2018**, *22*, 159–176. [CrossRef]
67. Hsiao, Y.; Chen, C.; Chien, C. Optimal capacitor placement in distribution systems using a combination fuzzy-GA method. *Int. J. Electr. Power Energy Syst.* **2004**, *26*, 501–508. [CrossRef]
68. Wolpert, D.H.; Macready, W.G. *No Free Lunch Theorems for Search*; Technical Report SFI-TR-95-02-010; Santa Fe Institute: Santa Fe, NM, USA, 1995.

 *energies*

Article

Distribution System Reconfiguration with Soft Open Point for Power Loss Reduction in Distribution Systems Based on Hybrid Water Cycle Algorithm

Shamam Alwash [1], **Sarmad Ibrahim** [1,*] **and Azher M. Abed** [2]

[1]   Department of Electrical Engineering, University of Babylon, Babylon 51001, Iraq
[2]   Air Conditioning and Refrigeration Techniques Engineering Department, Al-Mustaqbal University College, Babylon 51001, Iraq
*   Correspondence: sarmad.ibrahim@uobabylon.edu.iq

Abstract: In this paper, the role of soft open point (SOP) is investigated with and without system re-configuration (SR) in reducing overall system power losses and improving voltage profile, as well as the effect of increasing the number of SOPs connected to distribution systems under different scenarios using a proposed hybrid water cycle algorithm (HWCA). The HWCA is formulated to enhance the water cycle algorithm (WCA) search performance based on the genetic algorithm (GA) for a complex nonlinear problem with discrete and continuous variables represented in this paper by SOP installation and SR. The WCA is one of the most effective optimization algorithms, however, it may have difficulty striking a balance between exploration and exploitation due to the nature of the proposed nonlinear optimization problem, which mostly causes slow convergence and poor robustness. Consequently, the HWCA proposed in this paper is an efficient solution to improve the balance between exploration and exploitation, which in turn leads to improving the WCA's overall performance without the possibility of getting trapped in local minima. Several cases are studied and conducted on an IEEE 33-node and the IEEE 69-node to investigate the real benefit gained from using SOPs alone or simultaneously with the SR. Based on the obtained results, the proposed HWCA succeeds in enhancing the performance of the proposed test systems considerably in terms of loss reduction (e.g., 31.1–63.3% for IEEE 33-node and 55.7–82.1% for IEEE 69-node compared to the base case) and voltage profile when compared to the base case while maintaining acceptable voltage magnitudes in most cases. Furthermore, the superiority of the proposed method based on the HWCA is validated when compared with the GA and WCA separately for both test systems. The obtained results show the outperformance of the proposed HWCA in attaining the best optimal solution with the least number of iterations.

Keywords: distribution systems; hybrid water cycle algorithm (HWCA); loss reduction; soft open point (SOP); system reconfiguration; voltage profile

Citation: Alwash, S.; Ibrahim, S.; Abed, A.M. Distribution System Reconfiguration with Soft Open Point for Power Loss Reduction in Distribution Systems Based on Hybrid Water Cycle Algorithm. *Energies* **2023**, *16*, 199. https://doi.org/10.3390/en16010199

Academic Editor: Javier Contreras

Received: 26 November 2022
Revised: 19 December 2022
Accepted: 21 December 2022
Published: 24 December 2022

1. Introduction

Due to unanticipated load changes and poor configuration of distribution systems (DSs), the likelihood of voltage violations and increased system losses in traditional DSs has become a major challenge for system operators [1]. Because traditional voltage regulators often give a limited reaction and discrete voltage regulation, it is difficult to satisfy the demand for desired voltage regulation and loss reduction utilizing just traditional VAR devices such as on-load tap changers and switchable capacitor banks [2]. Several strategies have been presented by researchers to address these challenges. System reconfiguration (SR) is considered one of the most popular and cost-effective methods to enhance the performance of DSs, which has been proposed by many researchers over the past few decades. These researchers hypothesized that further techniques, such as capacitor installation and cable sizing upgrades, may exceed the DS utilities' financial constraints. Moreover, DSs are often equipped with switches that can be used to perform SR with the objective to balance loads and hence improve loss reduction. SR can be efficiently conducted by altering the state of sectionalized (closed) and tie (opened) switches while taking the system's radiality and operational constraints into account [3–8]. Regardless of the advantages of SR in terms of improving voltage profile and reducing system losses, further performance improvement in DSs is still restricted due to radial topology and construction limits.

With the evolution of power electronic technology, power electronic devices are nowadays playing a significant role in changing the traditional distribution network into a technologically sophisticated network. The soft open point (SOP), which is an innovative electronic device that generally uses back-to-back voltage source converters (VSC), has a significant influence on the design and functionality of the DS [9,10]. SOP can efficiently modify different operational set points by managing the active and reactive power flowing across nearby branches or feeders in real-time. Additionally, SOP has the ability to inject or absorb reactive power at their terminals to improve system loss reduction and the voltage profile of the network [10]. SOP is also distinguished by its capacity to isolate any voltage and current disturbances and regulate peak currents brought on by faults. As a result, the adoption of SOP will considerably improve the network's operating conditions, allowing for more operational flexibility and lower operating costs [11].

In this regard, several researchers have investigated the integration of SOP with DSs intending to minimize system losses, improve voltage profile, and increase affordability without affecting DS radiality [11–15]. For example, the taxi-cab and multi-objective particle swarm optimization methods were utilized to discover the optimal set-points of the SOP with the goal of minimizing power loss, improving feeder load balancing, and optimizing voltage profile [12]. The genetic algorithm (GA) was used to determine optimal SOP location and active/reactive power setting points in unbalanced distribution networks, considering the active power losses and voltage unbalance index as objective functions with correlated uncertain distributed generators [13]. In [14], the proposed nonlinear optimization problem was addressed using the mixed-integer second-order cone programming approach in order to save total operating and SOP expenses. Furthermore, the optimal SOP placement was selected utilizing the power flow betweenness index and voltage violation risk index. In [15], the SOP planning problem was handled based on a bi-level strategy to optimize DS resilience. The upper-level problem was solved using genetic algorithms to address the SOP location and capacity. The particle swarm optimization technique was also used to optimize SOP functions at the lower level.

To gain further improvement in the performance of the DS, numerous researchers have adopted simultaneous optimization of SR and SOP size and position to provide considerable advantages to DSs with a variety of objectives. For example, to minimize network power losses, a novel methodology of SR was provided based on the AC-SOP and the DC-SOP [16]. In [17], a novel optimization method known as the Archimedes optimization algorithm was applied to the multi-objective optimization problem in order to maximize DG penetration and decrease system losses through consecutive SR and SOP deployments. In [18], to discover the optimal radial topologies and minimize power loss in DSs, the discrete-continuous hyper-spherical search method was suggested to handle the SR approach while taking into account various distributed generations, SOPs, and SR strategies. In [19], a bi-level multi-objective optimization method was suggested to maximize hosting capacity and reduce overall active losses of DSs with simultaneous SR and SOP allocation, while ensuring operational constraint limitations. In [20], A modified version of the particle swarm optimization was created to solve SR and SOP integration problems in active DSs with the goal to lessen system power losses, improve system efficiency in steady-state operation, and enhance voltage profiles. Furthermore, in [20], choosing the optimal number, placement, and size of DGs was also considered with SR and SOP integration problems in DSs. In [21], with the goals of loss reduction, load balancing, and increasing DG penetration level, the SR and SOP operating problems were presented by a multi-objective framework utilizing the Pareto optimality.

In previously mentioned studies, various approaches were utilized to discover the appropriate allocation and size of SOP with SR by limiting SOPs' location to only tie switches. However, these may not achieve the optimal solution, particularly in large DSs. Unlike the other mentioned studies, this paper proposes that all sectionalized and tie switches could be efficient candidate locations for SOP installations. Additionally, for better improvement in DS performance, this paper proposes concurrent integration of SR and SOP by improving the search performance of the original water cycle algorithm (WCA) based on GA to handle the discrete–continuous search space. The WCA has been effectively employed in a variety of fields to solve constrained as well as unconstrained complex linear and nonlinear optimization problems with high accuracy and rapid convergence speed. The WCA is primarily designed to solve optimization problems with a continuous search space [22]. However, this conventional WCA often struggles to maintain an efficient balance between exploitation and exploration tendencies when it comes to specific optimization problems, such as a mixed-variables nonlinear optimization problem, which mostly leads to premature convergence. As a result, several researchers enhanced the conventional WCA to address the discrete and continuous search space in complex combinatorial optimization problems [23–27].

The main contribution of this paper is to propose a hybrid water cycle algorithm (HWCA) for concurrently solving SR and SOP installation problems to provide the best operation of DSs in terms of system loss and voltage profile. This hybridization eliminates the disadvantages of the original WCA, especially for the mixed-variables nonlinear optimization problem (i.e., limited exploration features due to the discrete–continuous nature of the proposed problem). It also emphasizes the benefits of combining the GA with WCA to deal with both discrete and continuous search spaces and improve the balance between exploration and exploitation phases. As a result, this efficient approach improves the search performance of the original WCA, allowing it to find the optimum solution without being trapped in local minima. The HWCA mainly aims to optimize the SR and SOP integration simultaneously for minimizing system power losses and improving the voltage profile in the DS while considering all operational constraints. Furthermore, unlike previous research, this paper provides a comprehensive evaluation of potential SOP locations that include not just tie switches, but also sectionalizing switches. Additionally, in this paper, the impact of increasing the number of SOPs coupled to distribution networks is evaluated with and without SR in lowering total active losses and improving voltage profiles under various scenarios. Finally, to validate the proposed method, the HWCA is compared to the WCA and the GA. In the comparison with the WCA, a sigmoid function is used to control the type of search space and its boundaries in order to avoid the continuous random movement of stream-to-river and stream-to-sea flows [28]. The main advantage of the sigmoid function is that it changes the values of the random movement of control variables selected by the original WCA to fall within the boundaries of the selective search space due to the discrete nature of SR (i.e., certain switches that are already present in the DS).

The remainder of this paper is organized as follows: Section 2 presents the formulation of the proposed problem. Section 3 presents a full description of the proposed HWCA. The numerical results and discussions are provided in Section 4. The conclusion is then provided in Section 5.

2. Problem Formulation

In this paper, a complex optimization problem with discrete and continuous variables represented by the SOP location and size along with SR is solved using the proposed HWCA to minimize system losses and improve the system voltage profile in radial DSs. The primary objective function taken into consideration in this paper is total power system losses, considering the operational and SOP constraints. To demonstrate the viability of the suggested method for reducing system losses as much as possible, eight cases, including the base case, will be used as examples. A description of the SOP modeling, objective function, and operational constraints is provided below:

2.1. SOP Modeling

SOP is a promising power electronic device recently used in radial distribution networks to replace sectionalizing switches as shown in Figure 1, or tie switches as shown in Figure 2. This cutting-edge power electronic device can effectively provide superior system performance in terms of reducing system losses, improving the system voltage profile, and balancing load flow [9]. This may be accomplished in an effective manner by providing quick, dynamic, and continuous active and reactive power flow control among nodes or feeders during normal network operating conditions, as well as fast fault isolation and supply restoration during abnormal events [10]. Figures 1 and 2 depict the simple structure of two terminals' VSCs placed in sectionalizing and tie switches as considered in this paper. The output reactive power of the two converters is independent since there is a DC bus connecting them [11]. The following equations can be used to simulate the proposed SOP topology:

$$P_n^{SOP} + P_m^{SOP} + P_n^{SOP_{loss}} + P_m^{SOP_{loss}} = 0, \tag{1}$$

where P_n^{SOP}, P_m^{SOP} are the injected active powers of SOP at the nth and m^{th} nodes, respectively, and $P_n^{SOP_{loss}}$ and $P_m^{SOP_{loss}}$ denote internal power losses of SOP converters at the nth and m^{th} nodes, respectively, which can be calculated by multiplying the loss coefficient and the injected apparent power of SOP at the nth and m^{th} nodes, respectively. Given that SOP has sufficient operational efficiency, created internal power losses from a small-scale power transfer can be ignored [18]. Lossless SOP is therefore taken into account in this paper. In the case of lossless SOP installation, the summation of the injected SOP powers to the mth and n^{th} nodes is equal to zero [12]. The active power boundaries for SOP are described in Equation (2):

$$P_n^{SOP} + P_m^{SOP} = 0, \tag{2}$$

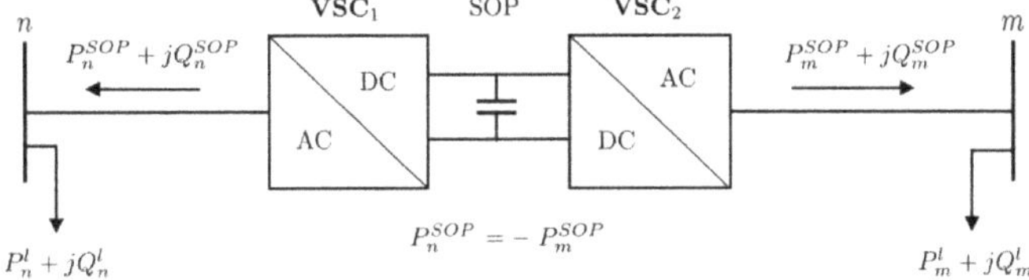

Figure 1. SOP model placed at the sectionalizing switch.

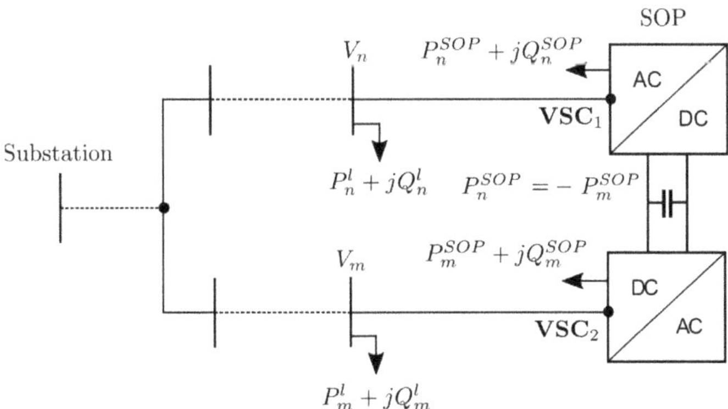

Figure 2. SOP model placed at tie switch.

The total reactive powers that SOP injects into the DS should not be more than the total reactive powers of the system loads, which may be described in Equation (3):

$$\sum_{k=1}^{N_{SOP}} \left(Q_n^{SOP}(k) + Q_m^{SOP}(k) \right) \leq \sum_{u=1}^{N_{load}} Q_u^l, \ \forall u \in N_{load}, \ \forall k \in N_{SOP}, \tag{3}$$

where Q_u^l represents the load reactive power at the uth node; N_{load} and N_{SOP} represent the total number of loads and SOPs, respectively; and Q_n^{SOP} and Q_m^{SOP} are the injected reactive powers of SOP at the nth and m^{th} nodes, respectively. The SOP capacity limits can be expressed in Equations (4) and (5):

$$\sqrt{\left(P_n^{SOP} \right)^2 + \left(Q_n^{SOP} \right)^2} \leq S_{rated}^{SOP}, \tag{4}$$

$$\sqrt{\left(P_m^{SOP} \right)^2 + \left(Q_m^{SOP} \right)^2} \leq S_{rated}^{SOP}, \tag{5}$$

where S_{rated}^{SOP} is the rated size of SOP.

2.2. Objective Function (OF)

The proposed OF is formulated in this paper based on the active power losses in a DS while maintaining all operational and SOP constraints within acceptable limits [12]. The OF is expressed in Equation (6):

$$\min_{\mathbf{X}} P_{loss} = \sum_{d=1}^{N_{br}} \alpha_d \left(\frac{P_d^2 + Q_d^2}{|V_d|^2} . r_d \right), \tag{6}$$

where $\mathbf{X}$ indicates the decision vector made up of the sectionalizing and tie switches status and SOP sizing and sitting; N_{br} represents the number of total branches; the value of α_d denotes the connection ($\alpha_d = 1$) or disconnection ($\alpha_d = 0$) of the dth branch; active power, reactive power, and the voltage at sending of the dth branch are represented by P_d, Q_d, V_d, respectively; and r_d is the resistance of the dth branch. To solve Equation (6), equality and inequality operational and SOP constraints of DSs are considered [18,19]. The problem constraints are given as follows:

- The voltage magnitude of each node in the DS must be maintained within the permitted boundaries for the inequality constraints specified in Equation (7):

$$V_{lower} \leq \left| \tilde{V}_n \right| \leq V_{upper}, \forall n \in N_{node}, \tag{7}$$

where $\tilde{V}_n$ is the voltage at the node n and V_{lower} and V_{upper} are lower and upper acceptable voltage limits (i.e., $V_{lower} = 0.95 \ p.u$ and $V_{upper} = 1.05 \ p.u$), respectively, and N_{node} represents the total number of system nodes.
- Another inequality constraint that is taken into account is the branch's current limitation, which is expressed in Equation (8):

$$|I_d| \leq I_d^{max}, \qquad \forall d \in N_{br}, \tag{8}$$

where I_d and I_d^{max} represent the current and the maximum current permitted for the dth branch.
- To ensure that all loads are kept connected and powered by the main substation during network reconfiguration, a radiality of the DS must be maintained. As a result, the radiality constraint is added as an equality constraint to the proposed objective function and is represented in Equation (9):

$$N_{br} = N_{node} - 1 \tag{9}$$

- SOP constraints can be expressed in Equations (2)–(5), as well as in Equations (10) and (11):

$$Q_{min}^{SOP-n} \leq Q_n^{SOP} \leq Q_{max}^{SOP-n},\tag{10}$$

$$Q_{min}^{SOP-m} \leq Q_m^{SOP} \leq Q_{max}^{SOP-m},\tag{11}$$

where Q_{min}^{SOP-n} and Q_{max}^{SOP-n} represent the minimum and maximum reactive power constraints of SOP injected to the nth node, respectively; Q_{min}^{SOP-m} and Q_{max}^{SOP-m} indicate the minimum and maximum reactive power constraints of SOP injected to the mth node, respectively; and Q_n^{SOP} and Q_m^{SOP} are the injected reactive powers of SOP at the nth and m^{th} nodes, respectively.

3. Description of Proposed HWCA

Many concepts of metaheuristic algorithms are inspired by nature. As a result, many researchers have drawn inspiration for their approaches from observations of these fascinating natural events, which are being controlled by a well-organized system. One of these natural systems that can be used to solve a wide range of difficult optimization problems is the water cycle process. The WCA was first developed by Eskandar et al. in 2012 [22] to solve constrained nonlinear optimization problems. Then, it was further improved by taking into account mixed-integer nonlinear optimization problems [29,30]. The WCA was created by researching how water circulates naturally, starting with raindrops produced by the evaporation process. Then raindrops gradually gathered to form streams, several of which were thereafter either moved toward rivers or sea. The following are the steps of the HWCA [22]:

3.1. Initial Population

An initial population matrix $\mathbf{M}$ of size $N_p \times N_v$ is represented in Equation (12). Each row in the matrix $\mathbf{M}$ consists of raindrops, taking into account the upper and lower constraints of the control variables (i.e., $\mathbf{UC}$ and $\mathbf{LC}$), and can be considered an initial candidate solution for the proposed optimization problem [22].

$$\mathbf{M} = \begin{bmatrix} R_{11} & \cdots & R_{1,Nv} \\ \vdots & \ddots & \vdots \\ R_{Np,1} & \cdots & R_{Np,Nv} \end{bmatrix} = \begin{bmatrix} \mathbf{R}_{D1} \\ \vdots \\ \mathbf{R}_{D,Np} \end{bmatrix},\tag{12}$$

where N_p and N_v indicate the population size and the total number of control variables, respectively, and $\mathbf{R}_{D1}$ represents a raindrop that contains a vector of each candidate solution.

3.2. Evaluation of Each Candidate's Solution

The cost value of each candidate solution represented by $\mathbf{R}_{Di}$ in the matrix $\mathbf{M}$ is evaluated using a fitness function F_i considering problem constraints, which can be expressed in Equation (13):

$$F_i = f(\mathbf{R}_{Di}), \quad i \in N_p\tag{13}$$

The initial population and evaluation of each candidate's solution are represented in block 1 of the main flowchart shown in Figure 3.

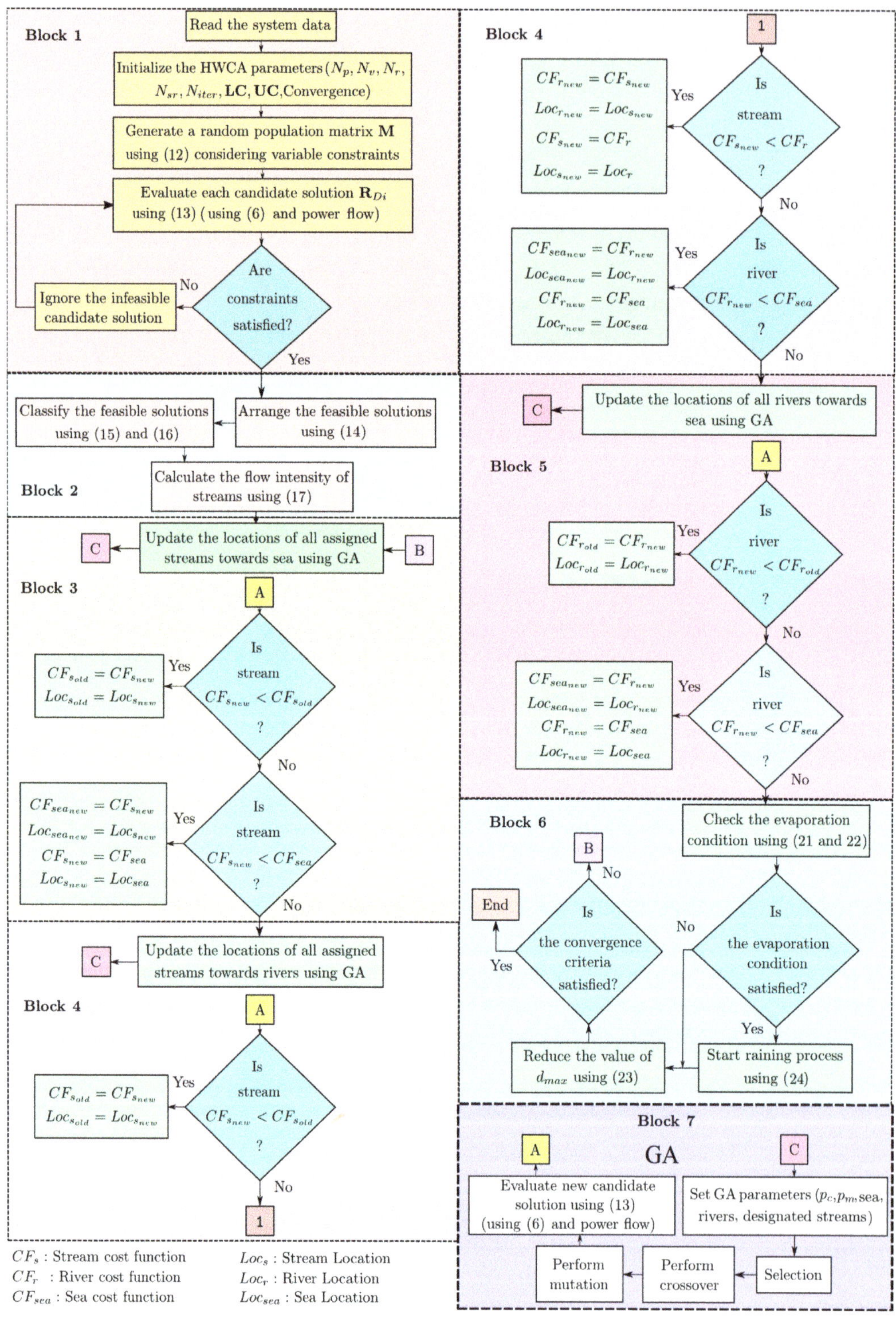

Figure 3. The flow chart of the proposed HWCA.

3.3. Raindrops Classification

In this step, after the cost value of each candidate solution has been evaluated using Equation (13), evaluated raindrops $\mathbf{R}_{Ds}$ are categorized in ascending order. The best $\mathbf{R}_{Di}$ with the lowest cost value is picked to be the sea and the best $\mathbf{R}_{Ds}$ number is chosen to be the rivers, N_r. Finally, the remaining $\mathbf{R}_{Ds}$ are assumed to be streams N_{sr} that may flow into the sea or the rivers [22]. The classification matrix $\mathbf{R}_{Ds}$ can be expressed in Equation (14):

$$\mathbf{R}_{Ds} = \begin{bmatrix} \mathbf{R}_{D,sea} \\ \mathbf{R}_{D,river_1} \\ \vdots \\ \mathbf{R}_{D,river_{N_r}} \\ \mathbf{R}_{D,stream_1} \\ \vdots \\ \mathbf{R}_{D,stream_{N_{sr}}} \end{bmatrix} \tag{14}$$

The total of rivers combined with a single sea can be expressed in Equation (15):

$$N_{sn} = Sea + N_r \tag{15}$$

The streams can also be calculated using Equation (16)

$$N_{sr} = N_p - N_{sn} \tag{16}$$

3.4. Determining the Stream's Maximum Flow Rate

Equation (17) is used to calculate the flow rate of streams that flow into rivers or directly into the sea [22]:

$$N_{sk} = round \left\{ \left| \frac{F_k}{\sum_{i=1}^{N_{sn}} F_i} \right| \times N_{sr} \right\}, \quad k \in N_{sn} \tag{17}$$

The raindrops classification and calculation of the flow intensity of streams are demonstrated in Block 2 of the main flowchart shown in Figure 3.

3.5. Update the Location of Streams and Rivers

In the original version of the WCA [22], the stream-to-river flow can be determined using a random distance given in Equation (18). Furthermore, the rivers can also flow to the sea with the same concept in Equation (18), so the new position for the streams and rivers can be expressed using Equations (19) and (20):

$$R \in (0, H \times d), \ H > 1 \tag{18}$$

$$\mathbf{R}_{D,stream_{i+1}} = \mathbf{R}_{D,stream_i} + rand \times H \times \left(\mathbf{R}_{D,river_i} - \mathbf{R}_{D,stream_i} \right) \tag{19}$$

$$\mathbf{R}_{D,river_{i+1}} = \mathbf{R}_{D,river_i} + rand \times H \times \left(\mathbf{R}_{D,sea_i} - \mathbf{R}_{D,river_i} \right), \tag{20}$$

where H is a constant and can be chosen as 2, d represents the distance between stream and river, and $rand$ is a random number between 0 and 1.

What can be deduced from Equations (19) and (20) is that the positions of streams and rivers are constantly adjusted in order to diversify the population and prevent becoming locked in a local optimum. These adjustments are based on a continuous search space. Therefore, a nonlinear optimization problem with either continuous or discrete control variables has been addressed by several modified WCAs [31–34] with different objectives. In this paper, the proposed problem aimed to be solved is also a nonlinear optimization problem with continuous and discrete control variables. These variables are categorized as discrete line numbers for the SR problem, discrete node numbers for SOP locations, and continuous SOP size limits. Therefore, to avoid becoming locked in a local optimum and handle continuous-discrete decision variables, the GA-based WCA is utilized to update the placement of intended streams and rivers.

As shown in Figure 3, the optimal new locations of assigned streams toward the sea and rivers as well as rivers toward the sea are illustrated in Blocks 3, 4, and 5. These optimal new locations can be efficiently discovered using GA, represented in Block 7.

Running WCA or GA optimizers, especially in big systems, may result in suboptimal results due to the nonlinear nature of the proposed problem. The proposed hybridization technique enhances the WCA generating and searching process, prevents becoming stuck at a local optimum point, and eventually solves this problem and yields the optimal solution. The GA is suggested in this paper to update the position of assigned streams and rivers because it is an evolutionary algorithm derived from the natural evolution and selection of human genetics and it can efficiently work in any search space [35]. Moreover, the GA increases the possibility of locating the best movement of streams and rivers because of its random nature. The GA procedure used in this paper can be summarized in the steps below [36]:

3.5.1. Initialization

In the general concept of the GA, the candidate solutions are randomly generated from an initial population. However, in this paper, candidate solutions are represented by designated streams and rivers, including the sea, selected by HWCA. The structure indicated in Figure 4 is categorized into two parts. The first part is integer control variables, which represent the discrete line numbers for the SR problem and SOP locations, and the second part is real control variables, which represent continuous SOP size limits. In this paper, when the GA is used to update the designated streams and river locations, we deal with each part of this initial structure separately in order to avoid selecting non-discrete numbers that could be located outside the range of a test system node, especially for SR and SOP location problems. As a result, each part of this structure (i.e., SOP location, SOP active power injection, SOP reactive power injection, and reconfiguration) has its own crossover and mutation operators used to combine candidate solutions to create new solutions.

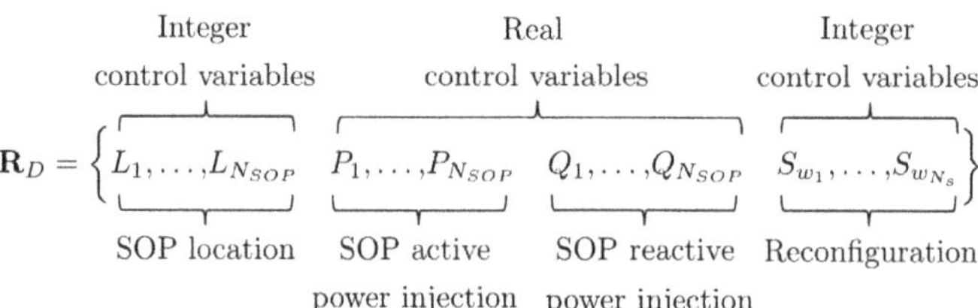

Figure 4. The structure of the initial individual HWCA.

The structure of each stream and river, including the sea, is shown in Figure 4:

Where L represents SOP location; $L \in (N_s, N_{tie})$, N_s and N_{tie} are the total number of sectionalizing and tie switches, respectively; P and Q are SOP active and reactive power injections, respectively; S_w represents sectionalizing (open) and tie (closed) switches; and $S_w \in (N_s, N_{tie})$.

3.5.2. Selection

In this stage, candidate individuals are chosen based on their cost functions, and individuals with better solutions have a higher chance of selection. In this paper, the best-evaluated individuals are already arranged by the HWCA in ascending order (i.e., minimum value) and categorized as sea, rivers, and streams.

3.5.3. Crossover

This operator combines two candidate individuals (parents) to create new individuals (offspring). If the new individual inherits the best features from both parents, it may outperform both. The simplest method used in this paper to perform crossover is to select a few crossover places based on a probability of crossover, which is selected in this paper as $p_c = 0.9$.

3.5.4. Mutation

After a crossover operator is performed, mutation takes place in the offspring generated from the crossover operation, with the probability of mutation selected in this paper to be $p_m = 0.01$. The use of mutation in the GA prevents falling into a local optimum and maintains the desired level of diversity in the population.

3.5.5. Evaluate the Candidate's Solutions

The offspring produced by crossover and mutation operators are evaluated using the proposed cost-function while taking operational restrictions into account in this step of the GA process. The position of streams and rivers that flow towards the sea will be changed to the one with the best cost-evaluation.

3.6. Check the Evaporation and Raining Process

The major purpose of the evaporation condition in Equations (21) and (22) is to lessen the chances of the algorithm getting stuck in local solutions and to identify whether or not streams or rivers have arrived, or are sufficiently close to the sea. If the evaporation requirement is met, it signifies that the distance between streams and rivers, or rivers and the sea [22], is less than d_{max}.

$$\|\mathbf{R}_{D,\text{sea}_i} - \mathbf{R}_{D,\text{river}_i}\| < d_{max} \tag{21}$$

$$\|\mathbf{R}_{D,\text{river}_i} - \mathbf{R}_{D,\text{stream}_i}\| < d_{max}, \tag{22}$$

where d_{max} is a preset number that is close to zero, $i \in N_{iter}$, and N_{iter} is the total number of iterations. In order to find the best solution, the value of d_{max} is used to adjust the search intensity close to the sea. When iterations occur, this predefined value is incrementally updated and is defined using Equation (23):

$$d_{max_{i+1}} = d_{max_i} - \frac{d_{max_i}}{N_{iter}} \tag{23}$$

The raining process begins immediately to generate new raindrops whenever either one, or both, of the evaporation requirements in Equations (21) and (22) are satisfied [22]. Once it rains, new streams are created in various places. The raining process in this paper is carried out using a mutation operator and is based on the following random probability defined in Equation (24):

$$\mathbf{R}_D^{new} = \mathbf{LC} + \text{rand} \times (\mathbf{UC} - \mathbf{LC}) \tag{24}$$

The evaporation condition and raining process are demonstrated in Block 6 of the main flowchart shown in Figure 3.

The main structure of the proposed HWCA used in this paper is demonstrated using the flow chart shown in Figure 3.

4. Numerical Results and Discussions

In this paper, eight cases are considered and applied to two modified versions of the IEEE 33-Node and IEEE 69-node DSs utilizing the proposed HWCA to show and assess the effectiveness of the HWCA in solving SR and installation of SOP unit problems [37,38]. All tie and sectionalizing switches are examined as candidate switches for solving the SR and

SOP allocation problem in both test systems. The maximum number of SOPs that can be placed on the provided test systems is two. However, the proposed approach may be used for any number of SOPs. Furthermore, in order to demonstrate that the HWCA is superior to other approaches in addressing the proposed problem with the goal of enhancing the voltage profile and reducing system loss, simulation results of the HWCA for case 7 are compared to those obtained by the WCA and GA, while the HWCA parameters that are initialized (e.g., population size of 80 and maximum iteration of 300) are shared with all cases and other compared optimization algorithms for both test systems. Seven different cases, including the base case, are simulated and categorized as follows:

- Base case: A power-flow solution without considering SOP installation and SR;
- Case 1: Only optimal SR without SOP installation;
- Case 2: One optimal SOP installation without SR;
- Case 3: One optimal SOP installation only at Tie switches without SR;
- Case 4: Two optimal SOP installations without SR.
- Case 5: Two optimal SOP installations only at Tie switches without SR.
- Case 6: Simultaneous optimal SR and one SOP installation (proposed method).
- Case 7: Simultaneous optimal SR and two SOP installations (proposed method).

In this paper, the proposed cost function indicated in Equation (6), operational limitations, and features of the system after SOP power injection and SR are evaluated using a modified version of the ladder-iterative power-flow technique [39]. Furthermore, evaluation of the proposed eight cases is performed using MATLAB and run on a personal computer with an Intel Core i7-3770 processor running at 3.40 GHz with 16 GB of RAM. The descriptions and optimal simulation results for IEEE 33-Node and 69-Node DSs are presented as follows:

4.1. Numerical Results of IEEE 33-Node Test System

The initial configuration system of the IEEE 33-bus test system includes 33 nodes, 37 branches, 32 normally-close sectionalized switches, and five normally-closed tie switches (e.g., T_{33} to T_{37}), with a 12.6 kV (or 1 p.u) base system voltage as shown in Figure 5. The total real and reactive power loads are 3.72 MW and 2.3 MVAR, respectively, and the base active power loss is 202.67 kW. The limits of real and reactive power injected by SOPs are 0 to 2.5 MW and 0 to 2.5 MVAR, respectively, and the voltage magnitude constraints of all buses are set at 0.95 and 1.05 p.u.

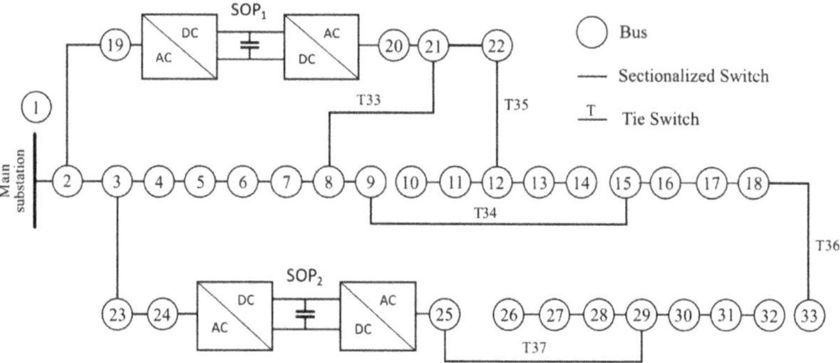

Figure 5. The optimal SR and two SOP locations for the IEEE 33-node radial DS.

The validity of the proposed HWCA in addressing SR and SOP installation problems is demonstrated in Table 1 with eight cases, including the base case. In this table, the base case illustrates the worst-case scenario, in which there is no performance improvement in the test system in terms of loss reduction and voltage limitations. Case 1 is also presented in Table 1. In this case, the optimal SR obtained by HWCA is to open and close some sectionalized and tie switches as shown in Table 1. This indicated that the system loss is improved by 31.144% compared with the base case, while the minimum voltage magnitude limit is still violated. This is expected because of the limitations of search space and radial topology. In case 2 as shown in Table 1, one SOP installation is optimally integrated to improve the system's performance without considering SR. In this case, it is observed that the HWCA made an optimal decision to select the best active and reactive setting and location for SOP as shown in Table 1. This decision resulted in a significant system loss reduction of nearly 43.45% compared with the base case with no voltage magnitude violation. Additionally, compared to scenario 1, the possibility of improving the system performance is increased by injecting reactive power along with active power into the test system. Case 3 is also included in Table 1. This case is similar to case 2, with the exception that the tie switch locations are the only viable options for the HWCA to place the SOP. In this case, it is clear that system performance is improved in terms of loss reduction, being roughly 41.1% compared to the base case, however, voltage magnitude limitation is violated. This happens because the search space of HWCA is restricted by only tie switch locations. Case 4 is also shown in Table 1. This case is similar to case 2, which suggests installing two SOPs rather than one. In this case, according to Table 1, the HWCA is successful in selecting the optimal SOP settings and locations, which considerably improves the system reduction by roughly 53.78% compared to the base case, with no voltage magnitude violations. In this case, the increase in loss reduction is expected compared to mentioned cases because the rising number of SOPs installed into the test system can bring further improvement to DSs in terms of loss reduction and voltage profile improvement. Table 1 also shows case 5. The only difference between this case and case 4 is that the tie switch locations are the only viable sites to install the SOPs. Consequently, as seen from Table 1, there is less improvement in loss reduction of about 53.27% compared to case 4, with no violation in voltage magnitude of all system nodes. Table 1 also includes case 6. Unlike cases 1 and 2, where the SOP and SR are integrated independently, in this case, the HWCA optimizes the SR and one SOP installation simultaneously. From Table 1, the HWCA is effective in determining the optimal SOP setting, position, as well as SR, resulting in a considerable system loss reduction of about 54.50% compared to the base case with no voltage magnitude violations. Finally, case 7 is also depicted in Table 1. This case is similar to case 6 except that it suggests installing two SOPs rather than one. As indicated in Table 1, the HWCA is capable of choosing the best settings for the two SOPs and the SR in this case, as shown in Figure 5, reducing system loss by around 63.33% compared to the base case with no voltage magnitude violations. Additionally, Figure 6 demonstrates the stages of loss reduction improvement from the base case to case 7. From this figure, it is obvious that case 7 is the best in system loss reduction.

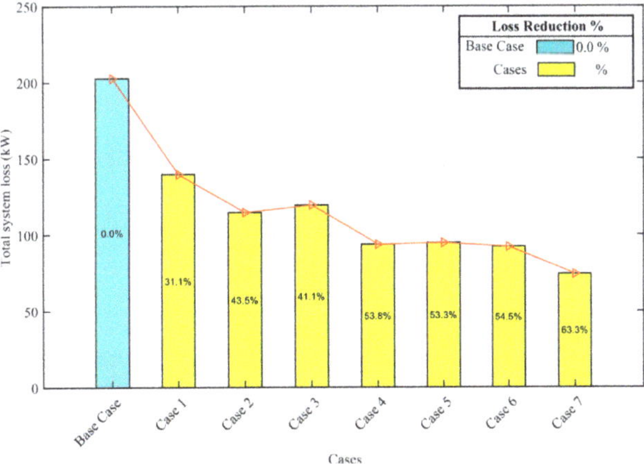

Figure 6. Loss reduction improvement of the 33-node test DS for all cases.

Table 1. Numerical results for IEEE 33-Node with and without SOP installation.

Case	No. of SOP	SOP Location (Node-Node)	Sec. and Tie Switch Status Opened Switch (OS) Closed Switch (CS)	Optimal SOP Active Power Injection (MW)	Optimal SOP Reactive Power Injection (MVAr)	Ploss (kW)	V_{min}/V_{max} (p.u)
Base Case	—	—	—	—	—	202.67	0.9131/0.9970
SR (Case 1)	—	—	OS: 7, 9, 14, 32 CS: 33, 34, 35, 36	—	—	139.55	0.9378/0.9971
SOP without SR (Case 2)	1	5–6	OS: 5 CS: 33	−1.3758/1.3758	0.3324/1.6926	114.59	0.9512/0.9977
SOP without SR (Tie-switches) (Case 3)	1	8–21	—	1.0813/−1.0813	1.3660/0.3268	119.39	0.9454/0.9976
SOP without SR (Case 4)	2	5–6 30–31	OS: 5,30 CS: 33, 36	−1.5554/1.5554, −0.5889/0.5889	0.3401/0.4996, 0.5000/0.4323	93.67	0.9599/0.9976
SOP without SR (Tie-switches) (Case 5)	2	25–29 12–22	—	−0.4190/0.4190 0.7439/−0.7439	0.4983/0.4956 0.4991/0.1397	94.71	0.9561/0.9976
SOP with SR (Case 6)	1	24–25	OS: 7, 9, 14, 17, 24 CS: 33, 34, 35, 36, 37	−0.9708/0.9708	0.3767/1.1359	92.22	0.9631/0.9976
SOP with SR (Case 7)	2	24–25 19–20	OS: 9, 14, 19, 24, 32 CS: 33, 34, 35, 36, 37	−0.8205/0.8205 −1.3458/1.3458	0.3424/1.1164 0.1742/0.5273	74.32	0.9670/0.9983

Figure 7 compares and presents the voltage profile curves for all cases, including the base case. This figure demonstrates that the voltage magnitudes at each test system node are within permissible bounds for all cases except the base case, case 1, and case 3. These cases are unable to find the best solution while keeping all system nodes within allowed limits because the search space is severely constrained by the test system's radial topology and tie switch positions.

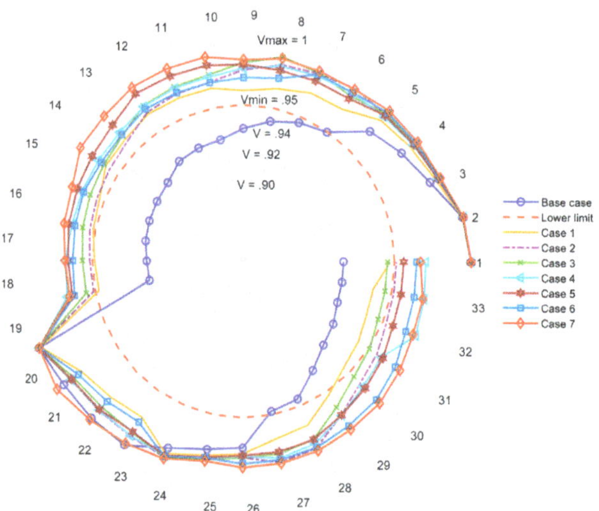

Figure 7. Voltage magnitude profile of the 33-node test DS for all cases.

4.2. Numerical Results of the IEEE 69-Node Test System

This test system consists of 69 nodes, 73 branches, 68 normally-close sectionalized switches, and five normally-closed tie switches (e.g., T_{69} to T_{73}), with a 12.6 kV (or 1 p.u) base system voltage, as shown in Figure 8. The total real and reactive power loads are 3.80 MW and 2.70 MVAR, respectively, and the base active power loss is 224.69 kW. The limits of real and reactive power injected by SOPs are 0 to 2.5 MW and 0 to 2.5 MVAR, respectively, and the voltage magnitude limits of all buses are 0.95 to 1.05 p.u.

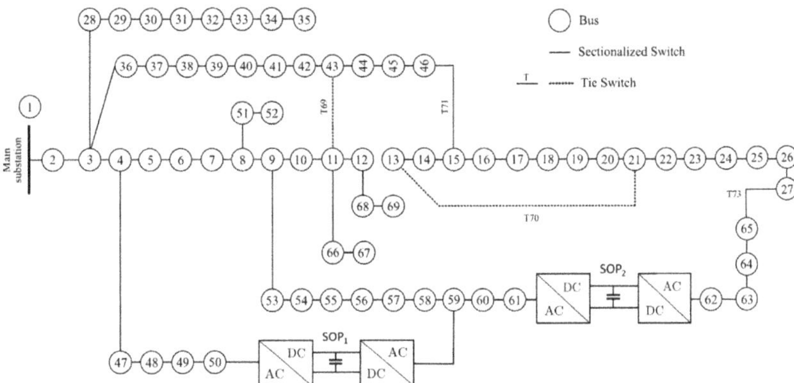

Figure 8. The optimal SR and two SOP locations for the IEEE 69-node radial DS.

Table 2 shows the performance of the proposed HWCA in handling SR and SOP installation problems in the 69-node system using eight cases, including the base case. According to Table 2, the worst-case scenario is the base case with no performance improvement in the test system with regards to loss reduction and voltage limit violation. Case 1 is also presented in Table 2. In this case, the HWCA determines an optimal sectionalized and tie switch as illustrated in Table 2. Additionally, the system loss reduction is improved by 55.7% compared to the base case, with a slight violation in the minimum voltage magnitude limit. This is expected because the search space of HWCA is constrained by the radial topology of the test system. Case 2 and case 3 are also shown in Table 2. In these cases, it is observed that the HWCA chooses an optimal SOP location at the same tie switch T_{72} with the best active and reactive settings without considering SR. This decision results in a significant system loss reduction of nearly 73.1% compared with the base case, with no voltage magnitude violation. Table 2 also depicts cases 4 and 5. In these cases, the results obtained by the HWCA exhibit high system performance, which can be explained by a loss reduction improvement of nearly 79.6% compared with the based case and no voltage magnitude violation. Furthermore, the HWCA selects the optimum SOP locations for both cases, which are T_{72} and T_{71} with the best active and reactive settings without taking SR into account, as shown in Table 2. Case 6 is also included in Table 2. In this case, the HWCA concurrently optimizes one SOP installation and SR. As shown in Table 2, the HWCA is effective in determining the optimal SOP setting and position, as well as the SR. This results in a considerable system loss reduction of around 78.7% with no voltage magnitude violations, as opposed to cases 1 and 2 where the SOP and SR are integrated individually. Case 7 is also shown in Table 2 as the last case. This case is comparable to case 6 with the exception that it recommends implementing two SOPs as opposed to one. Figure 8 illustrates how the HWCA is able to select the optimal configurations and placements for the two SOPs and the SR in this case, as given in Table 2. This results in a system loss reduction of around 82.1% compared to the base case, with no voltage magnitude violations. Furthermore, Table 2 shows that case 7 exhibits the greatest power loss reduction and voltage magnitude improvement when compared to the other cases. Finally, Figure 9 is used to visualize the improvement of loss reduction from the base case to case 7. This figure indicates that the HWCA offers the best system loss reduction in case 7.

Table 2. Numerical results for IEEE 69-Node with and without SOP installation.

Case	No. of SOP	SOP Location (Node-Node)	Sec. and Tie Switch Status Opened Switch (OS) Closed Switch (CS)	Optimal SOP Active Power Injection (MW)	Optimal SOP Reactive Power Injection (MVAr)	Ploss (kW)	V_{min}/V_{max} (p.u)
Base Case	—	—	—	—	—	224.96	0.9092/1.000
SR (Case 1)	—	—	OS: 14, 57, 61 CS: 71, 72, 73	$-\,-\,-$	$-\,-\,-$	99.60	0.9428/1.0000
SOP without SR (Case 2)	1	50–59	—	−1.5433/1.5433	0.5455/1.3852	60.43	0.9709/1.0000
SOP without SR (Tie-switches) (Case 3)	1	50–59	—	−1.5433/1.5433	0.5455/1.3852	60.43	0.9709/1.0000
SOP without SR (Case 4)	2	15–46 50–59	—	−1.5926/1.5926 0.4490/−0.4490	0.5503/1.2818 0.3574/0.0952	45.95	0.9794/1.0000
SOP without SR (Tie-switches) (Case 5)	2	15–46 50–59	—	−1.5926/1.5926 0.4490/−0.4490	0.5503/1.2818 0.3574/0.0952	45.95	0.9794/1.0000
SOP with SR (Case 6)	1	50–59	OS: 12, 64 CS: 71, 73	−1.5492/1.5492	0.5478/1.2654	47.89	0.9808/1.0000
SOP with SR (Case 7)	2	61–62 50–59	OS: 12, 61 CS: 71, 73	−1.4630/1.4630 −0.1812/0.1812	0.5549/0.2244 0.8557/0.4057	40.23	0.9852/1.0000

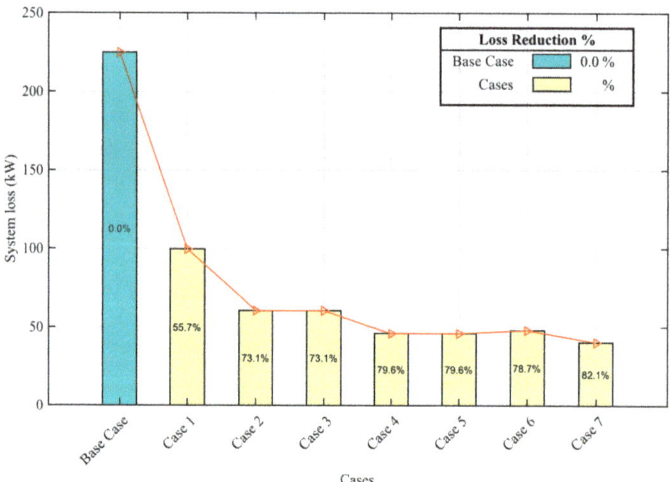

Figure 9. Loss reduction improvement profile of the 69-node test DS for all cases.

The voltage profile curve for each case, including the basic case, is compared and shown in Figure 10. Except for the base case and case 1, this figure shows that the voltage magnitudes for all system nodes are all within acceptable boundaries. Notwithstanding, as a result of the search space being heavily confined by the radial topology of the test system in case 1, the HWCA is unable to find the optimal solution while keeping all system nodes within permitted bounds.

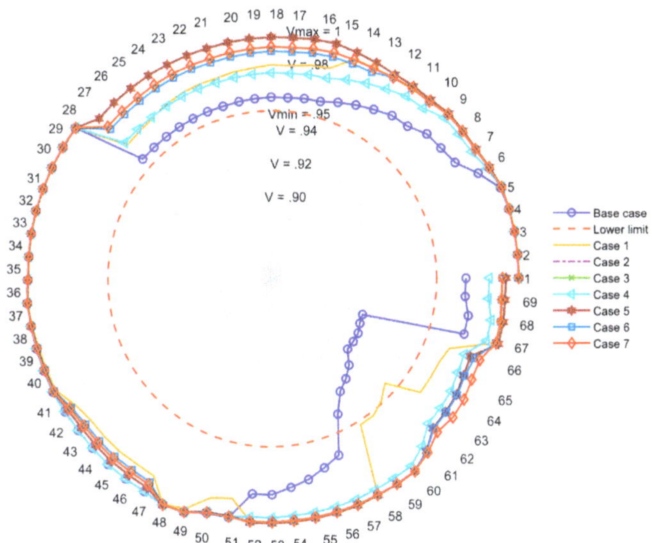

Figure 10. Voltage magnitude profile of the 69-node test DS for all cases.

4.3. Comparison of the HWCA Efficacy with Other Optimization Algorithms

The efficacy of the proposed method, represented by cases 6 and 7, based on the HWCA reducing system power loss and enhancing the voltage profile is further proven by comparison with two well-known optimization algorithms, GA and WCA, utilizing the same metrics from Tables 1 and 2 in addition to adding the number of iterations as indicated in Tables 3–6. Cases 6 and 7 are applied to these optimization algorithms with the same initial optimization settings to both the IEEE 33-node and 69-node DSs as shown in Table 7. Tables 3–6 demonstrate that the HWCA surpassed the other approaches in terms of loss reduction when compared to the base case. Moreover, the proposed HWCA shows fast convergence speed with the least iteration numbers as indicated in Figures 11–13 for both cases 6 and 7, except in Figure 14. In this figure, the proposed HWCA takes more iterations than the GA to attain the optimal solution; meanwhile, the GA failed to attain this optimal solution, as shown in Figure 14. The key reason for the HWCA's superiority is that the HWCA takes a distinctive indirect strategy to explore the most optimal solution based on updating the position of streams and rivers toward the sea by utilizing the GA, which is regarded as the temporary optimum solution. The HWCA can also efficiently prevent getting caught in a locally optimal solution or quick convergence thanks to the usage of the GA in updating the streams and rivers locations.

Table 3. Compared numerical results of HWCA for case 6 with WCA and GA for the 33-node system.

Case	SOP Location (Node-Node)	Sec. and Tie Switch Status Opened Switch (OS) Closed Switch (CS)	Optimal SOP Active Power Injection (MW)	Optimal SOP Reactive Power Injection (MVAr)	Ploss (kW)	V_{min}/V_{max} (p.u)	No. of iter.
Base Case	—	—	—	—	202.67	0.9131/0.9970	—
GA	24–25	OS: 7, 9, 14, 17, 24 CS: 33, 34, 35, 36, 37	−1.0800/1.088	0.477427/1.1129	93.02	0.9631/0.9976	172
WCA	25–29	OS: 7,9,14,17 CS: 33, 34, 35, 36	−0.4205/0.4205	0.3830/1.1411	93.47	0.9631/0.9976	291
HWCA	24–25	OS: 7, 9, 14, 17, 24 CS: 33, 34, 35, 36, 37	−0.9708/0.9708	0.3767/1.1359	92.24	0.9631/0.9976	122

Table 4. Compared numerical results of HWCA for case 6 with WCA and GA for the 69-node system.

Case	SOP Location (Node-Node)	Sec. and Tie Switch Status Opened Switch (OS) Closed Switch (CS)	Optimal SOP Active Power Injection (MW)	Optimal SOP Reactive Power Injection (MVAr)	Ploss (kW)	V_{min}/V_{max} (p.u)	No. of Iter.
Base Case	—	—	—	—	202.67	0.9131/0.9970	—
GA	50–59	OS: 12 CS: 71	−1.6418/1.6418	0.3670/1.4570	49.66	0.9820/1.0000	204
WCA	50–59	OS: 12, 64 CS: 71, 73	−1.5501/1.5501	0.5592/1.2864	47.90	0.9811/1.0000	253
HWCA	50–59	OS: 12, 64 CS: 71, 73	−1.5492/1.5492	0.5478/1.2654	47.89	0.9810/1.0000	81

Table 5. Compared numerical results of HWCA for case 7 with WCA and GA for the 33-node system.

Case	SOP Location (Node-Node)	Sec. and Tie Switch Status Opened Switch (OS) Closed Switch (CS)	Optimal SOP Active Power Injection (MW)	Optimal SOP Reactive Power Injection (MVAr)	Ploss (kW)	V_{min}/V_{max} (p.u)	No. of Iter.
Base Case	—	—	—	—	202.67	0.9131/ 0.9970	—
GA	19–20 25–29	OS: 9, 14, 19, 32 CS: 33, 34, 35, 36	−1.4914/1.4914 −0.2672/0.2672	0.4183/0.6429 0.4660/0.8353	76.37	0.9686/1.0043	202
WCA	19–20 23–24	OS: 9, 14, 19, 23 CS: 33, 34, 35, 37	−1.0528/1.0528 −1.5134/1.5134	1.0059/0.5078 0.5700/1.1248	79.54	0.9675/0.9981	266
HWCA	19–20 24–25	OS: 14, 32, 9, 19, 24 CS: 33, 34, 35, 36, 37	−0.8205/0.8205 −1.3458/1.3458	0.3424/1.1164 0.1742/0.5273	74.33	0.9670/0.9983	194

Table 6. Compared numerical results of HWCA for case 7 with WCA and GA for the 69-node system.

Case	SOP Location (Node-Node)	Sec. and Tie Switch Status Opened Switch (OS) Closed Switch (CS)	Optimal SOP Active Power Injection (MW)	Optimal SOP Reactive Power Injection (MVAr)	Ploss (kW)	V_{min}/V_{max} (p.u)	No. of Iter.
Base Case	—	—	—	—	224.96	0.9092/1.0000	—
GA	9–10 50–59	OS: 9, 13, 20, 63 CS: 69, 70, 71, 73	−1.4084/1.4084 −0.6483/0.6483	0.5526/0.9699 0.0811/0.7696	45.41	0.9755/1.0000	188
WCA	16–17 50–59	OS: 12, 16, 64 CS: 70, 71, 73	−0.2903/0.2903 −1.5520/1.5520	0.2029/0.1863 0.5508/1.1467	46.11	0.9795/1.0000	251
HWCA	61–62 50–59	OS: 12, 61 CS: 71, 73	−1.4630/1.4630 −0.1812/0.1812	0.5549/0.2244 0.8557/0.4057	40.23	0.9852/1.0000	205

Table 7. Optimization parameter settings of GA, WCA, and HWCA for cases 6 and 7.

Algorithm	33-Node and 69-Node Systems
GA	Population of chromosomes = 50, maximum iteration = 300, number of genes (number of variables) = 13, probability of crossover = 0.9, probability of mutation = 0.01.
WCA	Population of raindrops = 50, maximum iteration = 300, raindrops (number of variables) = 13, number of rivers = 6, a constant H = 2, $d_{max} = 10^{-16}$.
HWCA	Population of raindrops = 50, maximum iteration = 300, raindrops (number of variables) = 13, number of rivers = 6, $d_{max} = 10^{-16}$. probability of crossover = 0.9, probability of mutation = 0.01.

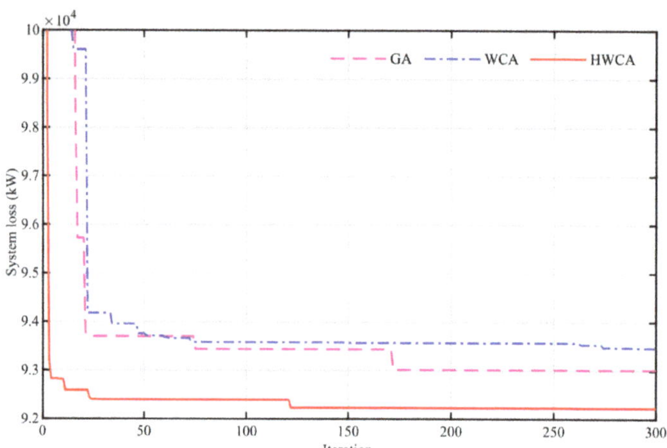

Figure 11. Convergence curve of the IEEE 33-node system using compared algorithms for case 6.

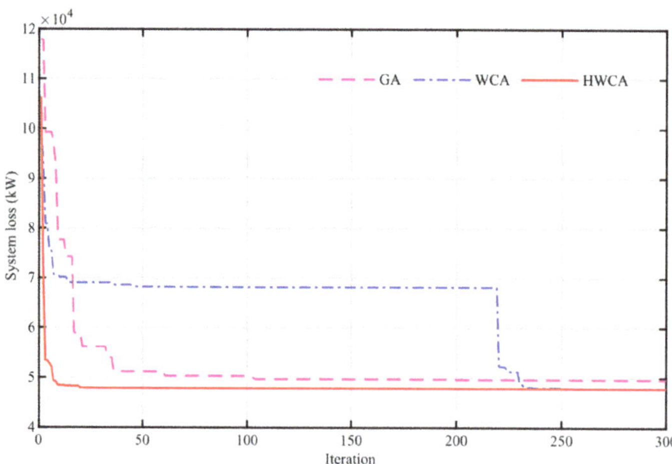

Figure 12. Convergence curve of the IEEE 69-node system using compared algorithms for case 6.

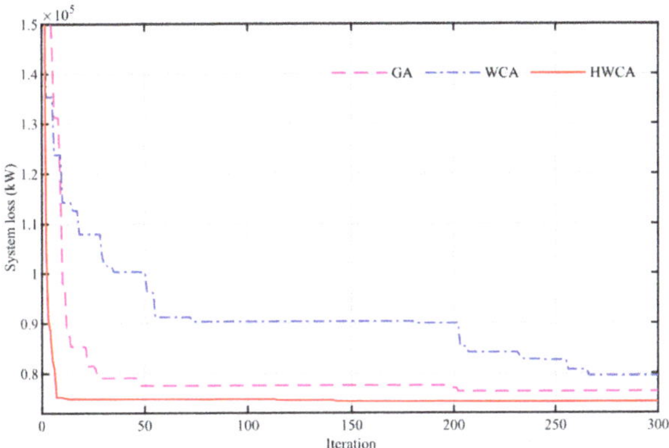

Figure 13. Convergence curve of the IEEE 33-node system using compared algorithms for cas.

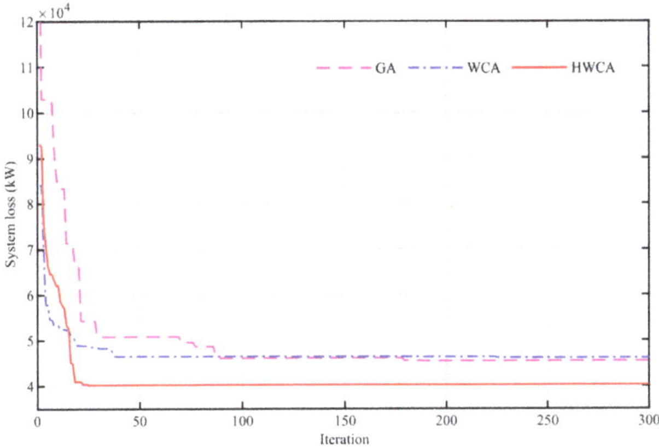

Figure 14. Convergence curve of the IEEE 69-node system using compared algorithms for case 7.

5. Conclusions

In this paper, the proposed hybrid water cycle algorithm (HWCA) was developed to address a complex nonlinear problem that is exemplified by simultaneously addressing the placement and size of the soft open point (SOP) and system reconfiguration (SR), with the objective to minimize system losses and improve the voltage profile while considering operational constraints in DSs. The water cycle algorithm (WCA) and genetic algorithm (GA) formed the basis of the proposed hybrid optimization algorithm to reduce the drawbacks of each compared method when utilized separately, deal with the discrete and continuous search space, and avoid getting trapped in local minima.

Eight cases, including the base case, were conducted on the IEEE 33-node and IEEE 69-node to evaluate the performance of the HWCA and investigate the real benefit gained from using SOPs alone or simultaneously with the SR, as well as show the effect of increasing the number of SOPs in improving system loss reduction and system voltage profiles in DSs under different scenarios. In each of these cases, the HWCA could efficiently improve the system loss reduction (e.g., 31.1–63.3% for IEEE 33-node and 55.7–82.1% for IEEE 69-node compared to the base case) while maintaining acceptable voltage magnitudes in most cases. Moreover, to demonstrate the efficacy of HWCA, case 7 was selected as a comparison case conducted on the modified IEEE 33-node and IEEE 69-node test systems with two other well-known metaheuristic optimization algorithms, WCA and GA. The simulation results show that the proposed method based on HWCA outperforms other optimization algorithms in terms of system loss reduction.

In future work, we will extend our method to simultaneously integrate the placement and size of the soft open point (SOP) and distributed generators (DGs) along with system reconfiguration (SR) to increase the DG hosting capacity and reduce system loss in DSs, while considering load uncertainty and system operational limits.

Author Contributions: Conceptualization, methodology and software, S.I., S.A. and A.M.A.; data collection, writing and original draft preparation, S.I.; visualization and investigation, S.I. and S.A.; supervision S.I.; software and validation S.I.; review and editing, S.I., S.A. and A.M.A. All authors have read and agreed to the published version of the manuscript.

Funding: This research received no external funding.

Data Availability Statement: The data supporting the reported results are available in the manuscript.

Acknowledgments: This work was supported by the Ministry of Higher Education and Scientific Research, the University of Babylon, and Al-Mustaqbal University College, Iraq.

Conflicts of Interest: The authors declare no conflict of interest.

References

1. Wang, X.; Wang, C.; Xu, T.; Guo, L.; Li, P.; Yu, L.; Meng, H. Optimal voltage regulation for distribution networks with multi-microgrids. *Appl. Energy* **2018**, *210*, 1027–1036. [CrossRef]
2. Gebru, Y.; Bitew, D.; Aberie, H.; Gizaw, K. Performance enhancement of radial distribution system using simultaneous network reconfiguration and switched capacitor bank placement. *Cogent Eng.* **2021**, *8*, 1897929. [CrossRef]
3. Chiang, H.-D.; Jean-Jumeau, R. Optimal network reconfigurations in distribution systems: Part 1: A new formulation and a solution methodology. *IEEE Trans. Power Deliv.* **1990**, *5*, 1902–1909. [CrossRef]
4. Civanlar, S.; Grainger, J.; Yin, H.; Lee, S. Distribution feeder reconfiguration for loss reduction. *IEEE Trans. Power Deliv.* **1988**, *3*, 1217–1223. [CrossRef]
5. Sahoo, N.; Prasad, K. A fuzzy genetic approach for network reconfiguration to enhance voltage stability in radial distribution systems. *Energy Convers. Manag.* **2006**, *47*, 3288–3306. [CrossRef]
6. Souifi, H.; Kahouli, O.; Abdallah, H.H. Multi-objective distribution network reconfiguration optimization problem. *Electr. Eng.* **2019**, *101*, 45–55. [CrossRef]
7. Truong, A.V.; Ton, T.N.; Nguyen, T.T.; Duong, T.L. Two States for Optimal Position and Capacity of Distributed Generators Considering Network Reconfiguration for Power Loss Minimization Based on Runner Root Algorithm. *Energies* **2018**, *12*, 106. [CrossRef]
8. Nguyen, T.T.; Nguyen, T.T. An improved cuckoo search algorithm for the problem of electric distribution network reconfiguration. *Appl. Soft Comput.* **2019**, *84*, 105720. [CrossRef]
9. Cao, W.; Wu, J.; Jenkins, N.; Wang, C.; Green, T. Operating principle of Soft Open Points for electrical distribution network operation. *Appl. Energy* **2016**, *164*, 245–257. [CrossRef]
10. Fuad, K.S.; Hafezi, H.; Kauhaniemi, K.; Laaksonen, H. Soft Open Point in Distribution Networks. *IEEE Access* **2020**, *8*, 210550–210565. [CrossRef]
11. Cao, W.; Wu, J.; Jenkins, N.; Wang, C.; Green, T. Benefits analysis of Soft Open Points for electrical distribution network operation. *Appl. Energy* **2016**, *165*, 36–47. [CrossRef]
12. Qi, Q.; Wu, J.; Long, C. Multi-objective operation optimization of an electrical distribution network with soft open point. *Appl. Energy* **2017**, *208*, 734–744. [CrossRef]
13. Farzamnia, A.; Marjani, S.; Galvani, S.; Kin, K.T.T. Optimal Allocation of Soft Open Point Devices in Renewable Energy Integrated Distribution Systems. *IEEE Access* **2022**, *10*, 9309–9320. [CrossRef]
14. Cong, P.; Hu, Z.; Tang, W.; Lou, C. Optimal allocation of soft open points in distribution networks based on candidate location opitimization. In Proceedings of the 8th Renewable Power Generation Conference (RPG 2019), Shanghai, China, 24–25 October 2019; pp. 1–6. [CrossRef]
15. Qin, Q.; Han, B.; Li, G.; Wang, K.; Xu, J.; Luo, L. Capacity allocations of SOPs considering distribution network resilience through elastic models. *Int. J. Electr. Power Energy Syst.* **2022**, *134*, 107371. [CrossRef]
16. Khan, M.; Wadood, A.; Abid, M.; Khurshaid, T.; Rhee, S. Minimization of Network Power Losses in the AC-DC Hybrid Distribution Network through Network Reconfiguration Using Soft Open Point. *Electronics* **2021**, *10*, 326. [CrossRef]
17. Ali, Z.M.; Diaaeldin, I.M.; El-Rafei, A.; Hasanien, H.M.; Aleem, S.H.A.; Abdelaziz, A.Y. A novel distributed generation planning algorithm via graphically-based network reconfiguration and soft open points placement using Archimedes optimization algorithm. *Ain Shams Eng. J.* **2021**, *12*, 1923–1941. [CrossRef]
18. Diaaeldin, I.; Aleem, S.A.; El-Rafei, A.; Abdelaziz, A.; Zobaa, A.F. Optimal Network Reconfiguration in Active Distribution Networks with Soft Open Points and Distributed Generation. *Energies* **2019**, *12*, 4172. [CrossRef]

19. Diaaeldin, I.M.; Aleem, S.H.E.A.; El-Rafei, A.; Abdelaziz, A.Y.; Zobaa, A.F. Enhancement of Hosting Capacity with Soft Open Points and Distribution System Reconfiguration: Multi-Objective Bilevel Stochastic Optimization. *Energies* **2020**, *13*, 5446. [CrossRef]
20. Shafik, M.; Rashed, G.; Chen, H.; Elkadeem, M.; Wang, S. Reconfiguration Strategy for Active Distribution Networks with Soft Open Points. In Proceedings of the 14th IEEE Conference on Industrial Electronics and Applications (ICIEA), Xi'an, China, 19–21 June 2019; pp. 330–334. [CrossRef]
21. Qi, Q.; Wu, J.; Zhang, L.; Cheng, M. Multi-Objective Optimization of Electrical Distribution Network Operation Considering Reconfiguration and Soft Open Points. *Energy Procedia* **2016**, *103*, 141–146. [CrossRef]
22. Eskandar, H.; Sadollah, A.; Bahreininejad, A.; Hamdi, M. Water cycle algorithm—A novel metaheuristic optimization method for solving constrained engineering optimization problems. *Comput. Struct.* **2012**, *110–111*, 151–166. [CrossRef]
23. El-Ela, A.A.A.; El-Sehiemy, R.A.; Abbas, A.S. Optimal Placement and Sizing of Distributed Generation and Capacitor Banks in Distribution Systems Using Water Cycle Algorithm. *IEEE Syst. J.* **2018**, *12*, 3629–3636. [CrossRef]
24. Sadollah, A.; Eskandar, H.; Kim, J.H. Water cycle algorithm for solving constrained multi-objective optimization problems. *Appl. Soft Comput.* **2015**, *27*, 279–298. [CrossRef]
25. Wang, J.; Liu, S. Novel binary encoding water cycle algorithm for solving Bayesian network structures learning problem. *Knowl.-Based Syst.* **2018**, *150*, 95–110. [CrossRef]
26. Ibrahim, S.; Alwash, S.; Liao, Y. A Binary Water Cycle Algorithm for Service Restoration Problem in Power Distribution Systems Considering Distributed Generation. *Electr. Power Compon. Syst.* **2020**, *48*, 844–857. [CrossRef]
27. Ibrahim, S.; Alwash, S.; Aldhahab, A. Optimal Network Reconfiguration and DG Integration in Power Distribution Systems Using Enhanced Water Cycle Algorithm. *Int. J. Intell. Eng. Syst.* **2020**, *13*, 379–389. [CrossRef]
28. Hizarci, H.; Demirel, O.; Turkay, B.E. Distribution network reconfiguration using time-varying acceleration coefficient assisted binary particle swarm optimization. *Eng. Sci. Technol. Int. J.* **2022**, *35*, 101230. [CrossRef]
29. Sadollah, A.; Eskandar, H.; Lee, H.M.; Yoo, D.G.; Kim, J.H. Water cycle algorithm: A detailed standard code. *SoftwareX* **2016**, *5*, 37–43. [CrossRef]
30. Sadollah, A.; Eskandar, H.; Bahreininejad, A.; Kim, J.H. Water cycle algorithm with evaporation rate for solving constrained and unconstrained optimization problems. *Appl. Soft Comput.* **2015**, *30*, 58–71. [CrossRef]
31. Xu, Y.; Mei, Y. A modified water cycle algorithm for long-term multi-reservoir optimization. *Appl. Soft Comput.* **2018**, *71*, 317–332. [CrossRef]
32. Barzegar, A.; Sadollah, A.; Su, R. A novel fully informed water cycle algorithm for solving optimal power flow problems in electric grids. *arXiv* **2019**, arXiv:1909.08800. [CrossRef]
33. Zhang, X.; Yuan, J.; Chen, X.; Zhang, X.; Zhan, C.; Fatollahi-Fard, A.M.; Wang, C.; Liu, Z.; Wu, J. Development of an Improved Water Cycle Algorithm for Solving an Energy-Efficient Disassembly-Line Balancing Problem. *Processes* **2022**, *10*, 1908. [CrossRef]
34. Veeramani, C.; Sharanya, S. An improved Evaporation Rate-Water Cycle Algorithm based Genetic Algorithm for solving generalized ratio problems. *RAIRO-Oper. Res.* **2021**, *55*, S461–S480. [CrossRef]
35. Chidanandappa, R.; Ananthapadmanabha, T.; Ranjith, H.C. Genetic Algorithm Based Network Reconfiguration in Distribution Systems with Multiple DGs for Time Varying Loads. *Procedia Technol.* **2015**, *21*, 460–467. [CrossRef]
36. Nara, K.; Shiose, A.; Kitagawa, M.; Ishihara, T. Implementation of genetic algorithm for distribution systems loss minimum re-configuration. *IEEE Trans. Power Syst.* **1992**, *7*, 1044–1051. [CrossRef]
37. Baran, M.E.; Wu, F.F. Network Reconfiguration in Distribution Systems for Loss Reduction and Load Balancing. *IEEE Power Eng. Rev.* **1989**, *9*, 101–102. [CrossRef]
38. Savier, J.S.; Das, D. Impact of Network Reconfiguration on Loss Allocation of Radial Distribution Systems. *IEEE Trans. Power Delin.* **2007**, *22*, 2473–2480. [CrossRef]
39. Kersting, W.H. *Distribution System Modeling and Analysis*, 3rd ed.; CRC Press: Boca Raton, NY, USA, 2012; pp. 154–196.

Article

Insuring a Small Retail Electric Provider's Procurement Cost Risk in Texas

Chi-Keung Woo [1], Jay Zarnikau [2], Asher Tishler [3] and Kang Hua Cao [4,*]

[1] Department of Asian and Policy Studies, Education University of Hong Kong, Hong Kong, China
[2] Department of Economics, University of Texas at Austin, Austin, TX 78712, USA
[3] Coller School of Management, Tel Aviv University, Tel Aviv 69978, Israel
[4] Department of Economics, Hong Kong Baptist University, Kowloon Tong, Hong Kong, China
* Correspondence: kanghuacao@hkbu.edu.hk

Abstract: Motivated by the relatively infrequent but very large price spikes in the day-ahead and real-time energy markets operated by the Electric Reliability Council of Texas, this paper proposes an insurance that a small and risk-averse retailer in Texas (i.e., a retail electric provider (REP)) may buy to prevent financial insolvency caused by inadequate risk management. It also demonstrates the insurance's practical design, pricing, and implementation. As participation in the REP's procurement auction is voluntary, the insurance is mutually beneficial for the REP and the insurance seller. Hence, the proposed insurance is a newly developed wholesale market product that deserves consideration by REPs in Texas and competitive retailers elsewhere.

Keywords: insurance; retail service provider; spot price spike; electricity markets; ERCOT

1. Introduction

The Electric Reliability Council of Texas (ERCOT) uses locational marginal pricing [1] to set the spot electricity prices in its day-ahead market (DAM) and real-time market (RTM) [2]. Varying with the day-ahead forecasts of fundamental drivers of natural gas price, system load, ancillary services requirements, nuclear generation, and wind generation, the volatile DAM prices move the even more volatile RTM prices shown in Figure 1, which is a reproduction from [3].

When ERCOT's retail market first opened in 2002, a qualified scheduling entity (QSE) must show that its projected aggregate supply had been procured through bilateral contracts and other means to meet its projected aggregate demand. A balancing energy market was created to resolve the mismatch between the QSE's projected and actual schedules of aggregate demand and supply. The QSE is responsible for commercial transactions involving multiple retail electric providers (REPs). Hence, the energy shortfall of one REP could be offset by the surplus energy of another REP within the QSE's portfolio of REPs (www.ercot.com/files/docs/2005/11/07/360prr_relaxed_balanced_schedules.doc; accessed on 27 May 2022).

The balanced schedule requirement was gradually relaxed, as ERCOT established a DAM in December 2010 that transformed the balancing energy market into today's RTM (Zarnikau et al., 2014). Hence, a REP in Texas can now decide whether and how to manage its procurement cost risk caused by spot price volatility and sales fluctuation, as exemplified by two case studies of a load serving entity in Florida [4,5].

A REP's procurement cost for serving the unhedged portion of the total load can explode during market conditions of scarcity, exacerbated by the various "adders" that apply to RTM prices (e.g., reliability deployment price adder) and operating reserve demand curve (ORDC) adder [6]. The reliability deployment price adder is designed to prevent prices from being depressed when ERCOT takes an out-of-market action, such as ordering a power plant to operate to maintain the reliability of the system or orders

Citation: Woo, C.-K.; Zarnikau, J.; Tishler, A.; Cao, K.H. Insuring a Small Retail Electric Provider's Procurement Cost Risk in Texas. *Energies* **2023**, *16*, 393. https://doi.org/10.3390/en16010393

Academic Editors: Yuan Liao and Ke Xu

Received: 28 November 2022
Revised: 16 December 2022
Accepted: 22 December 2022
Published: 29 December 2022

the deployment of a demand response program. See ERCOT Protocols, Section 6.5.7.3.1 at https://www.ercot.com/mktrules/nprotocols/current, (accessed on 27 May 2022). In addition, real-time ORDC price adders occur when ERCOT's physical operating reserve dips below a pre-set threshold. This phenomenon raises a substantive research question: can a small and risk-averse REP insure its procurement cost risk due to large spot price spikes in ERCOT's energy markets? This question accentuates our paper's primary focus of designing an insurance scheme to reduce the REP's procurement cost risk exposure to the price spikes during ERCOT's critical hours of low physical capacity reserve due to high system demand and/or generation plant outages.

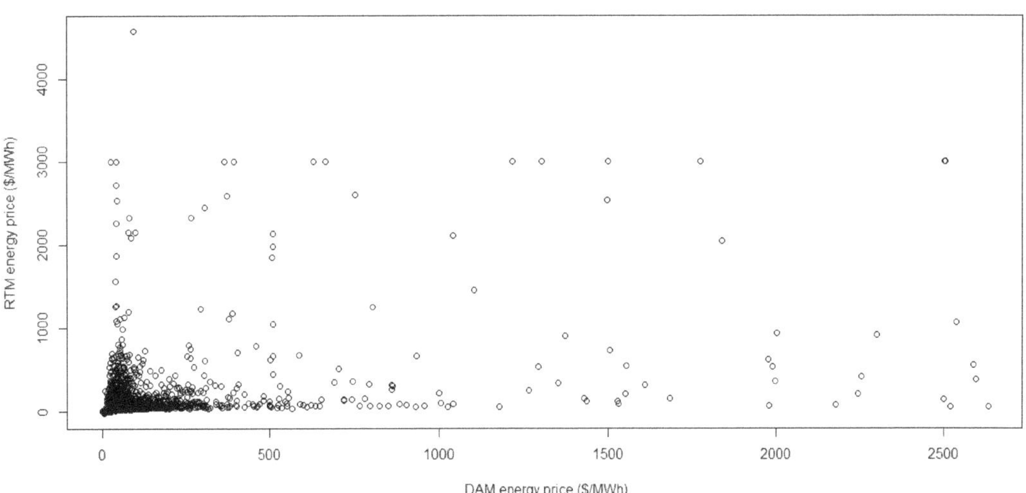

Figure 1. Scatter plot of hourly DAM energy price vs. hourly RTM energy price for the period of 01 January 2011 to 31 December 2017. The OLS regression based on the efficient market hypothesis is RTM price = $a + b \times$ DAM price + error, where $a \approx 0$ and $b \approx 1$ are the regression's coefficient estimates. Updating the figure with more recent data does not change its key message: DAM and RTM prices move in tandem and are highly volatile with infrequent but large spikes.

The preceding question's real-world relevance is best exemplified by Winter Storm Uri that caused the Texas deep freeze in February 2021, whose timeline and devastating effects of price spikes and blackouts are available from a report released in July 2022 by the University of Texas (Austin) (https://energy.utexas.edu/sites/default/files/UTAustin%20 %282021%29%20EventsFebruary2021TexasBlackout%2020210714.pdf; accessed on 27 May 2022). The large spot price spikes in that fateful month bankrupted several REPs, which had signed fixed price contracts with retail customers without adequately hedging their procurement cost risks (https://www.power-technology.com/news/industry-news/texas-snow-storm-bankrupt-fallout-energy-prices-ercot/; accessed on 27 May 2022). This kind of bankruptcy is not new, as underscored by the financial insolvency of two large electric utilities in California caused by the spot price spikes during the state's energy crisis in 2001 [7,8].

The same question is similarly important and relevant for other regions with volatile wholesale market prices that a competitive retailer inevitably faces. A partial list of these regions includes (a) the states served by the regional transmission organizations of PJM Interconnection, ISO New England, and New York ISO in the US; (b) the provinces of Alberta and Ontario in Canada; (c) Asia-Pacific countries such as Australia, New Zealand, and Singapore; (d) countries in the European Union (https://www.europarl.europa.eu/

factsheets/en/sheet/45/internal-energy-market; accessed on 27 May 2022); and (e) South American countries such as Brazil and Chile [9,10].

To answer the question in Texas's context, we propose an insurance that a small and risk-averse REP may buy to manage its procurement cost risk. Our proposed insurance comes from the first author's research funded by several electric utilities in North America. As such, it aims for practicality, rather than highly technical details often related to the pricing of electricity derivatives (e.g., [11,12]) and recently proposed insurance schemes (e.g., [13–16]). While aiding the REP to avoid financial insolvency, our proposed insurance is profitable for insurance sellers voluntarily participating in the REP's internet-based procurement auction described in Section 4.2. Hence, it is a newly developed wholesale market product that deserves consideration by REPs in Texas and competitive retailers elsewhere.

Complementing extant studies on electricity risk management (e.g., [5,11,12,17–20]), our proposed insurance is, to the best of our knowledge, a newly developed wholesale market product for use by a competitive retailer like those in Texas. Its practical pricing, design and implementation explained in Section 4 show that it differs from (a) the currently available electricity products described in Section 3 and (b) the insurance proposals for managing the risks related to system reliability [13], distributed generation [14], real-time pricing of energy consumption [15], power plant performance (https://www.munichre.com/hsb/en/products.html; accessed on 27 May 2022), and transmission and distribution [16].

The rest of this paper proceeds as follows. Section 2 states a REP's risk management problem. This section purposely omits a literature review of the voluminous studies on electricity risk management because (a) such a review is an unnecessary distraction from our narrowly focused paper; and (b) general overviews of electricity risk management are already available (e.g., [12,17]). Section 3 describes electricity products currently available for the REP's risk management. Section 4 explains our proposed insurance's design, pricing, and implementation. Section 5 is an indicative calculation of the insurance per MWh premium, whose empirics are reported in Section 6. Section 7 concludes.

2. Risk Management Problem of a Small REP in Texas

To provide a contextual background of our proposed insurance, consider the risk management problem in connection to a small REP's fixed price plan [21]. The plan's fixed price is \$$G$/MWh for generation, which inevitably differs from the wholesale spot price \$$P$/MWh after contract signing. For simplicity, we assume that G is mainly driven by P because the cost of ancillary services and the cost associated with other charges imposed by ERCOT are relatively small and fully passed through to the REP's customers. The REP earns ex post profit of \$$(G - P)$ for each MWh procured from the spot market for resale. Unfortunately, Figure 1 shows that P may surge above G, resulting in ex post loss of \$$(P - G)$/MWh. Thus, retail fixed pricing can cause the REP to face large financial risk exposure if it decides not to hedge adequately. Parenthetically, this outcome also applies to time-of-use and pre-pay plans with prices that do not closely track the fast-changing spot market prices in their delivery periods.

Figure 2 is an illustrative example of the REP's risk management problem under the assumption that the load duration curve (LDC) can be accurately forecasted [4,5]. The LDC forecast can be made using time series modeling of the REP's aggregate hourly load data. Alternatively, it can be based on a bottom-up approach that uses the data for (1) the average load profiles of customer segments differentiated by consumption size and residence type (e.g., apartment, town house, and single detached home) and (2) each segment's forecasted number of customers.

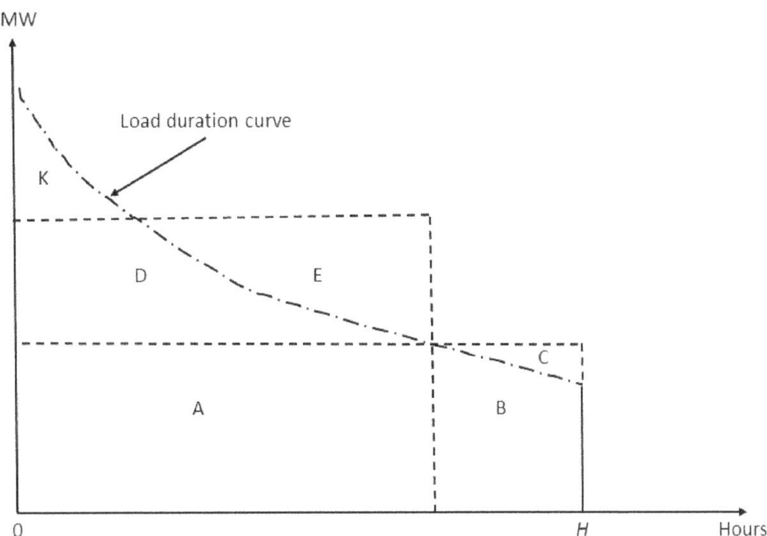

Figure 2. A hypothetical REP's load duration curve, procurement of electricity forward contracts, and residually unhedged peak loads.

Figure 2 assumes that the REP buys a forward contract at fixed price F/MWh for a baseload power block given by Areas A, B, and C. Area C is the contract's excess MWh sold by the REP in the spot market at P/MWh. As F exceeds $E(P)$ = expected value of P [22–25] due to the profitable forward premium required by generators and power marketers/traders, the contract's purchase causes a per MWh expected loss of $[F − E(P)]$ attributable to the REP's sale of excess MWh. The same line of reasoning applies to the REP's purchase of a forward contract for the shoulder power block given by Areas D and E to meet the load obligation given by Area D.

When the REP buys ERCOT's spot energy to meet the peak load obligation given by Area K, it has residual risk exposure. If the REP decides to reduce its procurement of forward contracts, it becomes increasingly vulnerable to large spot price spikes. When lasting multiple days, as in the case of the Texas deep freeze, such spikes can bankrupt the REP.

3. Demand for the Proposed Insurance
3.1. Electricity Products

To illustrate the market potential of the insurance proposed in Section 4, this section discusses the existing electricity products that a small REP may employ to manage its procurement cost risk. If these products can adequately meet the REP's risk management need, they obviate our proposed insurance's usefulness.

3.1.1. Wholesale Electricity Products

ERCOT operates a DAM for ancillary services and energy and a RTM for energy [3]. The main market participants are independent power producers (IPPs), power traders and marketers (PTMs), and REPs. While the DAM offers day-ahead forwards for hedging against the RTM's price risk [24], it cannot protect a small REP in a multiday financially ruinous event such as the Texas deep freeze. This is because DAM prices set on day *d-1* and RTM prices set on day *d* move on an almost dollar-for-dollar basis [3]. As DAM prices have forward premiums but are less volatile than RTM prices, buying energy from the DAM instead of the RTM does not reduce the REP's procurement cost expectation, notwithstanding that it can decrease the REP's procurement cost volatility.

Electricity derivatives aid a small REP's risk management [12,17]. A good example is the monthly 5-MW peak futures available from the Chicago Mercantile Exchange (CME) (https://www.cmegroup.com/markets/energy/electricity/ercot-houston-zone-mcpe-5-mw-peak-swap-futures.contractSpecs.html; accessed on 27 May 2022). However, these futures are thinly traded and have a delivery point (Houston 345 kV Hub) that does not geographically correspond to where the REP's customers reside. Moreover, their delivery hours (07:00 to 22:00, working weekdays) poorly match ERCOT's high price hours or the REP's peak load hours (e.g., hot summer afternoon hours of 12:00 to 16:00). Hence, the REP cannot easily use electricity futures to manage the procurement cost risk of the residually unhedged load shown in Figure 2.

Besides electricity futures, the REP's risk management may consider the bilaterally traded products listed below:

A full requirement contract with fixed price FR offered by an IPP or PTM eliminates the REP's price and quantity risks. Analogous to the contract for difference, it stipulates that if FR is above spot market price P, the REP pays $(FR - P)\,Q$ for its total retail sales of Q MWh to the contract seller; otherwise, the REP receives $(P - FR)\,Q$ from the contract seller. However, it may not be financially attractive to the REP because FR likely contains a large premium to compensate the contract seller for absorbing the wholesale market's spot price risk and the REP's quantity risk [4,5,7,23,25].

An electricity forward is a take-or-pay fixed price contract for daily delivery of a MW block to the REP by the seller in the contract period [7]. As illustrated in Figure 2, the REP purchases electricity forwards mainly for managing the price risk of its non-peak sales given by Areas A, B, and D [18].

A tolling agreement has a monthly capacity charge that gives the REP the right but not the obligation to dispatch the seller's natural-gas-fired generation unit (e.g., a combined cycle gas turbine) at per MWh fuel cost C = contracted heat rate (e.g., 7 MMBtu per MWh) $\times$ daily wholesale natural gas price [26]. Under least-cost dispatch, the REP's per MWh variable cost of energy is $C^* = \min(P, C)$. During ERCOT's high price hours, $C^* = C$ and the REP is immune to spot energy price spikes. Like an electricity forward, the agreement is typically used to manage the price risk related to the REP's non-peak sales.

A 1-MW capacity call option has a monthly premium that gives the REP the right but not the obligation to request MWh delivery from the seller at the option's strike price for a maximum duration (e.g., 6 h per call) and a maximum frequency (e.g., four calls per month) during the option's contract months (e.g., July and August) [27]. While the option can mitigate the REP's financial risk exposure in connection to the REP's peak sales during ERCOT's heat storm with extremely high spot prices, the REP needs to make the "right" calls in each month to maximize the option's monthly total payoff = monthly sum of max (spot price−strike price, 0) $\times$ hours per call. As will be seen in Section 4, our proposed insurance does not require the REP to know when to optimally exercise the option. Importantly, it can better protect the REP against spot price spikes because its contract specification does not have the option's duration and frequency restrictions.

3.1.2. Retail Electricity Products

There are retail electricity products that the REP may use to reduce its exposure to spot price risk. For example, the REP may employ real-time pricing (RTP) that shifts the spot price risk from the REP to its customers [28]. If the REP can apply RTP to all MWh sales under its LDC, it does not need to hedge to fully protect itself from the spot price risk's adverse financial impact. However, RTP is relatively unattractive to the REP's customers because of its complexity and bill instability, making fixed price plans the most popular in Texas [29]. Further, Texas has banned residential RTP (https://abc13.com/texas-legislature-ban-residential-wholesale-electricity-plans-house-bill-16-gov-greg-abbott/10633908/; accessed on 27 May 2022), a legislative response to the huge electricity bills for residential RTP customers in the aftermath of the February 2021 freeze (https://www.nytimes.com/2021/02/20/us/texas-storm-electric-bills.html; accessed on 27 May 2022).

The REP may employ demand response (DR) programs to reduce its procurement cost risk [28,30]. Roughly one million customers served by REPs in the ERCOT market are served on dynamic pricing plans or participate in load control programs, as shown by slide 18 of the 2020 Analysis of REP and NOIE Demand Response presentation by C.L. Raish to the ERCOT Demand Side Working Group, January 22, 2021 (https://www.ercot.com/files/docs/2021/02/04/15._RMS_2020_4CP__Retail_DR_Analysis_Raish.v3.pptx; accessed on 27 May 2022). However, many electricity consumers are not good candidates for such plans for reasons such as risk aversion, lack of understanding, and inability to respond to real-time price changes or comply with the REP's load reduction requests [28].

3.2. Possible Buyers

This section explains that the possible buyers of our proposed insurance are small REPs that do not own generation assets and are not subsidiaries of large holding companies. By buying the insurance, a small REP can transfer its procurement cost risk related to the residually unhedged load to an insurance seller that is less risk-averse than the REP.

Large REPs are unlikely buyers because they own generation assets or are subsidiaries of large holding companies [31]. Generation ownership implies that a large REP can self-generate to meet its retail sales when wholesale market prices surge. If this REP has excess generation capacity, it profits from its wholesale market sales during high price hours. If it is a subsidiary of a publicly traded holding company, sharing of its procurement risk among many shareholders implies risk neutrality that obviates its buying interest in the proposed insurance. Finally, the holding company has generation assets and retail customers dispersed across multiple states, resulting in geographic diversification that further reduces a large REP's buying interest in the proposed insurance.

4. Insurance's Design, Pricing, and Implementation

4.1. Design and Pricing

Our proposed insurance's focus is the REP's residual risk exposure related to the peak load obligation in Figure 2. This focus assumes that the REP uses an optimal portfolio of forward contacts and tolling agreements to meet its non-peak load obligations [4,5,18]. Removing this assumption implies that the REP mainly relies on the spot market purchases to meet its total load obligation measured by the entire area under the LDC, making the REP even more financially vulnerable to spot price spikes.

For easy reference and clarity, here are the key variables that characterize our proposed insurance scheme: (1) V = MWh volume that the REP wishes to insure; (2) S = per MWh insurance premium; (3) Y = per MWh payoff of buying insurance; (4) SV = total insurance premium; and (5) YV = total payoff.

To illustrate our proposed insurance's design, let V denote the MWh volume that the REP wishes to insure based on the REP's forecast of aggregate MWh sales. Beyond our paper's narrow scope, the complicated calculation of V is based on the forecast of Area K in Figure 2, the forecast's standard error, the spot price's expected level and volatility, the correlation between sales and spot prices, and the REP's risk preference [4,5]. Nevertheless, the REP can be almost fully insured if its chosen V is the forecast's MWh level + 1.65 $\times$ the forecast's standard error, which almost surely exceeds the REP's actual sales under normal circumstances with a 95% probability. Despite the presence of quantity risk, Section 5 shows that our indicative calculation of the per MWh insurance premium does not depend on the size of V. Further, Section 6 shows how the REP may determine V based on the impact of buying insurance on its retail price offer designed to attract and retain retail customers.

The insurance seller charges $\$SV$ for the insured MWh, where S = per MWh insurance premium. The payoff is $\$YV$, where Y = positive difference between the actual spot price and the threshold level T stipulated in the insurance contract. As a result, our proposed insurance looks like a capacity call option. However, there is an important difference. Our proposed insurance is simpler and more flexible than a multi-month capacity call option because it does not need to specify the maximum number of calls per month and maximum number of hours per call. Since T is unknown a priori, it presents an empirical challenge that our indicative calculation of S must overcome. Owing to the spot price spikes caused by Winter Storm Uri, this calculation assumes that T is the spot price level when ERCOT's physical reserve capability is above 5000 MW, as further detailed in Section 5 below.

Suppose we have data on $\mu = E(Y) =$ expected value of Y and $\sigma^2 = Var(Y) =$ variance of Y. We can use the forward contract pricing formula in Woo et al. (2001) to calculate S:

$$S = \mu + z\,\sigma, \tag{1}$$

where $z =$ standard normal variate. At $z = 1.65$, S is almost surely profitable for an insurance seller under normal circumstances with a 0.95 probability. However, the resulting S is likely cost-unreasonable for the REP, as demonstrated in Section 6 below. If fierce competition exists in a REP's procurement auction described in the next section, we expect $z \to 0$ and $S \to \mu$. To achieve profitability with a probability above 0.5, however, insurance sellers participating in the auction can submit S quotes that correspond to $z > 0$.

4.2. Implementation

Our proposed insurance differs from the insurances for uncorrelated events such as car accidents, illness, and fire, whose profitability is based on the law of large numbers. Rather, it resembles earthquake and flood insurances that provide buyer protection against rare but financially ruinous events. This is because extreme price spikes in ERCOT's RTM tend to occur in relatively few hours per year but can adversely affect small REPs, as evidenced by the bankruptcies in the aftermath of the Texas deep freeze.

As our proposed insurance is not a standardized wholesale product like the electricity futures traded on the CME, its implementation may occur via the REP's Anglo-Dutch internet-based procurement auction executed in two stages [27,32–34]. In Stage 1, all eligible participants submit open price offers that are continuously announced in a pre-set time window (e.g., one hour). A participant can then lower its offer after seeing the offers made by other participants within the window. After Stage 1's completion, the top three participants with low open offers are invited to participate in Stage 2 that entails a single submission of best and final closed offers. The winning participant is the one with the lowest closed offer.

Informed by the list of bidders underlying the auction results documented by [27,32], the likely insurance sellers are PTMs that are less risk-averse than the REP. Voluntary participation in the REP's procurement auction implies profitability of the insurance quotes submitted by the likely sellers.

Finally, IPPs and holding companies are unlikely participants in the small REP's insurance procurement auction, as their main business focus is profitable sales of the wholesale electricity products such as spot energy traded in ERCOT's DAM and RTM, forward contracts, and tolling agreements.

5. Indicative Calculation of the Insurance's per MWh Premium

While the values of μ and σ^2 may come from ERCOT's least-cost generation dispatch [3], they are difficult to forecast for determining a forward-looking value for S. As an illustrative alternative, we calculate μ and σ^2 based on the per MWh price adder set by ERCOT's ORDC [6].

Our indicative calculation presumes that the ORDC price adder is a reasonable approximation of Y, thus bypassing the need to know the threshold level T stipulated in the insurance contract. However, it does not mean that Y is caused by the ORDC adder. We use the ORDC adder solely for circumventing the data unavailability problem in our calculation of S. If more accurate values of μ and σ^2 are available from non-ORDC sources, they should replace those presented below.

Figure 3 portrays the ORDC adder during ERCOT's hours of low physical reserve capability [6]. Hence, these hours correspond to those containing the price spikes on the long right tail of ERCOT's skewed spot price distribution implied by Figure 1.

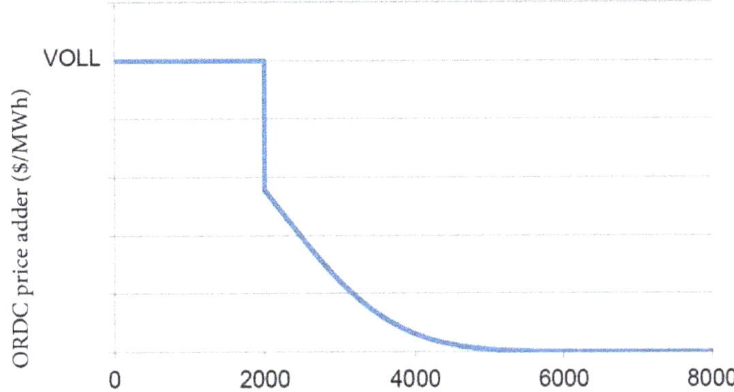

Figure 3. ERCOT's physical reserve capability vs. ORDC price adder [6].

When ERCOT's physical reserve capability is at or below 2000 MW, the ORDC adder is VOLL = $9000/MWh, which is ERCOT's value of lost load assumption prior to Winter Storm Uri. Technically, the ORDC adder is always below VOLL. If the system lambda is $X/MWh, the ORDC adder is $(VOLL–X)/MWh to bring the market price up to the systemwide offer cap of VOLL. As will be seen below, this technical adjustment shrinks the size of α used in our calculation of S. In contrast, the ORDC adder is $0/MWh when ERCOT's physical reserve capability exceeds 5000 MW. As the ORDC adder is used to measure the insurance's per MWh payoff Y, the insurance's threshold T is the spot price when ERCOT has over 5000 MW of physical reserve capability.

Replacing the VOLL number with the lower price caps adopted after the deep freeze does not change our calculation process. The price cap was $2000/MWh from March to December 2021 and had been increased to $5000/MWh in January 2022. Importantly, our methodology to estimate μ and σ^2 is applicable to any price cap in effect at the time of the per MWh premium's calculation.

Using Figure 3, we characterize Y as follows:

Case 1: $Y = 0$ if ERCOT's physical reserve capability is $R \geq R_U$, where $R_U = R$'s upper threshold at which the ORDC price adder is $0/MWh.

Case 2: $Y = \alpha + \beta R$ if $R_U > R > R_L$, where $R_L = R$'s lower threshold at which the ORDC price adder is strictly positive. The coefficient $\alpha > 0$ is the VOLL that can be replaced by a lower price cap. The coefficient $\beta = -(\alpha/R_U) < 0$ is the marginal effect of R on Y.

Case 3: $Y = \alpha$ if $R_L \geq R$.

Our linear characterization of Y is necessary to enable our per MWh premium's calculation based on the information readily available to the REP. Given the convex nature of the ORDC, linearization tends to overstate the ORDC price adder in Case 2. Hence, a nonlinear characterization of the ORDC can improve our premium calculation's accuracy by reducing the size of the per MWh premium. That said, it is a complication that does not materially enrich the qualitative understanding of our proposed insurance scheme, as further detailed in Section 6.2 below.

Based on Y's linear characterization, we can now calculate the conditional expectation and variance of Y for each case:

Case 1: As $Y = 0$, $E(Y \mid \text{Case 1}) = \theta_1 = 0$ and $\text{Var}(Y \mid \text{Case 1}) = 0$.

Case 2: As $Y > 0$, $E(Y \mid \text{Case 2}) = \theta_2 = \alpha + \beta \, \mu_R$, where $\mu_R = E(R \mid \text{Case 2})$. Let $\sigma_R^2 = \text{Var}(R \mid \text{Case 2})$ for $R_L < R < R_U$ so that $\text{Var}(Y \mid \text{Case 2}) = \beta^2 \, \sigma_R^2$.

Case 3: As $Y = \alpha$, $E(Y \mid \text{Case 3}) = \theta_3 = \alpha$ and $\text{Var}(Y \mid \text{Case 3}) = 0$.

The determination of θ_2 and $\beta^2 \, \sigma_R^2$ requires μ_R and σ_R^2 in Case 2. Hence, we find μ_R and σ_R^2 as follows. Let R = normally distributed reserve with mean η and variance λ^2, $\rho_L = (R_L - \mu)/\mu$ and $\rho_U = (R_U - \eta)/\lambda$. As a result,

$$\mu_R = \eta - \lambda M, \tag{2}$$

where $M = [\phi(\rho_U) - \phi(\rho_L)]/[\phi(\rho_U) - \phi(\rho_L)]$, $\phi(z)$ = normal density function, $\phi(z)$ = normal probability distribution function, and z = standard normal variate [35]. Further,

$$\sigma_R{}^2 = \lambda^2 \left(1 - M^2 - N\right), \tag{3}$$

where $N = [\rho_U \, \phi(\rho_U) - \rho_L \, \phi(\rho_L)]/[\phi(\rho_U) - \phi(\rho_L)]$.

Let $\pi_1 = \mathrm{Prob}(R \geq R_U)$, $\pi_2 = \mathrm{Prob}(R_U > R > R_L)$ and $\pi_3 = \mathrm{Prob}(R_L \geq R) = 1 - \pi_1 - \pi_2$. As indicated in Section 6, these probabilities can be estimated by an electric grid's generation reliability criterion and ERCOT's history of emergency hours.

The unconditional expectation of Y is:

$$\mu = \pi_2 \, \theta_2 + \pi_3 \, \theta_3. \tag{4}$$

Finally, the unconditional variance of Y is:

$$\sigma^2 = \pi_2 \left(\theta_2{}^2 + \beta^2 \sigma_R{}^2\right) + \pi_3 \, \alpha^2 - \left(\pi_2 \, \theta_2 + \pi_3 \, \theta_3\right)^2. \tag{5}$$

6. Empirics

6.1. Results

This section reports the empirics from our calculation of S based on Figure 3 that shows $\alpha = \$9000/\text{MWh}$, $R_L = 2000$ MW, $R_U = 5000$ MW, and $\beta = -9000 \div 5000 = -1.8$. Without invoking Equation (2), we use the midpoint between R_L and R_U as a simple estimate for $\mu_R = 3500$ in Case 2, which enables a REP's quick determination of the value of a competitively priced insurance premium. While this estimate for μ_R is less than the appropriately found estimate of 4475 MW based on Equation (2), our sensitivity check indicates that our calculated S is insensitive to the size of μ_R.

Assuming $\eta = 6000$ and $\lambda = 1000$ based on ERCOT's 2020 reserve data with mean = 5070 and standard deviation = 1163, we use Equation (3) to find $\sigma_R = 444.60$. This calculation of σ_R enables an insurer's determination of the per MWh insurance premium's probability of profitability.

To complete S's calculation, we assume $\pi_3 = 2.4\,\text{h}/8760\,\text{h} = 0.000274$ based on the loss-of-load-expectation criterion of 1 day in 10 years, which is commonly used to determine an electric grid's target of generation reserve margin. We further assume $\pi_2 = 12\,\text{h}/8760\,\text{h} = 0.00137$, which is five times π_3 and based on ERCOT's history of emergency hours of 10 to 20 h per year, excluding Winter Storm Uri's year of 2021.

We use Equations (4) and (5) to find $\mu = 6.16$ and $\sigma = 181.71$. Hence, the lower bound for S is $S_L \approx \$6.16/\text{MWh}$ at $z \approx 0$ when there is fierce competition in a REP's procurement auction. Thanks to Equation (4) that shows $\mu = \pi_2 (\alpha + \beta \, \mu_R) + \pi_3 a$; a REP can quickly determine S_L based on the simple estimate of $\mu_R = 0.5 \times (R_L + R_U)$, the price cap value of a, $b = - (a/R_U)$, and the readily available data for π_2 and π_3. The upper bound for S is $S_U = \$305.98/\text{MWh}$ at $z = 1.65$, reflecting S_U's profitability for an insurance seller with almost certainty.

Based on Equation (1), an insurance seller's per MWh profit with a 0.95 probability is $S_U - S_L = \$299.82$. However, making this large profit with almost certainty is unrealistic because a REP can use $S_L = \$6.16/\text{MWh}$ as the benchmark for selecting the winner of its procurement auction.

Finally, our sensitivity check shows that S_L and S_U are not materially affected by doubling or halving the values for η and λ. However, reducing α from 9000 to 5000, which is in effect as of 1 January 2022, leads to $S_L = \$3.42/\text{MWh}$ and $S_U = \$170.0/\text{MWh}$.

6.2. Discussion

Suppose a small REP considers buying insurance based on the insurance's impact on its fixed price offer. If this impact is deemed excessively large, buying the insurance can harm the REP's ability to attract customers, despite the insurance's benefit of pre-empting financial insolvency. This highlights the REP's trade-off between retail marketing and risk exposure. While offering a low fixed price plan made possible by not buying the insurance can increase customer sign-ups, it enlarges the REP's risk exposure to spot price spikes that apply to the REP's residually unhedged load. Underscoring this point are the REP bankruptcies observed in the wake of Texas's deep freeze in February 2021.

To assess the insurance's financial impact on the REP's fixed price offer, we use the average G value of ~\$60/MWh found by [21]. Table 1 reports our impact assessment results, which are the percentage changes in G by per MWh insurance premium and insured amount equal to γ percent of the REP's forecast of total fixed price sales.

Table 1. Percentage changes in G by per MWh insurance premium and insured amount γ.

Per MWh Insurance	$\gamma = 10\%$	$\gamma = 20\%$	$\gamma = 30\%$
S_L that results in ΔG_L	1.03%	2.05%	3.08%
S_U that results ΔG_U	51.0%	102.0%	153.0%

An example of the REP's hourly forecast for a given period (e.g., next month or quarter) is (a) the forecasted number of fixed price customers times (b) the estimated hourly kWh sales per customer. Using its marketing and customer data, the REP may find (a) based on n = number of existing customers + number of new customers − number of departed customers. The REP may also find (b) based on the hourly metered kWh sales per customer. The REP's total forecast can then be found as the sum of hourly forecasts. While there are alternative forecasting approaches (e.g., aggregate time series modeling and disaggregate panel data analysis), their elaboration is beyond our paper's intent and scope.

Table 1's calculation details are as follows. We first assume that the insured amount is $\gamma = 20\%$ of the REP's total fixed price sales. The price increase based on an insurer's highly competitive per MWh premium of S_L is $\Delta G_L = [0.8 \times 60 + 0.2 \times (60 + 6.16)] - 60 = \$1.23/\text{MWh} = 2.05\%$ of the average G value. The price increase based on an insurer's highly profitable per MWh premium of S_U is $\Delta G_U = [0.8 \times 60 + 0.2 \times (60 + 305.98)] - 60 = \$61.20/\text{MWh} = 102.0\%$ of the average G value. To complete Table 1, we alternatively assume $\gamma = 10\%$ (30%). The estimates for ΔG_L and ΔG_U are 1.03% and 51.00% (3.08% and 153.0%) of the average G value.

The ΔG estimates shown in Table 1 suggest that if Area K of the LDC in Figure 2 is ~30% of the REP's total fixed price sales, the amount of MWh insured may approach 100% of Area K when the per MWh insurance is highly competitive at S_L.

Informed by the range of ΔG_L and ΔG_U estimates in Table 1, the REP may consider buying the insurance when its procurement auction's winning S quote is close to S_L. However, it should reject S quotes that resemble S_U. Hence, our ΔG_L and ΔG_U estimates guide the REP's decision on insurance purchase, notwithstanding that such a decision is ultimately made by the REP's risk-averse management.

7. Conclusions

We conclude by recapping our key findings. First, large spot price spikes occasionally occur in ERCOT's DAM and RTM for energy. Second, inadequate risk management of these price spikes can financially ruin a REP that mainly offers fixed price plans. Third, even if the REP employs an optimal portfolio of electricity forward contracts, it faces residual risk exposure that may be eliminated by buying an insurance when other means are infeasible (e.g., RTP and DR) or financially unattractive (e.g., full requirement contract and peak forward contract). Fifth, a competitively priced per MWh insurance premium found by a REP's procurement auction does not substantially increase a REP's fixed price offer. Finally, the voluntary transaction between a REP and an insurance seller is mutually beneficial. In summary, our proposed insurance helps remove the risk exposure that can bankrupt a small REP. Hence, it deserves consideration by REPs in Texas and competitive retailers elsewhere.

Author Contributions: Conceptualization, C.-K.W. and J.Z.; methodology, C.-K.W., A.T. and K.H.C.; software, K.H.C.; data curation, J.Z.; writing—original draft preparation, C.-K.W. and A.T.; writing—review and editing, C.-K.W., J.Z., A.T. and K.H.C. All authors have read and agreed to the published version of the manuscript.

Funding: This paper extends C.K. Woo's research funded by Grant R3698 from the Education University of Hong Kong and Grant 04564 from the same university's Faculty of Liberal Arts & Social Sciences. Without implications, all errors are ours.

Informed Consent Statement: Not applicable.

Data Availability Statement: The data used in this paper are publicly available from the sources cited in text.

Conflicts of Interest: The authors declare no conflict of interest.

References

1. Stoft, S. *Power System Economics: Designing Markets for Electricity*; Wiley: New York, NY, USA, 2002.
2. Zarnikau, J.; Woo, C.K.; Baldick, R. Did the introduction of a nodal market structure impact wholesale electricity prices in the Texas (ERCOT) market? *J. Regul. Econ.* **2014**, *45*, 194–208. [CrossRef]
3. Zarnikau, J.; Woo, C.K.; Zhu, S.; Tsai, C.H. Market price behavior of wholesale electricity products: Texas. *Energy Policy* **2019**, *125*, 418–428. [CrossRef]
4. Woo, C.K.; Horowitz, I.; Horii, B.; Karimov, R. The efficient frontier for spot and forward purchases: An application to electricity. *J. Oper. Res. Soc.* **2004**, *55*, 1130–1136. [CrossRef]
5. Woo, C.K.; Karimov, R.; Horowitz, I. Managing electricity procurement cost and risk by a local distribution company. *Energy Policy* **2004**, *32*, 635–645. [CrossRef]
6. Zarnikau, J.; Zhu, S.; Woo, C.K.; Tsai, C.H. Texas's operating reserve demand curve's generation investment incentive. *Energy Policy* **2020**, *137*, 111143. [CrossRef]
7. Woo, C.K. What went wrong in California's electricity market? *Energy* **2001**, *26*, 747–758. [CrossRef]
8. Woo, C.K.; Lloyd, D.; Tishler, A. Electricity market reform failures: UK, Norway, Alberta and California. *Energy Policy* **2003**, *31*, 1103–1115. [CrossRef]
9. Joskow, P.L. Lessons learned from electricity market liberalization. *Energy J.* **2008**, *29*, 9–42. [CrossRef]
10. Sioshansi, F.P. *Evolution of Global Electricity Markets: New Paradigms, New Challenges, New Approaches*; Elsevier: San Diego, CA, USA, 2013.
11. Deng, S.J.; Johnson, B.; Sogomonian, A. Exotic electricity options and valuation of electricity generation and transmission. *Decis. Support Syst.* **2001**, *30*, 383–392. [CrossRef]
12. Eydeland, A.; Wolyniec, K. *Energy and Power Risk Management: New Development in Modeling, Pricing and Hedging*; John Wiley & Sons: Hoboken, NJ, USA, 2003.
13. Billimori, F.; Poudineh, R. Market design for resource adequacy: A reliability insurance overlay on energy-only electricity markets. *Util. Policy* **2019**, *60*, 100935. [CrossRef]
14. Fuentes, R.; Sengupta, A. Using insurance to manage reliability in the distributed electricity sector: Insights from an agent-based model. *Energy Policy* **2020**, *139*, 111251. [CrossRef]
15. Niromandfam, A.; Choboghloo, S.P.; Yazdankhah, A.S.; Kazemzadeh, R. Electricity customers' financial and reliability risk protection utilizing insurance mechanism. *Sustain. Energy Grids Netw.* **2020**, *24*, 100399. [CrossRef]
16. Lai, S.; Qiu, J.; Tao, Y.; Liu, Y. Risk hedging strategies for electricity retailers using insurance and strangle weather derivatives. *Int. J. Electr. Power Energy Syst.* **2022**, *134*, 107372. [CrossRef]
17. Deng, S.J.; Oren, S.S. Electricity derivatives and risk management. *Energy* **2006**, *31*, 940–953. [CrossRef]
18. Woo, C.K.; Horowitz, I.; Olson, A.; Horii, B.; Baskette, C. Efficient frontiers for electricity procurement by an LDC with multiple purchase options. *OMEGA* **2006**, *34*, 70–80. [CrossRef]
19. Kleindorfer, P.R.; Li, L. Multi-period VaR-constrained portfolio optimization with applications to the electric power sector. *Energy J.* **2005**, *26*, 1–26. [CrossRef]
20. Iria, J.; Soares, F. Real-time provision of multiple electricity market products by an aggregator of prosumers. *Appl. Energy* **2019**, *255*, 113792. [CrossRef]
21. Brown, D.P.; Tsai, C.H.; Woo, C.K.; Zarnikau, J.; Zhu, S. Residential electricity pricing in Texas's competitive retail market. *Energy Econ.* **2020**, *92*, 104953. [CrossRef]
22. Woo, C.K.; Horowitz, I.; Hoang, K. Cross hedging and forward-contract pricing of electricity. *Energy Econ.* **2001**, *23*, 1–15. [CrossRef]
23. DeBenedictis, A.; Miller, D.; Moore, J.; Olson, A.; Woo, C.K. How big is the risk premium in an electricity forward price? Evidence from the Pacific Northwest. *Electr. J.* **2011**, *24*, 72–76. [CrossRef]
24. Zarnikau, J.; Woo, C.K.; Gillett, C.; Ho, T.; Zhu, S.; Leung, E. Day-ahead forward premium in the Texas electricity market. *J. Energy Mark.* **2015**, *8*, 1–20. [CrossRef]

25. Woo, C.K.; Horowitz, I.; Olson ADeBenedictis, A.; Miller, D.; Moore, J. Cross-hedging and forward-contract pricing of electricity in the Pacific Northwest. *Manag. Decis. Econ.* **2011**, *32*, 265–279. [CrossRef]
26. Woo, C.K.; Horowitz, I.; Zarnikau, J.; Moore, J.; Schneiderman, B.; Ho, T.; Leung, E. What moves the ex post variable profit of natural-gas-fired generation in California? *Energy J.* **2016**, *37*, 29–57.
27. Lloyd, D.; Woo, C.K.; Borden, M.; Warrington, R.; Baskette, C. Competitive procurement and internet-based auction: Electricity capacity option. *Electr. J.* **2004**, *17*, 74–78. [CrossRef]
28. Woo, C.K.; Sreedharan, P.; Hargreaves, J.; Kahrl, F.; Wang, J.; Horowitz, I. A review of electricity product differentiation. *Appl. Energy* **2014**, *114*, 262–272. [CrossRef]
29. Brown, D.P.; Zarnikau, J.; Adib, P.; Tsai, C.H.; Woo, C.K. Rising market concentration in Texas's retail electricity market. *Electr. J.* **2020**, *33*, 106848. [CrossRef]
30. Moore, J.; Woo, C.K.; Horii, B.; Price, S.; Olson, A. Estimating the option value of a non-firm electricity tariff. *Energy* **2010**, *35*, 1609–1614. [CrossRef]
31. Brown, D.B.; Tsai, C.H.; Zarnikau, J. Do gentailers charge higher residential electricity prices in the Texas retail electricity market? *Electr. J.* **2022**, *35*, 107109. [CrossRef]
32. Woo, C.K.; Lloyd, D.; Borden, M.; Warrington, R.; Baskette, C. A robust internet-based auction to procure electricity forwards. *Energy* **2004**, *29*, 1–11. [CrossRef]
33. Woo, C.K.; Borden, M.; Warrington, R.; Cheng, W. Avoiding overpriced risk management: Exploring the cyber auction alternative. *Public Util. Fortn.* **2003**, *141*, 30–37.
34. Klemperer, P. What really matters in auction design. *J. Econ. Perspect.* **2002**, *16*, 169–189. [CrossRef]
35. Johnson, N.L.; Kotz, S.; Balakrishnan, N. *Continuous Univariate Distributions*; Wiley: New York, NY, USA, 1994.

Article

Study of Short Circuit and Inrush Current Impact on the Current-Limiting Reactor Operation in an Industrial Grid

Yuriy Varetsky [1,2,*] and Michal Gajdzica [1]

[1] Department of Energy and Fuels, AGH University of Science and Technology, 30-059 Kraków, Poland
[2] Institute of Power Engineering and Control Systems, Lviv Polytechnic National University, 79013 Lviv, Ukraine
* Correspondence: jwarecki@agh.edu.pl

Abstract: Current-limiting reactors are widely used in industrial electrical grids to reduce the current amplitude in the equipment and stabilize the voltage on the busbar during short circuits. Their application is distinguished by high technical and economic efficiency. However, mechanical damage to the reactors has been observed within extensive industrial grids with many induction motors and internal synchronous generators. The article analyses a case study of the reactor damage in the true industrial grid during a short circuit. An analysis of the damaged reactor's previous operation had shown that there was a weakening of the fastenings in the reactor design, caused by the repeated starting currents of the grid motors and generators. A study of grid transients during short circuits was carried out by Matlab/Simulink software. The simulation results showed that the reactor could be damaged by a critical peak current in an unfavourable combination of the grid configuration and the short circuit location. The results of the study prove that, for industrial networks containing powerful induction motors and internal synchronous generators, the standardized procedure for selecting current-limiting reactors should additionally consider such factors as the localization in the grid, the effect of equipment-starting currents and possible grid configurations.

Keywords: current-limiting reactor; industrial grid; modeling transients; mechanical stress; short circuit current peak; generator energizing

Citation: Varetsky, Y.; Gajdzica, M. Study of Short Circuit and Inrush Current Impact on the Current-Limiting Reactor Operation in an Industrial Grid. *Energies* **2023**, *16*, 811. https://doi.org/10.3390/en16020811

Academic Editors: Yuan Liao and Ke Xu

Received: 26 November 2022
Revised: 25 December 2022
Accepted: 4 January 2023
Published: 10 January 2023

1. Introduction

The high levels of short circuit currents and high transient overvoltages in medium voltage grids require more and more knowledge of the methods for limiting short circuit currents and their effects on switchgear and other electrical equipment. Developers split the grid, introduce higher impedance transformers and series current-limiting reactors, or use complex strategies such as sequential grid tripping. However, these solutions may create other problems such as the loss of the security and reliability of the power system, high costs and increased power losses.

The issues of the application of current-limiting reactors, their design and their location in power industrial grids and transmission systems are still relevant and discussed in monographs, IEEE standards, technical handbooks and manuals as well as numerous publications. The use of current-limiting reactors (CLRs) in various connections and locations is the most popular solution for industrial power grids [1]. Their application is inexpensive and they significantly reduce short circuits. The IEEE Standard [2] provides general information about dry-type air-core reactors that are installed in transmission and distribution systems, their effect on steady-state power flow and short circuit current limitation. The IEEE [3] standard provides ANSI/IEEE recommendations for power grid designers on short circuit current calculation to select the appropriately sized air-core reactors for industrial and power grid applications. The IEC Standard [4] describes the methodology for the technical specification and testing of reactors for manufacturers. The available literature [5–9] presents a wide range of issues on the design and testing of air-core current-limiting reactors for various industrial applications in accordance with the above standards. The methods for optimizing the configuration of a grid equipped with reactors and reducing the cost of the power grid due to the appropriate limitation of short circuit currents for the used electrical equipment are discussed.

The location of the current-limiting reactor at the power grid must be analysed as well in terms of a total lumped inductance in an equivalent electric circuit of the power grid, which could lead to an increase in the severity of the transient recovery voltage across the circuit breaker contacts, associated with the interruption of the circuit current [10–12]. Thus, for each of these grids, it is necessary to carry out the proper design of the reactor. The optimization of a reactor mainly focuses on solving the contradiction between the fundamental properties, such as inductance, current carrying ability and the metal conductor materials' (copper or aluminium) cost. Several computational methods are used to improve the utilization ratio of the metal conductors of reactor coils and their electromagnetic and thermal characteristics. Several papers [13–17] present the calculations of electromagnetic and thermal fields for reactors to improve their thermal efficiency.

The basic installation of CLR units is a series reactor connected between source and loads and is designed mainly to protect feeders and electrical equipment and devices against the high magnitude of short circuit currents and their thermal action. The more complex applications with series current-limiting reactors are designed as circuits where these units are installed between busbar sections of each power supply system, in series with switchboards equipped with a large number of induction motors with rated power above 250 kW and characterized by direct start or in series with industrial small synchronous generators that are connected to the main industrial grid bus. In these applications, the CLR allows the main busbar voltage to be stabilized and, when there is a fault, reduces the amplitude of the short circuit current to ensure the correct operation of sensitive feeders.

Nevertheless, operational experience indicates that faults within complicated industrial power grids have caused damage to the current-limiting reactors due to the incorrect estimating of their sizing. Such cases are provoked by a change in the operating configuration of an industrial power grid with internal generators when several of them can be connected in parallel regarding the possible short circuit location. The transient analysis of the possible operating configurations of the MV industrial power grid under fault conditions can support the selection of appropriate CLR parameters to ensure its reliable and sustained operation. The literature review has shown little information on the impact of short circuit and inrush currents on the mechanical strength of current-limiting reactors in industrial electrical grids. The authors of this study carried out a detailed technical analysis of the operation of current-limiting reactors affected by multiple current pulses caused by short circuits and the switching on of motors and generators in a specific industrial grid. The analysis of the damage to one of the grid reactors and the results of the following grid simulations have supported the supposed thesis on declining the rated mechanical strength of the current-limiting reactor due to the specific operating conditions.

The article presents studies that were prompted by analysing CLR damage in the true industrial grid with the intensive use of powerful induction motors and local synchronous generators. This paper intends to analyse the case and causes of a current-limiting reactor failure due to a three-phase short circuit within the industrial grid. The CLR was connected via a cable line between a busbar fed from the power system and internal generators and a busbar where eleven induction motors with ratings of 320 kW up to 2000 kW were operated. As a result of energizing one of the induction motors, a three-phase short circuit occurred due to damage to the insulation of the cable supplying it.

The article has been essentially divided into two parts. The first part of the article discusses the operational issues of CLRs in various configurations of MV industrial power grids. The second part presents the results of experimental studies carried out by modelling transients under short circuits in the described industrial grid. To analyse transients occurring under short circuits and equipment switching, the authors developed a model of the grid using the Matlab/Simulink software package, which provides simulating transients for potential switching events and network configurations.

2. Current-Limiting Reactor Specification

2.1. Design

The current-limiting reactor unit, which is not equipped with an iron core magnetic material within the windings, is called an air-core current-limiting reactor. It consists of a circular coil wound around a nonmagnetic material for greater mechanical strength. Usually, in an industrial environment, the windings of current-limiting reactors are the air-cooled dry type or immersed in oil or a similar cooling fluid. Hence, there are different applications in the electrical field, where reactors improve the performance of a system or play a role in protecting it [1].

During a fault condition, the reactance of the current-limiting reactor should not diminish due to the saturation effect. This is an essential requirement to limit the short-time fault currents. Ideally, the current-limiting reactor must have a no-iron circuit and must be of the air core or coreless type. The iron core type provides non-linear saturating type characteristics, and during overcurrents tends to diminish their reactance due to the saturation effect, while the reactors are required to offer high impedances to limit the fault currents. The CLR type without a core will provide a near-constant reactance at all currents due to the absence of an iron core, hence their preference over other types for such applications. These reactors can be made of copper or aluminium winding without any core, similar to an air-cored solenoid. They are merely coils of wire wrapped around a non-metallic core. The cores are usually made of ceramics, concrete, fibreglass or glass polyester. In the absence of an iron core, there is a large amount of leakage flux (stray magnetic field) in the space, which may also infringe on the metallic tank housing of the reactor, and affect the reactance of the coil, in addition to heating the tank itself. It is therefore important to provide some kind of shielding between the winding of the reactor and the tank.

In cases without an iron core, there is no saturation of the core, so these reactors are more useful when they are required to be used as current-limiting devices for limiting the inrush currents during the switching of large inductive motors.

In the MV industrial power grid, the current-limiting reactors usually consist of three-phase coils stacked one above or next to the other with support insulators between them as shown in Figure 1.

In the first solution, which is a stacked three-phase reactor (Figure 1a), the gap between the two coils is maintained (>300 mm) such that each of them would fall nearly out of the inductive interference of the other. Moreover, that kind of construction of CLR unit will exhibit time-varying axial forces between the phase coils during a two-phase or three-phase short circuit. As the short circuit force depends on the distance between the coils, the force from the centre to the outward coils is much higher than between the two outward coils. Reversing the winding direction of the centre coil concerning the outward coils causes the neighbouring coils to be attracted to each other when the time-varying force reaches its maximum during a three-phase short circuit. This contraction force is absorbed by the supporting insulators between the coils. Further, the magnetic coupling between the coils increases the effective inductance of the centre coil. This effect can be compensated by reducing the self-inductance of the central phase due to a corresponding decrease in the number of turns. The second type of construction of the CLR unit is the location of three-phase coils next to each other (Figure 1b) with the dimension between the two coils as in the first solution to avoid the high inductive forces between them. The presented solution of the three-phase reactors, which are wound in the same direction and are placed with their axes parallel, causes the magnetic forces to be attractive if the current in one flows in the opposite direction to the current in the other, and repulsive if the current is in the same direction. In addition, setting three-phase reactor coils placed with axes parallel with braces between phases ensures better cooling conditions. Nevertheless, more busbars and operational areas are required in that case, which is associated with higher costs of preparing the electric room. In both cases, the foundation of the inductor coils must also be ensured to have no closed loops in the reinforcement concrete steel (RCS) [1].

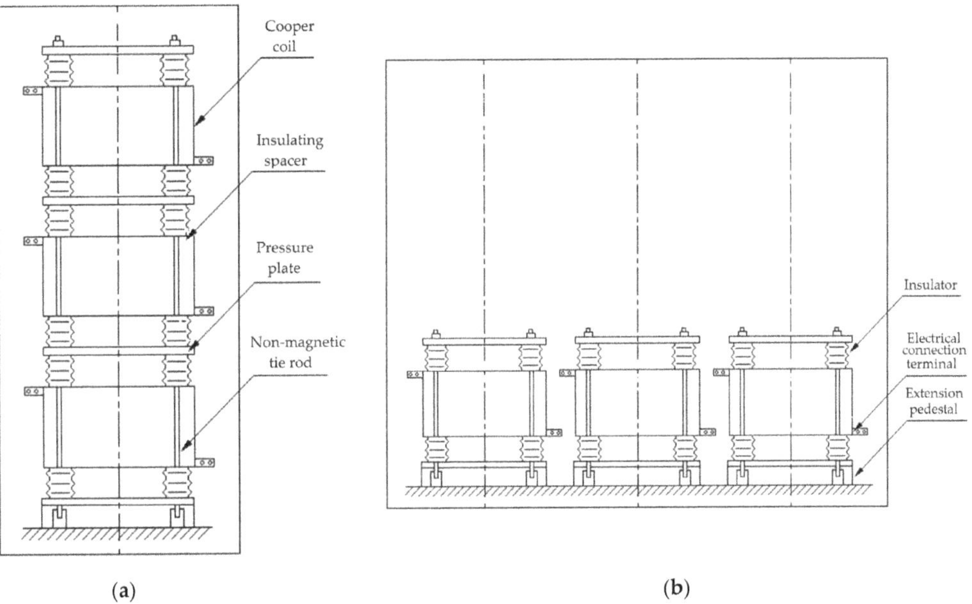

Figure 1. A typical solution for a three-phase unit of air-core current-limiting reactors for an MV industrial grid with non-magnetic shielding placed: (**a**) coaxially with provision for bracing against the compartment wall (stacked); (**b**) parallel with braces between phases.

The magnitude of the force between the adjacent reactors of a three-phase circuit is greater during a two-phase short circuit than it is during a three-phase short circuit, (although a two-phase short circuit current is only 86.6 per cent of the three-phase current) and therefore stresses caused by these forces should be based on two-phase short circuits. Forces due to two-phase short circuits are higher because the fluxes from adjacent reactors are in phase or 180 deg. out of phase during two-phase short circuits, while they are 120 deg. or 60 deg. out of phase during three-phase short circuits.

In the current-limiting reactor without an iron core a cylindrical shield of non-magnetic material, such as aluminium or copper, is provided around the inductor coil instead of a magnetic material. There is no iron path for the magnetic field, and the coil may not maintain a constant inductive reactance, as in the case of magnetic shielding. Instead of that, it can become reduced with an increase in the current due to a counter-field generated in the coil by the non-magnetic shielding. The effect is that there is no saturation of the non-magnetic core (inductor coil) and the reactor's V-I characteristic remains almost linear [1].

2.2. Short Circuit Current Rating

Referring to Standard [2], the short circuit current rating of the air-core current-limiting reactor should describe two impacts: mechanical and thermal. The mechanical short circuit current rating is based on the worst-case assumption of a simultaneous three-phase fault and the resulting offset peak current. The degree of offset is a function of system damping. The magnitude of the thermal short circuit current rating is obtained from system fault calculations and the duration is a function of the system operating policy including several "autorecloses" of the breaker, etc. (Section 5.6 in [2]).

The mechanical short circuit current rating is defined through the maximum peak value of the transient current during a three-phase short circuit, the so-called maximum asymmetrical current peak [3]. The maximum asymmetrical current peak can be obtained from industrial grid information and by calculating the reactor impedance. Unless specified, grid parameters estimating the maximum peak asymmetrical current I_{pk} shall be described as 2.55 times the RMS value of the symmetrical fault current:

$$I_{pk} = 2.55 I_{SC}. \tag{1}$$

The symmetrical RMS fault current of the reactor is calculated from a fault occurring at the load side terminals by using the system equivalent series reactance of an industrial grid:

$$I_{SC} = \frac{U_S}{\sqrt{3}(X_R + X_S)}, \tag{2}$$

where
  I_{SC} - is the RMS short circuit current, [kA];
  X_R - is the reactance of the series current-limiting reactor, [Ω];
  X_S - is the total system equivalent series reactance, [Ω];
  U_S - is the system line-to-line RMS voltage, [kV].
  The reactances in Formulae (2) are calculated as follows:

$$X_S = \frac{U_s^2}{S_{SCs}}, \tag{3}$$

and

$$X_R = U_S^2 \left(\frac{1}{S_{SCl}} - \frac{1}{S_{SCs}} \right), \tag{4}$$

where
  S_{SCs} - is the three-phase fault level on the source side of the reactor, [MVA];
  S_{SCl} - is the three-phase fault level on the load side of the reactor, [MVA].
  The thermal short circuit current rating of a current-limiting reactor is defined as the RMS symmetrical fault current specified by (2). The duration should be specified by the user and should typically take into consideration system protection practices such as breaker interrupting time, breaker auto reclose sequence, etc. Typical values for the duration of the thermal short circuit current rating are 1, 2 or 3 s. If not specified by the user, the duration of the thermal short circuit current shall be considered to be 3 s [2,4].

2.3. Concern about Repetitive High Inrush Currents

Short circuit currents and motor inrush currents cause the electromagnetic forces in the reactor windings and auxiliary components that can be tens and hundreds of times higher than the values in steady-state operation. These forces are proportional to the square of the short circuit current amplitude value. The windings are subjected to the simultaneous action of circumferential and compressive forces. The mechanical clamping structure and the supporting elements of the reactor are subject to direct action forces and reaction forces. Current-limiting reactors in MV industrial grids containing powerful induction motors and local synchronous generators are exposed to repetitive inrush currents [5]. Hence, the CLR unit should be designed and verified by calculation so that its coils do not have mechanical resonances close to twice the inrush current frequency [2,13,14]. On the other hand, these repetitive currents cause the mechanical fatigue of the CLR construction.

The problem of considering the impact of repetitive transient currents when choosing the parameters of air reactors for medium voltage harmonic filters is well-known and described in various publications [18–21]. Its essence lies in the requirements to analyse the operating conditions of the air-core filter reactor within a specific power supply system, considering such impact factors as the transient current amplitudes and their repetition rate over a specific time interval. Depending on these data, the requirements for the mechanical strengthening of the air-core filter reactor design are established. Energizing the filter circuit or a powerful transformer is a typical repetitive event in industrial grids that causes high current magnitudes in the filter harmonic reactor. The number of such dynamic overcurrents per year is critical for the design of the filter reactor. The IEEE Standard [2] establishes that, if transient currents of a magnitude close to the mechanical short circuit current rating are expected to be frequent more than 10 times a year, the number of such transient current peaks per year shall be specified.

Our experience has shown that the problem of repetitive inrush currents should also be taken into account for air-core limiting reactors installed in industrial grids. The designer and the manufacturer must verify the location and function of the reactor in the industrial network every time, especially when exposed to the grid synchronous generators and induction motors since they affect the operating conditions of the CLR during transient currents. Therefore, before installing CLR in the MV grid, it is necessary to carry out appropriate tests and calculations to confirm that each coil of the reactors can withstand the mechanical effect of short circuit currents after the repetitive operational inrush currents of the equipment that occurred during the previous operation of the reactor.

2.4. Typical Applications

In medium and high-voltage industrial grids, the amplitude of the short circuit current is a function of the nominal voltage and reactance of the power grid. To limit its value to the appropriate level of electrical devices and equipment at the same voltage rating, the primary method is to increase the inductance of the supply system seen at the fault location. This can be done by either increasing the reactance of the circuit or by removing grid elements from the fault path. The first is done by adding current-limiting reactors and the second by current limiters [10].

Current-limiting reactor units can be installed anywhere in the industrial grid to limit the short circuit currents. Since they are essentially a linear inductive reactance, their impedance will add arithmetically to the power system impedance, reducing the fault current's magnitude.

The current-limiting reactors may be introduced in the outgoing feeders, incoming feeders or between the bus sections, so there are different applications in power engineering where CLRs improve the performance of a system or play a role in protecting it [6,7,12,22]. Besides the aspects mentioned above, the position of the reactor at the substation busbar to meet better the needs of the specific case should be analysed [8]. Various methods are proposed to develop the reactor's optimal design in each specific case [9,23,24]. The main applications of CLR installation diagrams in industrial power systems are illustrated in the figures below: the connection of the bus sections (see Figure 2a), the series connection with the incoming feeder (see Figure 2b) and the series connection with the outgoing feeder (see Figure 2c).

In the industrial power grid, which is characterized by the high value of the system fault level or based on more than one section of buses, the current-limiting reactor is installed between the sections to reduce the total short circuit current, Figure 2a. In case of a fault, the CLR reduces the total short circuit current in each section, allows the better sharing of transformer loading and reduces the required equipment short circuit rating of the industrial grid. In transient conditions, the CLR mechanically withstands the short circuit current peak amplitudes from incoming and outgoing feeders. Therefore, its rating parameters should be calculated considering the operational characteristics of the loads that are connected to each section of the substation between which it was installed. Moreover, in that application, CLRs are responsible for limiting fault currents from internal synchronous generators by effectively increasing their resultant impedance relative to the fault location through the bus tie. To the disadvantages of that kind of diagram, the CLR does not impact incoming feeder contribution and transients.

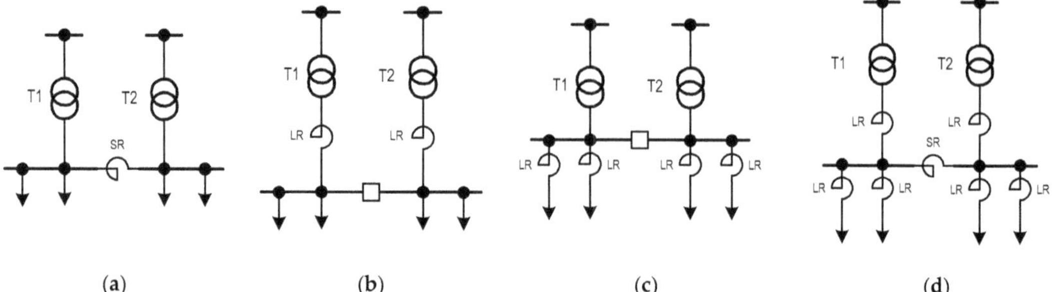

(a)   (b)   (c)   (d)

Figure 2. Typical diagrams of air-core current-limiting reactor applications for an MV industrial grid: (**a**) Connection of the bus sections; (**b**) Series connection with the incoming feeder; (**c**) Series connection with the outgoing feeder; (**d**) Mixed type of connection with bus sections and feeders.

In industrial power grids containing a large number of high-capacity induction motors, the series current-limiting reactors are installed in the incoming feeders, Figure 2b. Also, if a fault occurs, each CLR reduces the short circuit current amplitude at the first current rise. The reactor implementation in series with incoming feeders makes it possible to individually reduce the short circuit currents in the grid, but at the same time, in such a diagram, power losses increase, and there is a problem with voltage control on each busbar section.

Further, CLR can be installed in each outgoing feeder, Figure 2c. In this application, in the event of a fault, each current-limiting reactor reduces the amplitude of the short circuit current at the very first current increase, but with low power losses and better voltage control at each load at the end of the power cables during normal operation. The disadvantage of this case is the switching conditions of asynchronous motors. The amplitude of the starting current is 2–10 of the rated current, so the voltage drop across the reactor is in phase with the load voltage and causes large voltage drops, which can cause serious problems when starting large drives directly. As practice shows, to avoid a large voltage drop when starting motors, the main busbars of switchboards should be operated with a voltage higher than the nominal voltage, with a voltage factor from 1.05 to 1.1 of the rated voltage of the power system. In such a scheme, the reactors are not installed directly in the inrush current path of large feeders or loads with heavy starting. The motor starting reactor is connected in series with a motor to limit the inrush current during the motor starting operation. After the motor is started, the reactor is typically bypassed to limit power losses in continuous operation. That solution is frequently used in industrial power grids where cable lines are connected in series with high-capacity loads: at steel, coking and mining plants. In this case, a CLR limits the current under system fault conditions to a level that is compatible with the system equipment ratings.

In some MV industrial electrical grids, current-limiting reactors are installed in various parts of the network to improve their operational functionality and reliability of the power supply and prevent equipment damage at high short circuit transient currents, Figure 2d.

That type of application is a very cost-effective solution, as it eliminates the need for upgrading the entire switching and protection system when the short circuit power of the system is increased. Such a reactor is designed to withstand the rated and short-period transient fault currents, giving the total power supply system a specified impedance value.

The examined industrial grid in the presented article uses a mixed configuration of CLR units as in Figure 2d. In a further study, we describe the peculiarities of the CLR operation within true grid conditions.

3. Case Study

3.1. The Analysed Industrial Grid

The following methodology was adopted in the studies:

- the detailed analysis of the examined reactor failure;
- the analysis of the repeatability of the short circuits and switching currents of motors and generators in the industrial grid, which the reactor was subjected to during operation;
- the development of the grid model for studying transients affecting the reactor;
- the discussion of the simulation results to confirm the thesis about the decrease in the mechanical strength of the reactor due to the impact of multiple transient current pulses;
- suggesting a possible solution to the problem.

The analysis of short circuit currents and their impact on the current-limiting reactor was carried out on the example of an MV industrial cable grid shown in Figure 3. The 6 kV power grid under consideration is powered by an external 110/6 kV substation and contains two 6 kV synchronous generators, G1 and G2, with a rating of 25 MW.

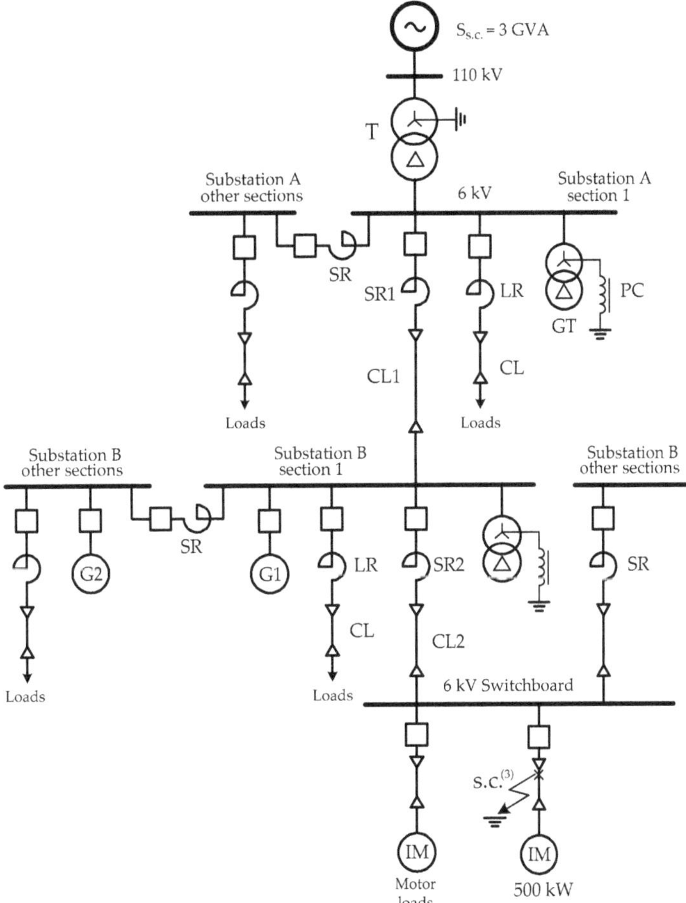

Figure 3. Single-line diagram of the industrial grid.

The industrial grid under study contains two substations, A and B, connected by cable CL1 and reactor SR1. The sections of these substations are connected by intersectional reactors SR. The feeders in all the sections are supplied through cables CL and current-limiting reactors LR. Two Petersen coils (PC) to compensate for capacitive single-line-to-ground fault currents are connected to the busbars of substations A and B by grounding transformers (GT) with a total rating of 690 kVA.

The examined grid is a highly extensive cable industrial grid with various powering configurations of substation B and thereby all loads are connected to the bus sections of the substation. The power supply of substation B in the typical configuration of the presented grid is provided by synchronous generators and by substation A. However, several topologies are strictly dedicated to the emergency operation of the industrial grid or have been applied during the technical inspection or routine maintenance of the electrical equipment and substation cell devices. Table 1, respectively, presents the rating data of the electrical equipment and the synchronous generators of the examined industrial grid.

Table 1. Catalogue data of the industrial grid.

Grid Component	Parameter	Unit	Value
Supply system	Transformer rated voltage, U_n	kV/kV	110/6
	Transformer capacity, S_n	MVA	31.50
	Short circuit power, S_{sc}	GVA	3.00
Cable lines, CL and SCL	Rated voltage, U_n	kV	6.00
	Specific resistance, R_0	Ω/km	0.16
	Specific inductance, L_0	mH/km	0.34
	Specific capacitance, C_0	µF/km	0.44
Line reactor, LR	Rated current, I_n	kA	0.60
	Resistance, R	Ω	0.35
	Inductance, L	mH	0.45
	Rated short circuit current peak, I_p	kA	32.00
Intersection reactor, SR1	Rated current, I_n	kA	1.00
	Resistance, R	Ω	0.27
	Inductance, L	mH	1.10
	Rated short circuit current peak, I_p	kA	37.00
Intersection reactor, SR2	Rated current, I_n	kA	1.60
	Resistance, R	Ω	0.13
	Inductance, L	mH	0.22
	Rated short circuit current peak, I_p	kA	52.00
Synchronous generators, G1 and G2	Rated active power, P_n	MW	25.00
	Rated voltage, U_n	kV	6.30
	Rated current, I_n	kA	2.29
	Power factor, $\cos \varphi$	-	0.80
	Stator resistance, R_s	mΩ	5.22
	Subtransient reactance, X''_d	%	14.30
	Transient reactance, X'_d	%	24.00
	Synchronous reactance: X_d	%	245.00
	Mechanical time constant, T_m	s	9.32

In the examined industrial grid, the studies focus on the SR2 unit located in the electric room of substation B, which was connected in series with an aluminium cable line CL2 to power a 6 kV switchboard. It supplies the eleven MV induction motors with ratings from 320 kW to 2000 kW. The analysed current-limiting reactor SR2 was designed for rated current 1600 A to limit the current amplitude during the frequent starting of MV induction motors. The repetitive high current amplitudes during MV motor switching operations have similar negative impacts on the SR2 coils and their flexible connections to substation busbars, such as transient conditions that occur during short circuits. Therefore, the analysed current-limiting reactor is operated under the most severe conditions compared to the other reactors (SR or LR) installed in the industrial grid. The rated data of the current-limiting reactor SR2 are presented in Table 1.

The 6 kV switchboard is implemented as 14 bays that supply the auxiliaries for the industrial grid and is equipped with an Automatic Transfer Switch unit (ATS) with digital protections on each feeder. The automatic switching on of the reserve power supply of the MV switchboard is realized between two sections of substation B. Moreover, its electric devices and equipment have been designed for a voltage rating of 12 kV and transient overvoltage equal to 75 kV. The 6 kV bus can withstand a 1-s RMS short circuit current of 31.5 kA. The switchboard busbars feed eight induction boiler motors, four with a power rating of 320 kW, the rest 500 kW and three induction pump motors with power ratings of 630 kW, 900 kW and 2000 kW, respectively. The total power rating installed and supplied by the MV switchboard amounts to 6810 kW.

3.2. System Configuration Analysis during Fault

The deep analysis of the current-limiting reactor SR2 mechanical damage proved that the event was a result of the high peak current impact due to a three-phase short circuit (S.C.[3]) on the cable connection of a 500 kW induction motor to the 6 kV switchboard (see Figure 3). At the time of the failure, the power industrial grid was supplied by two synchronous generators, G1 and G2, and additionally there was a power supply from the substation of 110/6 kV. The settings of the digital protection devices that were located in the relay section of the switchboard bays are presented in Table 2.

Table 2. Protection settings of digital protection devices.

Localization of Digital Protection Devices	Protection Settings			
	Overcurrent Protection		Ground Fault Protection	
	I [kA]	t [s]	I [kA]	t [s]
Bays of 6 kV Switchboard	30.50	0.12	-	signalling
Bay of 500 kW induction motor	0.71	0.08	2.18	0.02

Due to long service life (about 30 years), the considered industrial power grid and its electrical equipment are characterized by a significant accident rate. All the current-limiting reactors of the grid were subjected to short circuit currents and inrush currents from internal generators and motors. The average estimations summarized by operational personnel showed that each of the current-limiting reactors in this grid was subject to about 1–4 short circuits per year. At the same time, the analysed reactor, which was operated for about 15 years, was subjected to about 30 boiler motor starts and 5–10 pump motor starts per quarter. Before the failure, it withstood (without any damage) an average of about 5–8 short circuits per year. The damaged current-limiting reactor was regularly and properly serviced following the manufacturer's instructions and manual. Periodic inspections were carried out at least once a year and included the following operational activities: the inspection of reactor corrosion and dust, checking the screw connections and the condition of the terminals, as well as checking protection operation and the resistance insulation measurement of the reactor coils.

During the switching-on operation of the induction motor, the arcing single-line-to-ground fault on the cable connection was initiated. The fault lasted a time because the ground fault protection module for the current after 80 ms was not activated and the circuit breaker of the induction motor was still closed under the fault. Next, the fault turned into a three-phase current with high amplitude and the time delay overcurrent protection tripped the circuit breaker of the primary supply of the 6 kV switchboard after 120 ms from Section 1 of substation B. At the same time, the ATS system switched on the circuit breaker of the reserve power supply from the other section of substation B to prevent a shutdown of all the feeders and loads that were operated from the 6 kV switchboard. The detailed analysis showed that the main cause of the non-activated time delay over-current protection module of the induction motor was a too high measurement error from the input current transformers and analogue-to-digital converter of digital protection [25]. The detailed analysis of the protection settings showed that, during ground fault tests, the digital criterion of protection detected the disturbances only 40 ms after their occurrence. Therefore, after the failure, the settings of the ground fault protection module of the digital protection of the induction motor were changed, reducing the time delays of the decision to 5 ms and thus increasing the signal sampling frequency to 200 Hz.

The inspection of the current-limiting reactor SR2 in the electric room of substation B by operating personnel showed that its coils were faulty in all phases. Figure 4 presents a damaged coil of the current-limiting reactor SR2.

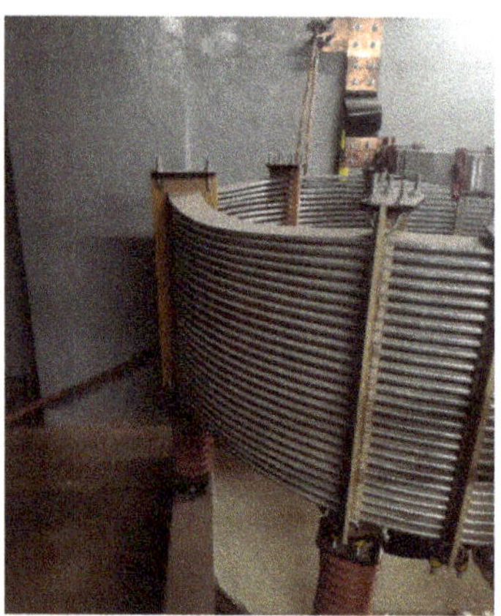

Figure 4. A faulty coil of the intersection current-limiting reactor SR2.

The detailed analysis of the current-limiting reactor failure showed that each phase coil was damaged due to the loosening of the screws of the elastomer springs on each turn of the reactor coil and the inappropriate number of epoxy isolate spacers between the turns of the reactor coil. Reducing the thickness of the insulating spacers led to the bending and indentation of the coil turns between themselves as a result of severe forces due to the high pulse of the short circuit transient current. Thus, it can be stated that the main reason for the failure of the considered current-limiting reactor was the incorrect mechanical calculation of its design and the subsequent choice of epoxy isolate spacers between the reactor coil turns, which led to the mentioned consequences.

3.3. Modelling the Industrial Grid

Figure 5 shows a block diagram of the industrial grid model developed by the SimPowerSystem and Electric Drives libraries of the Matlab software package. The following model blocks are used here: 110 kV external power supply system and step-down transformer T; G1 and G2—synchronous generators; CL and SCL—cables; arc suppression devices on the base of transformers GT and reactors PC, SR and CLR—short circuit current-limiting reactors; IM—induction motors; CB—power circuit breakers. The synchronous generator equivalent was implemented as a three-phase machine block, modelled in the d-q rotor reference frame. The external power supply system operates with a direct grounded neutral, and the examined industrial grid with an ungrounded one. Limiting the high ground fault current, arising due to the significant equivalent ground capacitance of the grid cables, is carried out by the arcing devices. The equivalent parameters of all the grid reactors are modelled using R-L elements. The effect of increasing cable resistance due to escalating transient frequency under short circuit faults in the industrial cable grid has negligible impacts on the transient behaviours. Therefore, it is reasonable to present the grid cable equivalent as a π-model with equivalent resistance and inductance series-connected and shunt capacitances. The three-phase induction motor in the discussion was implemented by a three-phase induction machine model with a squirrel cage and was loaded by mechanical torque modelled by a Step block. The remaining three-phase induction motors were implemented as one block with appropriate equivalent parameters.

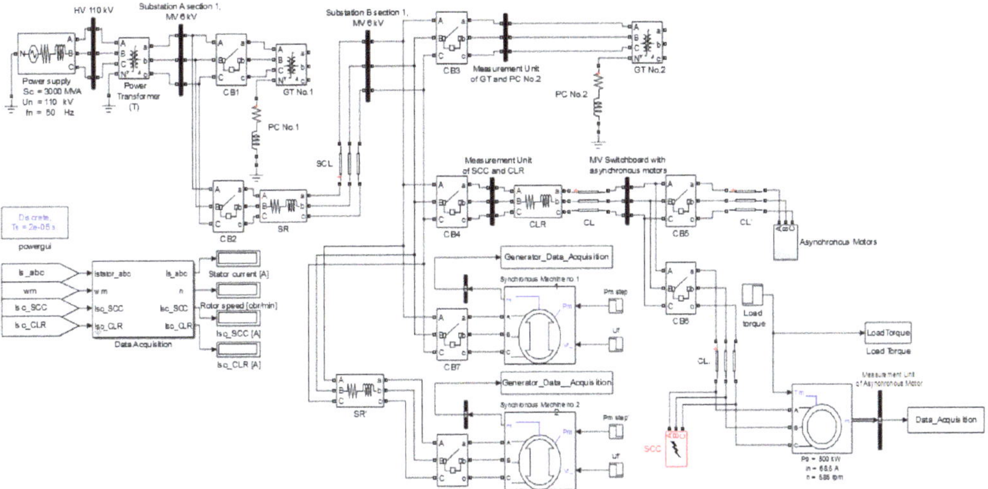

Figure 5. Block diagram of the industrial grid model.

During a short circuit that occurred on the cable feeder of the 500 kW induction motor, the total operational active power of the other induction motors was equal to 2910 kW. The three-phase short circuit on the cable connection was implemented by a three-phase fault block SCC from the SimPowerSystem library by two parameters: fault resistance $R_{on} = 0.04\ \Omega$ and ground resistance $R_g = 1.0\ \Omega$.

Transients caused by switching on motors and short circuits in the industrial grid under study are characterized by significant variations in the time constants and natural frequencies of the equivalent circuit loops. Therefore, automatic step recalculation by the ode23t solver was implemented to solve the model differential equations. The initial value of the step was taken as 20 μs. Using this solver makes it possible to provide the required accuracy and optimize the time for solving model equations. The equivalent circuit of the examined grid was built following the diagram in Figure 3, and the parameters of the model blocks are calculated by equipment catalogue data from Tables 1 and 2.

3.4. Results of Simulations and Discussion

The damage to the examined reactor selected according to the standard methodology initiated detailed simulations to determine the causes of such an event. Therefore, the simulations can give precise answers to the points about the actual short circuit current peak at the time of the accident, and the short circuit current peaks and motor starting currents that occurred in other grid configurations.

Several operation configurations can be used during the examined industrial grid's operating conditions. In the basic configuration, the power supply of the substation B sections is provided by generators G1, G2 and the external power system. Nevertheless, several configurations are predicted for technical inspection or emergency conditions where the substation sections are powered by one generator or just an external substation.

Hence, the article discusses only configurations of the industrial power grid that have a practical interest considering the analysed current-limiting reactor SR2 failure. The analysis was carried out for two types of industrial grid configurations. The first describes only one source to supply the substation B sections: the power supply system or synchronous generator G1 or G2. The second type describes a combination of the power sources connected to the industrial grid.

The main task of simulations is to obtain the short circuit current peaks in the examined reactor SR2 coils and compare their values under the considered industrial grid configurations. Figure 6 shows the transient voltages and currents in the grid components of interest during a three-phase short circuit at the motor cable feeder of the 6 kV switchboard (see Figure 3) when various power sources are connected to Section 1 of substation B, following the second type of configuration described above. Within these industrial network configurations, the highest current peaks in the coils of the SR2 reactor will occur during the considered short circuit.

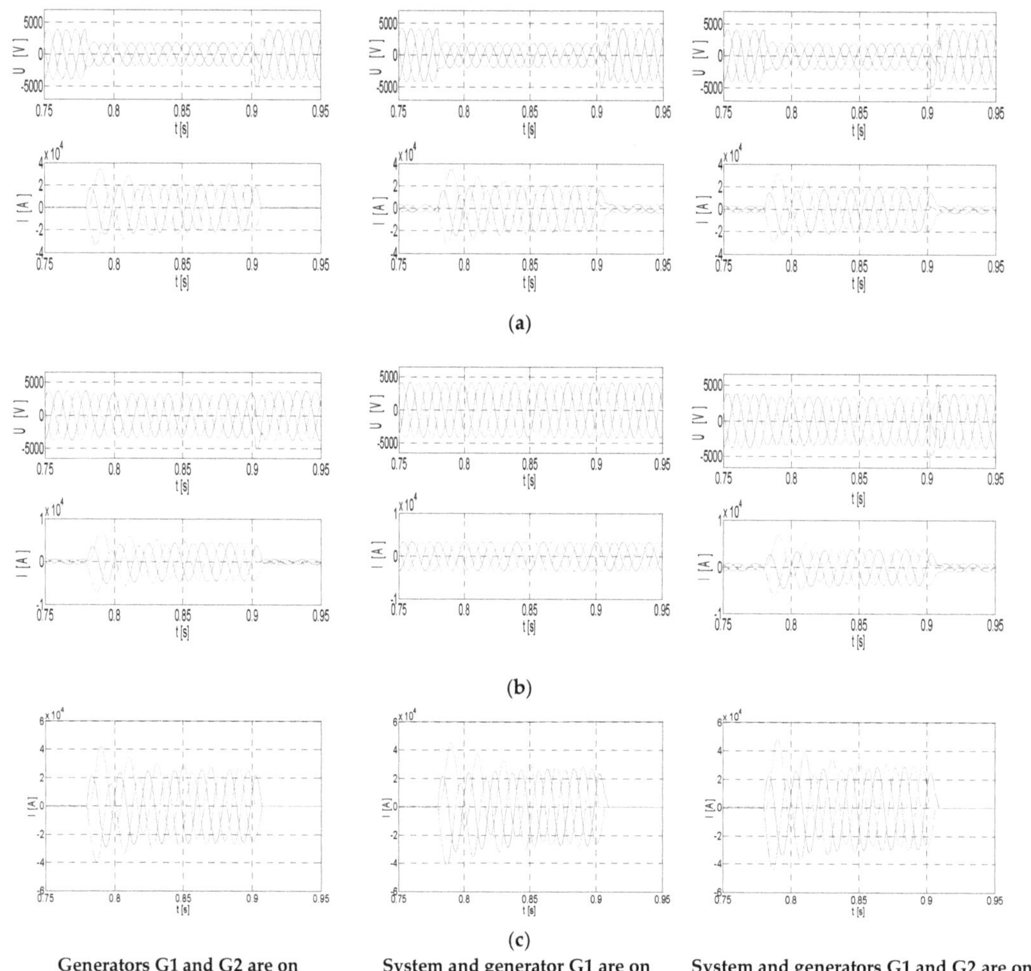

Generators G1 and G2 are on   System and generator G1 are on   System and generators G1 and G2 are on

Figure 6. The transients in the grid components (phase A - blue, phase B - green, phase C - red): (**a**) Bus phase voltages and currents of synchronous generator G1; (**b**) Bus phase voltages and currents of synchronous generator G2; (**c**) Currents of the reactor SR2.

The comparison of short circuit current peaks on the fault point and in the reactor SR2 coil was carried out for all studied industrial grid configurations, and the results are presented in Table 3. Comparing the simulation results, one can note the following: the most impact on the transient short circuit current peak in the reactor winding is exerted by the synchronous generator G1, located "electrically closest" to the short circuit point.

Table 3. Maximum current peaks in the reactor SR2 coils (I_{Rp}) and at the short circuit point I_{SCp}.

Connected Power Sources	I_{Rp} [kA]	I_{SCp} [kA]
The first type of the grid configurations		
System	17.50	24.86
Synchronous generator G1	38.01	53.88
Synchronous generator G2	12.74	18.22
The second type of the grid configurations		
Synchronous generators G1 and G2	42.00	59.49
System and synchronous generator G1	45.17	63.98
System and synchronous generators G1 and G2	48.14	68.28

During the examined industrial grid short circuit, the induction motors operating on the 6 kV switchboard do not have an impact on the transient current of the SR2 reactor. They only increase the total transient currents in the fault location.

In the case of configurations of the first type, where only one power source is connected to Section 1 of substation B, the fault currents are not dangerous for this reactor.

The most severe fault conditions for the SR2 current-limiting reactor are observed for the second type of grid configurations, while different combinations of power sources are connected to Section 1 of substation B, and high current peaks through the analysed SR2 reactor cause significant mechanical stress on its windings.

The worst-case scenario is when all power sources are connected to Section 1 of Substation B, both the internal synchronous generators and the external power system. Analysing the values of current peaks in the SR2 current-limiting reactor coils, obtained by simulating for this type of grid configuration and presented in Table 3 and Figure 6, it can be seen that these values are close to the reactor rating short circuit current peak value. Such current peaks, as shown by operating experience, can cause damage to the reactor structure due to its mechanical fatigue caused by inrush currents of lower amplitude through the repeated starts of the induction motors and generators in the industrial grid during the previous operation of the reactor.

The example given in the article is intended to draw the attention of operating and design institutions to the problem of sizing current-limiting reactors in specific operating conditions. The requirement for transient analysis aiming for the correct definition of critical current peaks is especially relevant for industrial grids containing powerful asynchronous motors and generators. In our opinion, at the design stage calculations, it is also necessary to account for the number of repeated peaks of inrush currents of a given amplitude to ensure the reliable operation of the reactor at the declared rating values. Such an approach has been adopted, for example, for air-core filter reactors, as discussed in detail in [18,19,21]. Once the values of the transient peaks and their expected number are determined, the methodology recommended in the Standard [2] can be used to calculate the required mechanical short circuit current rating. The essence of this methodology is the use of weighting factors [2] when calculating the mechanical strength of a reactor designed to operate under conditions of repetitive inrush currents. The result is a more robust reactor design, electrically represented as a higher short circuit current rating than that derived from standard requirements for limiting short circuit current and busbar voltage drop.

4. Conclusions

The article describes the concern of operating current-limiting reactors in industrial power grids containing powerful induction motors and internal synchronous generators. During long-term operation, certain reactors of the grids are subjected to frequently repeated inrush currents under energizing induction motors and generators. Although these inrush currents have a lower amplitude than the reactor mechanical short circuit current rating, they cause a gradual weakening of the mechanical design. Repetitive electrodynamic blows during long-term operations cause a well-known phenomenon, which is commonly called "mechanical fatigue" in the reactor design. As a result of the following short circuit in the grid with a transient current peak amplitude close to the rating one, the reactor can be damaged. It can be claimed that the previous operation of the reactor under the described operating conditions caused a certain derating (decreasing) of such a parameter as the mechanical short circuit current rating.

To support the described statement on the possible impact of the previous operation of the air-core limiting reactor, the authors carried out a detailed simulation of a true industrial grid and its operating conditions during a short circuit that caused damage to the reactor. Compared with other reactors of the power system, the analysed reactor has run under specific (the most severe) operating conditions, as it was subjected to the frequent starting currents of powerful induction motors. The simulation results showed that, under these conditions, the short circuit current peak that caused damage to the reactor was lower than the nameplate rating one, which confirms the thesis about the mechanical fatigue of the reactor design. In other words, damage to the reactor by a short circuit current with a peak current lower than specified on the nameplate denotes that, at the time of the accident, the mechanical strength of the reactor was significantly weakened due to the previous inrush current blows and did not meet the declared guarantees.

The studies in this paper indicate that, for industrial grids containing powerful induction motors and internal synchronous generators, the sizing of current-limiting reactors should be carried out not only by the standard requirements for limiting short circuit currents and ensuring residual voltage on busbars but also considering their location in the network, the number of motor and generator starts and the possible number of faults and grid configurations that provoke the highest amplitudes of short circuit currents. Transient analysis should be used to identify critical current peaks in such industrial grids correctly.

Author Contributions: Conceptualization, Y.V. and M.G.; methodology, Y.V. and M.G.; investigation, Y.V. and M.G.; validation, Y.V. and M.G.; writing—original draft, Y.V. and M.G.; writing—review and editing, Y.V. and M.G.; supervision, Y.V.; project administration, Y.V.; funding acquisition, Y.V. All authors have read and agreed to the published version of the manuscript.

Funding: The present study was financially supported by AGH University of Science and Technology (Grant AGH No. 16.16.210.476) and by the program "Excellence Initiative—Research University" for the AGH University of Science and Technology.

Data Availability Statement: Not applicable.

Conflicts of Interest: The authors declare no conflict of interest.

References

1. Agrawal, K.C. Selection of power reactors. In *Electrical Power Engineering. Reference and Applications Handbook*, 1st ed.; CRC Press: Boca Raton, FL, USA, 2007; pp. 971–980.
2. *IEEE Std C57.16™2011*; IEEE Standard for Requirements, Terminology, and Test Code for Dry-Type Air-Core Series-Connected Reactors. IEEE Power & Energy Society: Piscataway, NJ, USA, 2012. [CrossRef]
3. *IEEE Std 551™-2006*; IEEE Recommended Practice for Calculating Short-Circuit Currents in Industrial and Commercial Power Systems. IEEE-SA Standards Board: Piscataway, NJ, USA, 2006.
4. *Int. Std IEC 60076-6*; Power Transformers—Part 6 Reactors. International Electrotechnical Commission: Geneva, Switzerland, 2007.
5. Swati Devabhaktuni, H.S.; Jain, K.K.P. Optimum selection of inductor as a current limiter in AC power systems. In Proceedings of the 2016 IEEE Region 10 Symposium (TENSYMP), Bali, Indonesia, 9–11 May 2016; pp. 416–419. [CrossRef]
6. Amon, J.F.; Fernandez, P.C.; Rose, E.H.; D'Ajuz, A.; Castanheira, A. Brazilian Successful Experience in the Usage of Current Limiting Reactors for Short-Circuit Limitation. In Proceedings of the International Conference on Power Systems Transients (IPST'05), Montreal, QC, Canada, 19–23 June 2005. Paper No. IPST05-215.
7. Mehta, J.; Chudasama, P.; Sangdot, K.; Patel, H.; Patel, P. Reactor: A Vital Part of Electrical System. *Int. J. Eng. Sci. Comput.* **2017**, *7*, 4556–4561.

8. Caverly, D.; Pointner, K.; Presta, R.; Griebler, P.; Reisinger, H.; Haslehner, O. Air Core Reactors: Magnetic Clearances, Electrical Connection and Grounding of Their Supports. 2017. pp. 1–24. Available online: https://pdfslide.net/documents/air-core-reactors-magnetic-clearances-electrical-air-core-reactors-magnetic.html? (accessed on 25 November 2022).
9. Liu, Z.; Ouyang, S.; Geng, Y.; Wang, J. Study of genetic algorithm in the optimum design of air-core reactor. *Adv. Technol. Electr. Eng. Energy* **2003**, *22*, 45–49.
10. Shuiming, C.; Pengcheng, Y.; Wei, X.; Xiao, B.; Wei, W. Side effects of Current-Limiting Reactors on power system. *Int. J. Electr. Power Energy Syst.* **2013**, *45*, 340–345. [CrossRef]
11. Shoup, D.; Paserba, J.; Colclaser, R.G., Jr.; Rosenberger, T.; Ganatra, L.; Isaac, C. Transient recovery voltage requirements associated with the application of current-limiting series reactors. *Electr. Power Syst. Res.* **2007**, *77*, 1466–1474. [CrossRef]
12. Gegeng, Z.; Lin, X.; Xu, J.; Tian, C. Effects of Series Reactor on Short-circuit Current and Transient Recovery Voltage. In Proceedings of the 2008 International Conference on High Voltage Engineering and Application, Chongqing, China, 9–13 November 2008; pp. 524–526. [CrossRef]
13. Fating, Y.; Zhao, Y.; Lixue, C.; Wang, Y.; Junxiang, L.; Junjia, H.; Yuan, P. Thermal and Electromagnetic Combined Optimization Design of Dry Type Air Core Reactor. *Energies* **2017**, *10*, 1989. [CrossRef]
14. Antonov, A.S.; Glushkov, D.A. Research on Electromagnetic and Heat Processes in Dry Type Air Core Current-Limiting Reactors. In Proceedings of the IEEE Conference of Russian Young Researchers in Electrical and Electronic Engineering (ElConRus), St. Petersburg, Russia, 1–3 February 2017; pp. 1471–1476. [CrossRef]
15. Fating, Y.; Zhao, Y.; Jun, X.L.; Yong, W.; Wen, X.M.; Jun, J.H. Research on temperature field simulation of dry type air core reactor. In Proceedings of the 20th International Conference on Electrical Machines and Systems (ICEMS), Sydney, NSW, Australia, 1–14 August 2017; pp. 1–5. [CrossRef]
16. Xinlao, W.; Chunlai, Y.; Jianquan, L.; Heqian, L.; Hongda, Z. Effect of Thermal Aging on Interturn Insulation Properties of Dry-type Air-core Power Reactors. In Proceedings of the 5th Asia Conference on Power and Electrical Engineering (ACPEE), Chengdu, China, 4–7 June 2020; pp. 1590–1594. [CrossRef]
17. Xiuke, Y.; Zhongbin, D.; Yanli, Z.; Cunzhan, Y. Fluid-Thermal Field Coupled Analysis of Air Core Power Reactor. In Proceedings of the 6th International Conference on Electromagnetic Field Problems and Applications, Dalian, China, 19–21 June 2012; pp. 1–4. [CrossRef]
18. Bonner, J.A.; Hurst, W.M.; Rocamora, R.G.; Dudley, R.F.; Sharp, M.R.; Twiss, J.A. Selecting ratings for capacitors and reactors in applications involving multiple single-tuned filters. *IEEE Trans. Power Deliv.* **1995**, *10*, 547–555. [CrossRef]
19. Dudley, R.F.; Fellers, C.L. Special Design Considerations for Filter Banks in Arc Furnace Installations. *IEEE Trans. Ind. Appl.* **1997**, *33*, 226–233. [CrossRef]
20. Warecki, J.; Gajdzica, M. Energizing arc furnace transformer in power grid involving harmonic filter installation. *Prz. Elektrotechniczny* **2015**, *4*, 64–69. [CrossRef]
21. Varetsky, J.; Gajdzica, M. The procedure for selecting the ratings of capacitor banks and reactors of the filtering systems. *Prz. Elektrotechniczny* **2020**, *3*, 77–81. [CrossRef]
22. Lubośny, Z.; Klucznik, J.; Dobrzyński, K. Problems of Selecting Protection for Shunt Reactors Working in Extra-High Voltage Grids. *Acta Energetica* **2016**, *2/27*, 139–143. [CrossRef]
23. Zhao, Y.; Chen, F.; Kang, B.; Ma, X. Optimum design of dry-type air-core reactor based on the additional constraints balance and hybrid genetic algorithm. *Int. J. Appl. Electromagn. Mech.* **2010**, *33*, 279–284. [CrossRef]
24. Yuan, Z.; He, J.; Pan, Y.; Yin, X.; Ding, C.; Ning, S.; Li, H. Thermal analysis of air-core power reactors. *Int. Sch. Res. Not.* **2013**, *2013*, 865015. [CrossRef]
25. Przygrodzki, M.; Rzepka, P.; Szablicki, M. Criteria and decisive algorithm of a zero-sequence current directional ground fault protection on HV and LV overhead lines. *Energetyka* **2017**, *7*, 442–447.

 *energies*

Article

Hardware-in-the-Loop Testing for Protective Relays Using Real Time Digital Simulator (RTDS)

Gaurav Yadav [1], Yuan Liao [1,*] and Austin D. Burfield [2]

[1] Department of Electrical and Computer Engineering, University of Kentucky, Lexington, KY 40506, USA
[2] Schweitzer Engineering Laboratories, Inc., Columbus, OH 43035, USA
* Correspondence: yuan.liao@uky.edu

Abstract: With the increasing size and complexity of power systems, it is crucial to have an effective protection system in place to ensure its reliability. One of the important components of the protection system are relays. It is important for a relay to operate dependably and securely so that any fault can be cleared in time to minimize damages to the power network. However, it is important to test a relay in a realistic environment before commissioning it to the network. Testing a relay in the actual network can be expensive with limited fault scenarios. Hence, Hardware-in-the-Loop (HIL) testing is an efficient method to perform closed-loop testing of a relay since numerous fault cases can be simulated to provide a realistic operating environment for the relay under test. This paper sheds light on the HIL testing done for protective relays using a sample distribution system using RTDS. Two SEL-351 relays have been used in this experiment, and proper settings for the relays are calculated for coordination. The paper also describes the procedure of configuring the relay and other RTDS components crucial for interfacing of the relay with RTDS. After test setup, a pre-fault, fault, and post-fault analysis was done for the system. The results obtained from these analyses are cross-checked with the event history of the two SEL-351 relays, obtained with AcSELerator Quickset software. This paper provides thorough information for researchers to replicate the presented study or to develop new HIL experiments. It can also help in developing a fundamental understanding of the HIL testing setup that can be further applied to a more complex power system.

Keywords: Hardware-in-the-Loop test; real time digital simulator; protective relay; overcurrent protection

Citation: Yadav, G.; Liao, Y.; Burfield, A.D. Hardware-in-the-Loop Testing for Protective Relays Using Real Time Digital Simulator (RTDS). *Energies* **2023**, *16*, 1039. https://doi.org/10.3390/en16031039

Academic Editor: J. C. Hernandez

Received: 29 November 2022
Revised: 11 January 2023
Accepted: 13 January 2023
Published: 17 January 2023

1. Introduction

The power system in the modern era has evolved into large and complex interconnected systems which is composed of nonlinear and dynamic elements. The development of smart grids has led to wide-scale integration of numerous elements at different operation levels of the grid. This has created complexities in the system and operational challenges that can be resolved using real-time digital computation, and advanced sensors and communication technologies [1–3]. With the increased sophistication in the methods and equipment for testing, power system simulation tools can play a major role in the testing and analysis of the system.

Power system simulation methods are mainly classified in two types: offline simulation and real-time simulation. Offline simulation methods have limited ability to provide an accurate simulation for the behavior of real-life equipment and practical systems [4]. The real-time simulation technology can perform numerous complex operations on the power system as they consist of high-speed processors and can provide higher efficiency and performance. Recently, Hardware-in-the-Loop (HIL) simulation has gained popularity, as it can wedge the gap between the behavior of the simulated and real system. In HIL testing, the power system under study is represented by a simulation model on a real time simulation platform, and the equipment under test is outside of the model but is linked to the simulation model through special Digital to Analog and Analog to Digital cards and necessary sensors and amplifiers [4,5]. One of the biggest advantages of HIL is that it can reduce the risk of field errors in actual scenarios [6]. HIL testing allows for simulating a real scenario of the power system, such as a fault situation, and provides accurate data on handling such situations in the real world [7]. HIL simulation platform performs time domain simulation to generate transient waveforms that are encountered in many protection applications [8–10].

With the increased energy demand, it has become crucial to procure the demand in an economically and environmentally sustainable way. Due to this, the stability and safety limit of modern power grids are being pushed. To handle the surge and the pressure, utilities are adding sophisticated new schemes and devices to their networks [11]. Whenever a new equipment and controller is added to the system, it needs to be tested and tuned before adding it to the network. In this situation, HIL simulation can be used for analyzing the new equipment in the simulated model before commissioning in the field [4,5,12].

Over the years, real-time simulation has become popular among manufacturers and utilities to assist them in the study of behavior and operation of the power system, performing the closed-loop testing of new equipment, and developing new protection and control functions [11].

Out of all the functions performed by a real-time simulation system, this paper is concerned with performing closed-loop testing of directional overcurrent relays using RTDS. Closed-loop testing of equipment in a model power system can mimic the characteristics of the system in a realistic scenario. Hence, readily decipherable information can be obtained on the performance of the equipment. In closed-loop testing of the relay, the RTDS outputs voltage and currents signals that are fed to the relay, and the relay provides an output, i.e., a tripping signal to the RTDS to open the circuit breaker in the RTDS model to disconnect the faulted section, and the simulation continues with the updated model [13]. The closed loop setting is ideal for testing the coordination between multiple relays.

A directional overcurrent relay is one of the primary safety equipment used in a networked, non-radial electrical distribution system to ensure the reliable and efficient operation of the power system. This relay monitors the current flowing through the protected element and generates a trip signal to the circuit breaker in the case of an occurrence of a fault, i.e., when the current exceeds the value of pickup current in the relay [14].

Numerous studies have been conducted in the field of HIL testing. These studies have been performed for various power system equipment using different real-time simulation software. F. R. Adegbohun et al. conducted a HIL testing for a grid-connected PV system consisting of SEL-421 relay, to test the hardware compatibility and model reliability [7]. A performance analysis of relays was done by M. Afshar et al. for detection of high impedance faults using a RTDS. A 15-bus system was modeled in RSCAD with an arc model for the high impedance fault [15]. The authors of [16] proposed an improved numerical algorithm based on PSCAD/EMTDC simulation to eliminate the overreach issue in distance relay caused due to transients of the coupling capacitor voltage transformer. The authors proposed a prototype for distance relay that was tested using RTDS. The proposed relay had a faster operation speed and higher reliability. Y. S. Cho et al. designed a RSCAD model containing multi-function relay and mimic board to improve the understanding of the technology associated with the protection and operation of the power system [17]. K. H. So et al. performed a closed-loop testing for differential and distance relay using EMTP [18]. O. Dias et al. [19] and T. Ledwaba et al. [20] focused on the HIL testing of transmission network. In [19], a 500 kV Brazilian transmission network was protected using SEL-421 distance relays. The simulation was performed for two fault cases: permanent and transient faults. The results were also cross-checked with the available field data. The HIL testing using ABB RED 670, a differential relay for a 400 kV, 50 Hz transmission network was performed in [20]. They also studied the system dynamics with additional equipment, such as series capacitor. A non-linear programming based adaptive protection scheme for coordination of directional overcurrent relays in a 7-bus system was proposed in [21]. The results were validated using RTDS. Reference [22] proposed a coordination algorithm for advanced distance, and an over current relay was proposed that was validated using RTDS and Avera hardware relays. In [23], a flywheel energy storage system (FESS), a primary energy storage component of an energy magazine, was tested using a Power Hardware-in-the Loop (PHIL) simulation. A 3.5MJ, 320kW FESS was virtually interfaced with a simulated model of a shipboard. M. Stifter et al. presented a PMU-RTS-HIL testbed for validation and testing of multiple PMUs. Various HIL platforms exist, such as Opal-RT, Typhoon, and RTDS. Opal-RT was used for the aforesaid task [24]. Reference [25] illustrated recent best practices in the field of development of real-time simulator, from the perspective of RTDS.

This paper uses RTDS to perform HIL tests, and it is expected that the method is readily extendable to other real time simulation platforms. The paper provides a thorough illustration of calculating scaling factors of GTAO card of RTDS, cable connection between digital I/O cards of RTDS and low-level interface of relay, connection with power amplifiers, and consideration of amplification factor of amplifiers in calculating GTAO scaling factors. It is crucial that these tasks are performed correctly to have a successful experiment.

The rest of the paper is organized as follows: Section 2 describes the RSCAD model developed in RTDS for a five-bus system selected for this experiment. Section 3 describes the procedure to configure the SEL-351 relays before they are used for HIL testing. Section 4 explains the different relay schemes used in the HIL testing. Section 5 discusses the key components and procedures in interfacing of the relays with the RTDS. Section 6 presents the findings achieved for the work carried out in Sections 2–5. Section 7 concludes the methodology used for this research.

2. Development of a RTDS Model for the Five-Bus System

This section describes the development of an RTDS model of a five-bus system. The model is developed in power mode. Additionally, crucial components for interfacing relays under testing conditions with the model simulated with the RTDS will be shown.

2.1. Five-Bus System

Figure 1 depicts a five-bus system considered for protection. R1 and R2 are the two SEL-351 relays (i.e., directional overcurrent relays under study). SEL-351 relays can protect lines and equipment using phase, negative sequence, residual-ground, and neutral-ground overcurrent elements with directional control. The modeled system is a sample 12.47 kV (L-L RMS) overhead power distribution system. The lengths of feeders are in miles, and the load ratings are given in kVA. A power factor of 0.9 is assumed for all the loads. A three-phase fault (F) has been introduced in the system for the relays under study.

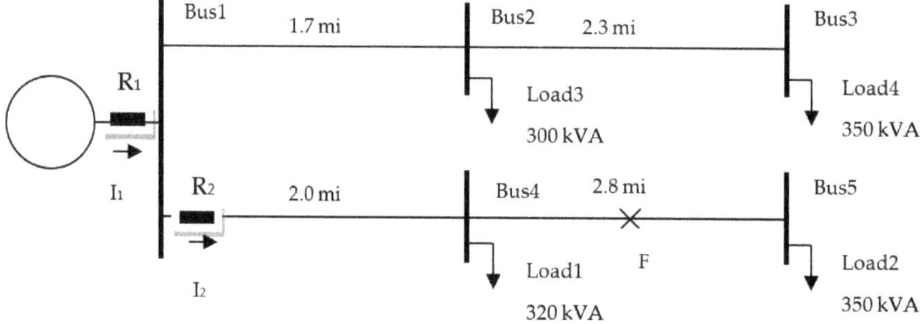

Figure 1. System considered for protection.

2.2. RTDS Model

The five-bus system shown in Figure 1 has been modeled for RTDS using the RSCAD software, as shown in Figures 2 and 3. The model has been divided into two modules for illustration purpose. Module 1 shows the RSCAD model developed for the five-bus system, and module 2 shows how the RTDS model interfaces with the relays. In reality, it is a single model containing both modules to perform the HIL testing.

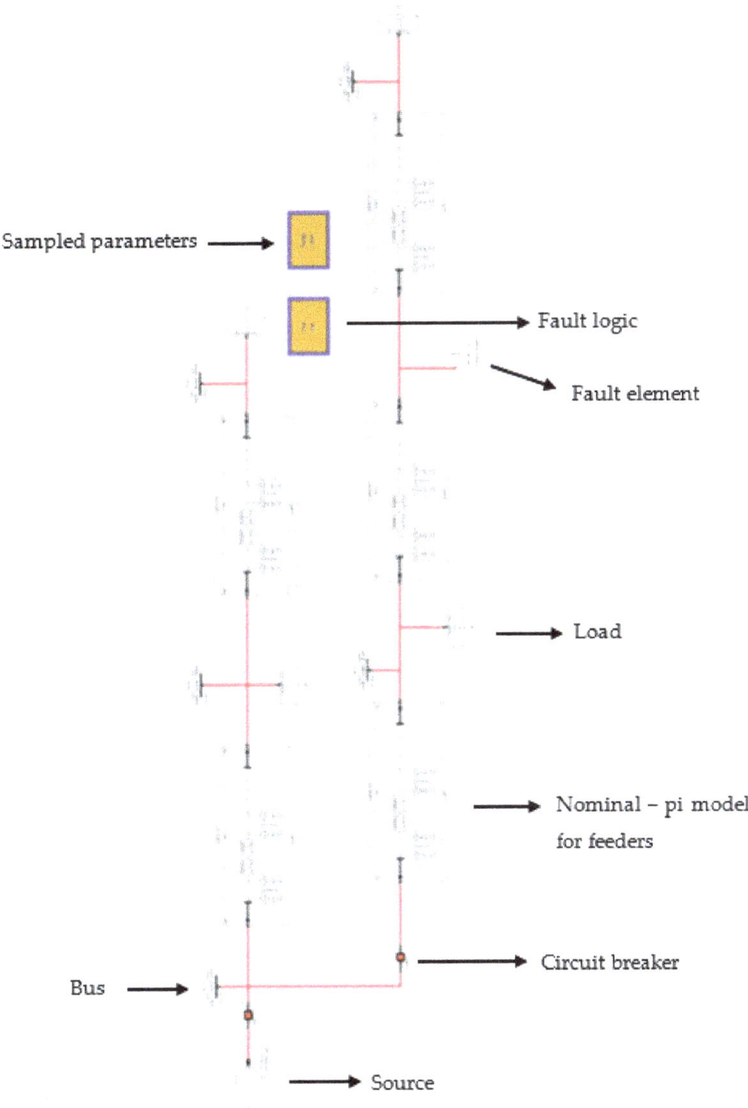

Figure 2. Module 1 of the RSCAD model.

The feeder lines shown in Figure 2 are labelled using the pi-section element in RSCAD. The feeder line impedances between each bus have been provided as input to the relevant pi-section model connecting them. A three-phase sinusoidal source of rating 12.47 kV line-to-line RMS voltage has been selected for the model. Four grounded R-L loads were connected to buses 2, 3, 4, and 5. A circuit breaker was added between source and bus 1 and another between bus 1 and bus 4. The feeder impedance matrix in ohms/mile was as follows:

$[0.3465 + \text{j}1.0179 \; 0.1560 + \text{j}0.5017 \; 0.1580 + \text{j}0.4236$

$0.1560 + \text{j}0.5017 \; 0.3375 + \text{j}1.0478 \; 0.1535 + \text{j}0.3849$

$0.1580 + \text{j}0.4236 \; 0.1535 + \text{j}0.3849 \; 0.3414 + \text{j}1.0348]$.

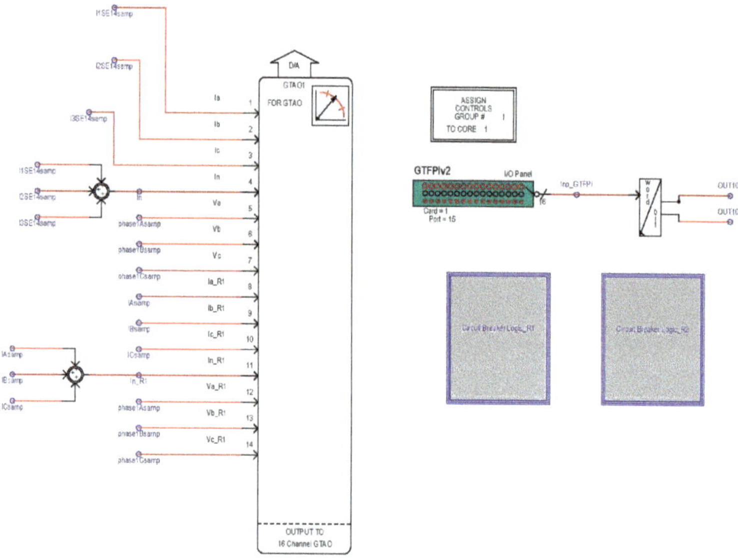

Figure 3. Module 2 of the RSCAD model.

The pre-fault rms current at R1 was 60.2 A, and fault current was 1770.0 A. The pre-fault rms current at R2 was 30.6 A, and the fault current was 1758.7 A.

The instantaneous overcurrent relay settings were calculated as follows. At R1, the tripping current was set as three times the pre-fault current: $60.2 \times 3 \; / \; 120 = 1.51$ A secondary, where 120:1 was the CT ratio when using a CT tapped at 600:5. The fault current will be $1770 \; / \; 120 = 14.75$ A secondary.

At R2, the tripping current was set as three times the pre-fault current: $30.6 \times 3 \; / \; 120 = 0.77$ A secondary, where 120:1 was the CT ratio for a CT tapped at 600:5. The fault current will be $1758.7 \; / \; 120 = 14.66$ A secondary, which will be used in setting the time dial for the inverse-time overcurrent protection elements, to achieve a proper coordination.

3. SEL-351 Relay Configuration

To configure the SEL-351 relay, SEL's AcSELerator Quickset software was used. In this study, two SEL-351 relays were chosen. The configuration was done as follows. First, the communication between the computer and the relay was established. Then the as-found relay settings were read to the computer. In the 'General settings', the CT and PT ratio were set to 120 and 60 (12,470 / 60 = 207 V secondary L-L), respectively, for both the relays. The parameters in 'Line Settings and Fault locator' section, as shown in Figure 4, needed to be calculated as follows:

$$Self\ impedance\ (Z_s)$$
$$= average\ of\ the\ diagonal\ elements\ of\ feeder\ impedance\ matrix \qquad (1)$$
$$= \frac{Z_{11} + Z_{22} + Z_{33}}{3}$$

$$Mutual\ impedance\ (Z_m)$$
$$= average\ of\ non$$
$$- diagonal\ elements\ of\ feeder\ impedance\ matrix \qquad (2)$$
$$= \frac{Z_{12} + Z_{21} + Z_{13} + Z_{31} + Z_{23} + Z_{32}}{6}$$

$$\text{Zero sequence primary impedance } (Z_{0p}) = Z_s + (2 * Z_m) \tag{3}$$

$$\text{Positive sequence primary impedance } (Z_{1p}) = Z_s - Z_m \tag{4}$$

$$\text{Zero sequence secondary impedance } (Z_{0s}) = Z_{0p} * \frac{CT\ ratio}{PT\ ratio} \tag{5}$$

$$\text{Positive sequence secondary impedance } (Z_{1s}) = Z_{1p} * \frac{CT\ ratio}{PT\ ratio} \tag{6}$$

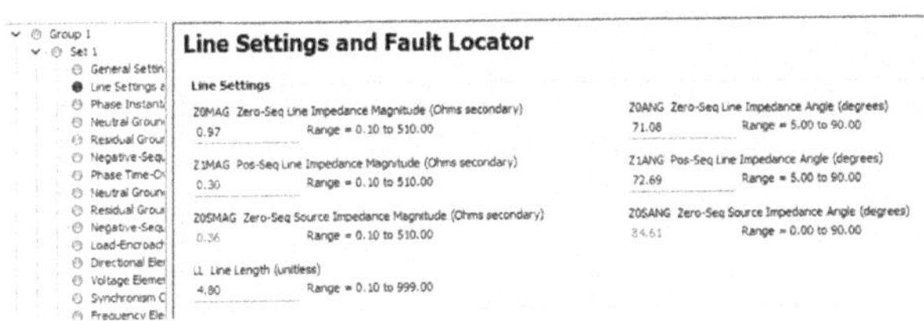

Figure 4. Entering parameters for 'Line settings and Fault locator' section.

The pickup current of 50P1P level 1 in the 'Phase Instantaneous Overcurrent Elements' and the value of '67P1D level 1' in the 'Phase Definite-Time Overcurrent Elements' can also be set as shown in Figure 5. Settings have been provided for R1, shown in Figure 5. R2 can be set in a similar way. The 50P1P level for R2 will be less than R1 (0.77 in this case) as R2 is supposed to act first. 67P1D level for R2 will be set to zero as R2 is operating under instantaneous overcurrent scheme.

Phase Instantaneous Overcurrent

E50P Enable Phase Overcurrent Elements

E50P Enable Phase Overcurrent Elements

1 Select: N, 1-6

Phase Instantaneous Overcurrent Elements

50P1P Level 1 (Amps secondary)
1.51 Range = 0.25 to 100.00, OFF

50P2P Level 2 (Amps secondary)
OFF Range = 0.25 to 100.00, OFF

50P3P Level 3 (Amps secondary)
OFF Range = 0.25 to 100.00, OFF

50P4P Level 4 (Amps secondary)
OFF Range = 0.25 to 100.00, OFF

50P5P Level 5 (Amps secondary)
OFF Range = 0.25 to 100.00, OFF

50P6P Level 6 (Amps secondary)
OFF Range = 0.25 to 100.00, OFF

Phase Definite-Time Overcurrent Elements

67P1D Level 1 (cycles in 0.25 increments)
4.00 Range = 0.00 to 16000.00

Figure 5. Setting 50P1P and 67P1D level for R1.

Another setting is the Trip logic equation, which will be 67P1T, 50P1, 51PT—to be explained in later sections. The Trip logic equation for R1 and R2 was set to 51PT under the IDMT overcurrent scheme in Figure 6. For R2, 50P1 was used under the instantaneous overcurrent scheme, and 67P1T was used for R1 under the definite-time overcurrent scheme. After entering the desired relay settings, the settings were written to the relay.

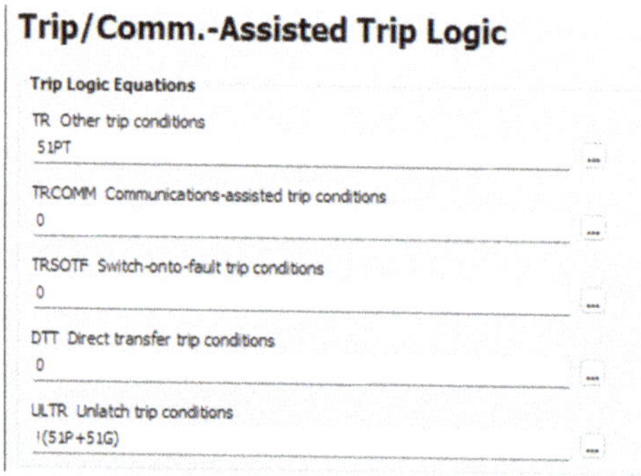

Trip/Comm.-Assisted Trip Logic

Trip Logic Equations

TR Other trip conditions
  51PT

TRCOMM Communications-assisted trip conditions
  0

TRSOTF Switch-onto-fault trip conditions
  0

DTT Direct transfer trip conditions
  0

ULTR Unlatch trip conditions
  !(51P+51G)

Figure 6. Trip logic equation for R1 and R2 under IDMT scheme.

4. Relay Schemes

This paper focuses on the following three different relay schemes.

4.1. Instantaneous Overcurrent Scheme

Under this scheme, there was no intentional time delay in operation of the relay. The relay acts as soon as the current exceeds the pickup current setting of the relay. Relay R2 in this experiment works on this scheme. The results shown for relay R2 in case 1 of Section 6.2 of this paper were obtained using the aforesaid scheme. The 50P1P level 1 element in the Phase Instantaneous Overcurrent Elements was set to three times the value of the pre-fault current as calculated in Section 2.2.

The Trip Logic Equation was set to 50P1.

4.2. Definite Time Overcurrent Scheme

The relay R1 used the definite time scheme. Under this scheme, a pick-up setting was selected for the relay R1 as done in the instantaneous overcurrent scheme. In addition to the pickup setting, a time-setting was also provided for R1. When the current exceeded the pickup setting, R1 was supposed to wait for the specified time setting before it could issue the trip signal. A time delay of 4 cycles was set for R1, i.e., 67P1D level 1 in the Phase Definite-Time Overcurrent Elements was set to 4.0 cycles. The results for R1 using this scheme can be seen in case 2 of Section 6.2 of this paper.

The Trip Logic Equation was set to 67P1T.

4.3. Inverse Definite Minimum Time (IDMT) Overcurrent Scheme

In inverse time scheme, the relay operation time was inversely proportional to the fault current, i.e., a faster operation for a bigger fault current. In this paper, this scheme was applied to both R1 and R2 in addition to the aforementioned two schemes. A coordination was done between the two relays, where the protected feeder had a primary protection (R2) and secondary protection (R1).

The following steps were required to set up a coordination scheme for the relays under study:

Step 1—When setting up coordination, start with the device that will be closest to the fault within the zone of protection, i.e., relay R2 in this study.

Step 2—Select the appropriate relay settings for both relays as shown in Table 1.

Table 1. Relay settings for coordination scheme.

Parameters	Relay R1	Relay R2
Pickup current setting (51PP)	1	0.5
IDMT curve (51PC)	U3	U3

Step 3—Calculate the operating time and time dial setting (51PTD) for both relay R1 and R2. Initially, the operating time of R2 will be calculated. The equation used to calculate the operating time is selected as per the IDMT curve (51PC) chosen for the relay. Since curve U3 has been selected for R2, the following formula (given in SEL-351's instruction manual) will be used to calculate the operating time:

$$t_{operate_{R2}} = 51PTD_{R_2} \times \left(0.0963 + \frac{3.88}{(M^2 - 1)} \right) \tag{7}$$

where, M is a multiple of the pickup current setting, and $51PTD_{R_2}$ is the time dial setting of R2. In this experiment, for R2, the value of M is 29.32 (14.66 / 0.5, where 0.5 is the pickup current setting of R2) (see Section 2.2), and the value of 51PTD is 1. Hence, the operating time of relay R2 will be as follows:

$$t_{operate_{R2}} = 1 \times \left(0.0963 + \frac{3.88}{(29.32^2 - 1)} \right) = 0.101 \; seconds \tag{8}$$

Step 4—To calculate the operating time of R1, determine the operating time of R2 and a Coordination Interval (CI) between both relays. CI is required to ensure proper coordination with upstream devices (R1 in this case). It ensures that R2 has enough time to operate before R1 begins to operate.

A CI typically consists of relay/element operate time (pickup/tolerance), relay output operate time, breaker operate time, and a safety margin. Generally, CI ranges between 0.2 and 0.5 s for a distribution system. For this experiment, CI was taken as 0.3 s. Thus, the operating time of R1 was given as follows:

$$t_{operate_{R1}} = t_{operate_{R2}} + CI = 0.101 + 0.3 = 0.401 \; seconds \tag{9}$$

Step 5—Using the parameters in step 3 and 4, calculate 51PTD of R1 as follows:

$$51PTD_{R_1} = \frac{t_{operate_{R1}}}{\left(0.0963 + \frac{3.88}{(M^2-1)} \right)} = \frac{0.401}{\left(0.0963 + \frac{3.88}{(14.75^2-1)} \right)} = 3.51 \tag{10}$$

where M is 14.75 (14.75/1, where 1 is the pickup current setting of R1), as calculated in Section 2.2.

The IDMT curve (i.e., operating time v/s fault current for relay R1 and R2 for a U3 curve setting) can be plotted as shown in Figure 7.

Different curve types, pickups, or time dials can be chosen to adjust the curves shapes to achieve different coordination margins.

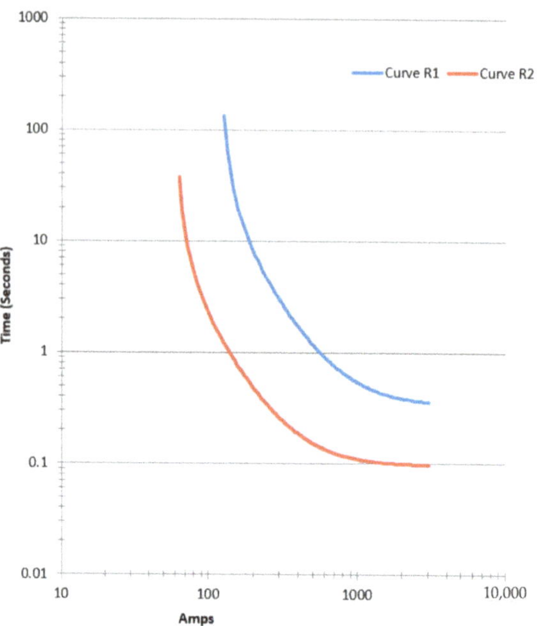

Figure 7. Relay operating time v/s Fault current for the inverse time scheme.

5. Interfacing Relays with RTDS

Figure 8 illustrates the setup for Hardware-in-the-Loop testing of SEL-351 relay using RTDS for the five-bus system. Figure 9 shows the functional block diagram for the HIL setup.

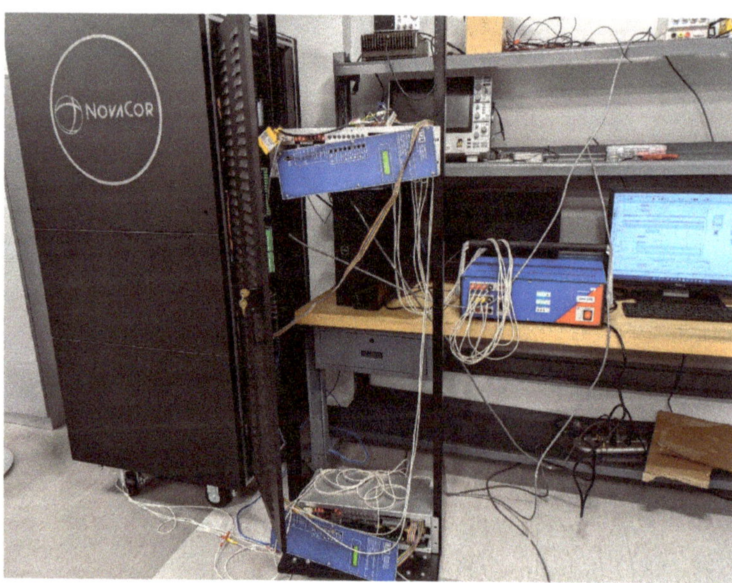

Figure 8. HIL setup for both SEL-351 relays, using RTDS.

As it can be observed from Figure 9, two circuit breakers, CB1 and CB2, were used in the power system under study. CB1 was used for relay R1, and CB2 was used for R2. To connect both relays, R1 and R2, with RTDS, we required the Giga-Transceiver Analogue Output (GTAO) card and Gigabit-Transceiver Front Panel Interface (GTFPI) card provided in the RTDS. The GTAO provided analog signals, which could then be fed to a relay through an amplifier or through the low- level interface of the relay. In this study, we performed experiments using both the low-level interface method and CMS 356 (i.e., the amplifier). For successful implementation, one must consider the relay's low-level interface scales, the current and potential transformer turns ratio, and the GTAO voltage and current scales. The GTAO provided the modeled system output to the relay, and when a fault was simulated, the relay sent a trip signal to the GTFPI card that in turn signaled the circuit breaker, which modified the simulation model, thus forming a closed-loop system. R1 here was to serve as a backup relay for faults within the protection zone of R2. When the fault occurs, R2 was supposed to act first to clear the fault. However, in case of failure of R2, R1 must act after a specified coordination delay, e.g., 4 cycles.

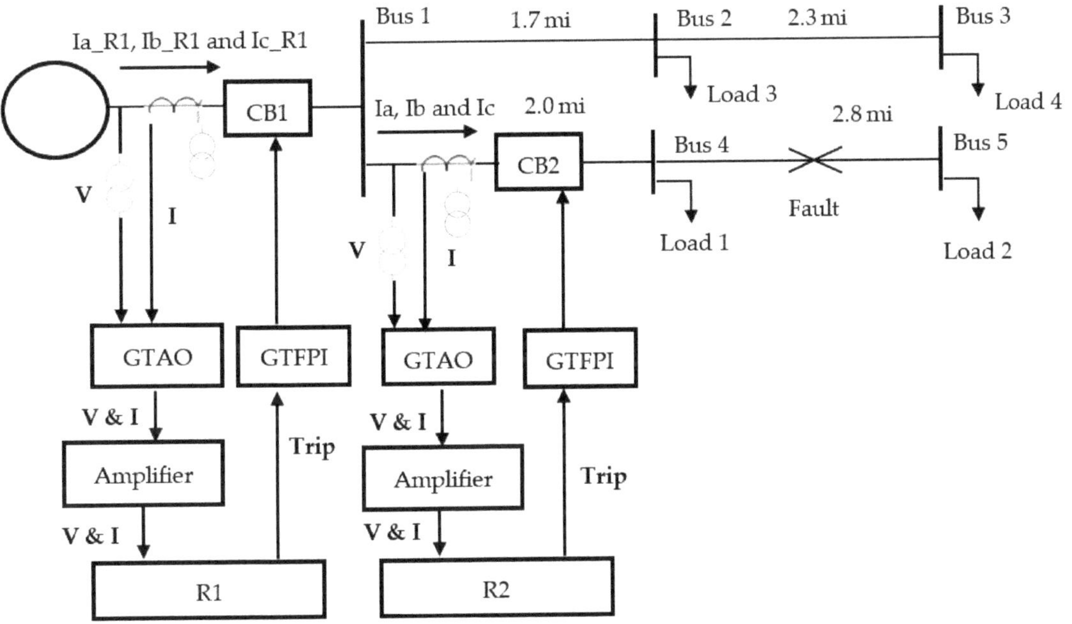

Figure 9. Block diagram for HIL setup.

In Figure 9, the Amplifier refers to external devices, such as the Omicron CMS 356, which can convert RTDS low current and voltage signals (0–5 V rms) to higher current and voltage signals (0–5 A, 0–120 V rms) that are at the same level of actual CT and VT signals that can be fed to the tested relays' current and voltage input terminals on the backside without using the relay's low-level signal interface.

Sections 5.1 and 5.2 provide a detailed description about the configuration of GTAO and GTFPI card, respectively, to perform their assigned operations shown in Figure 9.

5.1. GTAO Card

The GTAO card can be modelled in the RSCAD software using the GTAO element as shown in Figure 10. A GTAO card has 16 input ports. For our experiment, we needed 14 ports. In Figure 10, port 1 to 7 were the current and voltage inputs for relay R2, and port 8 to 14 were the current and voltage inputs for relay R1. The GTAO card took these inputs and provided analogue outputs to the relay input port. In Figures 9 and 10, Ia, Ib, and Ic were the input phase currents for R2, and Ia_R1, Ib_R1, and Ic_R1 were the input phase currents for R1. In and In_R1 were the neutral currents for R2 and R1, respectively. At bus 1, 'phase1Asamp' (Va), 'phase2Asamp' (Vb) and 'phase3Asamp' (Vc) were the phase voltages.

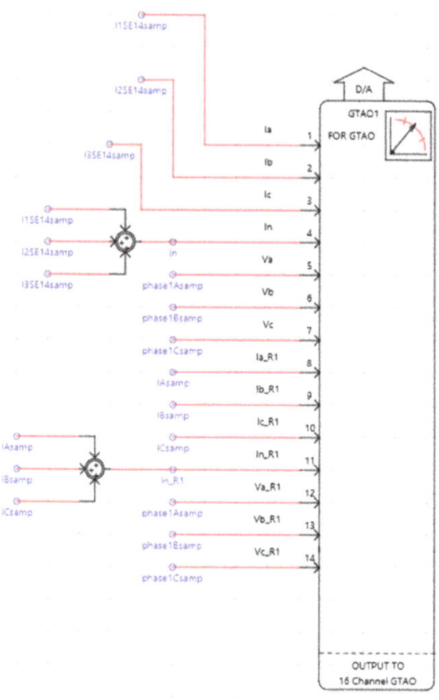

Figure 10. GTAO element in the RSCAD model.

The connections shown in Figure 10 were determined as per the highlighted connections given on the relay low-level test interface as shown in Figure 11. The rightmost parameter 'IA' given on the relay panel was considered as the connection on the input port 1 on the GTAO element shown in Figure 10 for relay R2 and so on. As observed from highlighted area in Figure 11, the first four parameters were for currents, and the rest were for voltages. Not following the given connection pattern will result in errors in values on the relay front panel.

Figure 11. Low–level test interface for SEL–351 relay.

The above-mentioned parameters were sampled using a sampling frequency of 7.68 kHz before being provided as an input to the GTAO card. This can be done using the sampler element provided in the RSCAD library as shown in Figure 12.

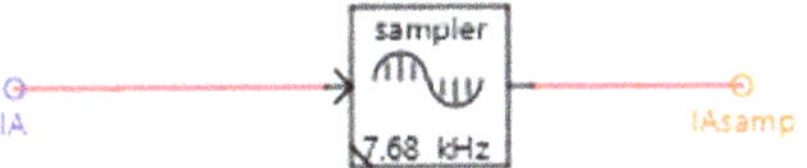

Figure 12. Sampling of the parameters.

5.1.1. GTAO Scaling When Using Relay's Low-Level Interface

The input parameters shown in Figure 10 needed to be scaled to a value within ± 5 V before being sent as input to the GTAO card. These values were internally scaled in the GTAO element. This section discusses the process of obtaining the scaling factor for input current and voltage parameters for GTAO. When providing the output of the GTAO card to the relay, it is also important to consider the scaling factors of the relay. The scaling factors available for SEL-351 relay can be seen in Table 2. Initially, the required GTAO output voltages were calculated using the relay scaling factors, Current Transformer Ratio (CTR), and Potential Transformer Ratio (PTR). Then, these voltages were used to calculate the current and voltage scaling factors for the GTAO card.

Figure 13 shows the connection on the relay end between the GTAO card of the RTDS and relay. J1 was the terminal connector for input module of the relay, and J10 was the terminal for the processing module of the relay. J10 received the input secondary current and voltage parameters from the GTAO. After receiving the input secondary values, the processing module converted these values to primary values as per the relay scaling factor, CTR and PTR. These primary values could be verified on the relay front panel. Figure 14 shows the connection made on the GTAO card to provide output current and voltage values to the relay. The bottom most port of the card was the first output port of the card. The connections shown here were made as per the connection diagram shown in Figures 10 and 11. Port 1 to 7 were the output values provided to relay R2, and port 8 to 14 were the output values provided to relay R1.

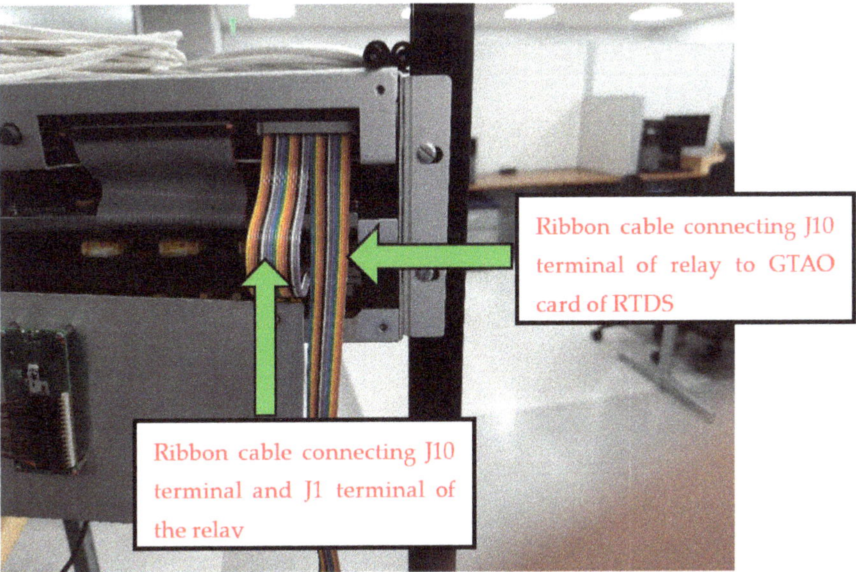

Figure 13. Connection on the relay end between relay and GTAO card.

Figure 14. Connection on the GTAO end between the relay and the GTAO card.

Connections shown in Figures 13 and 14 were done for both relays shown in Figure 15. We also tested using the CMS 356 to power Relay R2, as shown in Figure 8.

Table 2. Scaling factors for SEL-351 relay.

Input Channels (Relay Rear Panel)	Input Channel Nominal Rating	Input Value	Corresponding J1 Output Value	Scale Factor (Input/Output) (A/V or V/V)
IA, IB, IC, IN	1 A	1 A	100 mV	10
IA, IB, IC, IN	5 A	5 A	100 mV	50
IN	0.2 A	0.2 A	114.1 mV	1.753
IN	0.05 A	0.05 A	50 mV	1
VA, VB, VC, VS	150 V	67 VLN	1313.7 mV	51
VA, VB, VC, VS	300 V	134 VLN	1313.7 mV	102
Power (+, −)	48/125 Vdc or 125/120 Vdc	125 Vdc	1.25 Vdc	100

For the relays used in this study, a scaling factor of 50 for current and 102 for voltage in the relay was adopted.

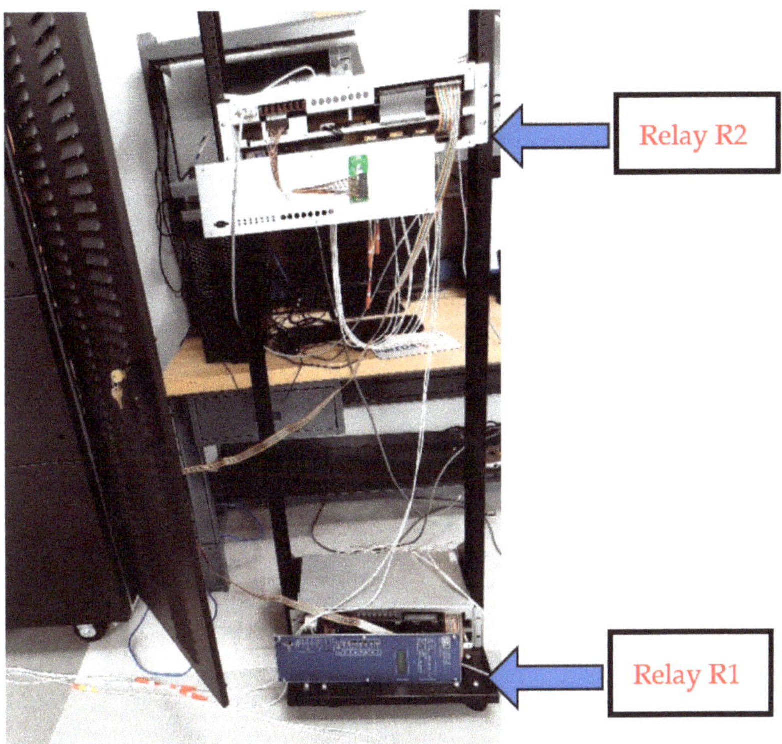

Figure 15. Relay R1 and R2 used for experiment.

Calculating Required Voltage Output from GTAO Card for Input Current Parameters

As an example, assume a fault current of 4000 A and follow the steps given below to calculate the required output from the GTAO card.

Step 1—Calculate the secondary side value of the fault current. In this case, the CTR is 120 (i.e., 120:1). Therefore,

$$Secondary\ relay\ current = \frac{Primary\ Current}{CTR} = \frac{4000A}{120} = 33.33\ A \tag{11}$$

This secondary relay current is then stepped down within the relay and converted to a mV signal used by the relay's processor based upon the scale factors shown in Table 2.

Step 2—To calculate the required output voltage from the GTAO card, divide the secondary current by the scaling factor of the relay. In our case, the scale was 50, as shown in Table 2. Therefore,

$$GTAO\ output\ for\ current\ (V_{GOUT1}) = \frac{Secondary\ relay\ current}{Relay\ current\ scale\ factor} = \frac{33.33}{50} = 0.67\ V \tag{12}$$

Calculating Required Output from GTAO Card for Input Voltage Parameters

For a pre-fault voltage of 10 kV, the required GTAO voltage output can be calculated as shown below.

Step 1—Calculate the value on the secondary side of the relay by dividing fault current by PTR. In our case, the PTR was 60 (i.e., 60:1). Therefore,

$$Secondary\ relay\ voltage = \frac{10000V}{60} = 166.67\ V \tag{13}$$

This secondary relay voltage is then stepped down within the relay to a mV signal used by the relay's processor based upon the scale factors shown in Table 2.

Step 2—To calculate the required output voltage from the GTAO card, divide the secondary voltage by the scaling factor of the relay. In this case, the scale factor was 102, as shown in Table 2. Therefore,

$$GTAO\ output\ for\ current\ (V_{GOUTI}) = \frac{Secondary\ relay\ voltage}{Relay\ voltage\ scale\ factor} = \frac{166.67}{102} = 1.63V \quad (14)$$

After obtaining the required GTAO outputs, calculate the scaling factor for the GTAO card to ensure that the above calculated GTAO outputs are obtained at the output of the GTAO card. The following formulas can be used to calculate the scaling factor for the voltage and current signal:

$$GTAO\ output\ for\ voltage\ (V_{GOUTV}) = \frac{5}{V_{sf}*1000} \times V_{actual}\ (in\ volts) \quad (15)$$

$$\Rightarrow V_{sf} = \frac{5}{V_{GOUTV} \times 1000} \times V_{actual}\ (in\ volts) = \frac{5}{1.63*1000} \times 10000 = 30.6$$

or a more general equation is shown below that can be used to calculate V_{sf}:

$$\frac{5}{V_{sf}*1000} \times V_{actual}\ (in\ volts) \times S_V \times PTR = V_{actual}\ (in\ volts) \quad (16)$$

where S_V is the relay voltage scale factor, which is 102. Therefore, we have

$$\frac{5}{V_{sf} \times 1000} \times V_{actual}\ (in\ volts) \times 102 \times 60 = V_{actual}\ (in\ volts)$$

from which V_{sf} can be derived

$$GTAO\ output\ for\ current\ (V_{GOUTI}) = \frac{5}{I_{sf} \times 1000} \times I_{actual}\ (in\ amperes) \quad (17)$$

$$\Rightarrow I_{sf} = \frac{5}{V_{GOUTI} \times 1000} \times I_{actual}\ (in\ amperes) = \frac{5}{0.67*1000} \times 4000 = 30$$

or a more general equation is shown below that can be used to calculate I_{sf}:

$$\frac{5}{I_{sf} \times 1000} \times I_{actual}\ (in\ amperes) \times S_I \times CTR = I_{actual}\ (in\ amperes) \quad (18)$$

where S_I is the relay current scale factor, which is 50. Therefore, we have

$$\frac{5}{I_{sf} \times 1000} \times I_{actual}\ (in\ amperes) \times 50 \times 120 = I_{actual}\ (in\ amperes)$$

from which I_{sf} can be derived.

Here, V_{sf} and I_{sf} are the scaling factors in the GTAO card for the voltage and current, respectively.

5.1.2. GTAO Scaling When Using CMS 356

The GTAO scaling was calculated differently when using CMS 356 in contrast with when using relay's low-level interface.

In this study, the voltage amplification factor was 60 V/V, and the current amplification factor was 6.4 A/V for the CMS 356.

The GTAO scaling was calculated as follows. We have the following relationship between the GTAO output and the actual voltage and current:

$$GTAO\ output\ for\ voltage\ (V_{GOUTV}) = \frac{5}{V_{sf} \times 1000} \times V_{actual} \tag{19}$$

$$GTAO\ output\ for\ current\ (V_{GOUTI}) = \frac{5}{I_{sf} \times 1000} \times I_{actual} \tag{20}$$

The actual voltage is also expressed as:

$$V_{actual} = V_{GOUTV} \times AMP_V \times PTR \tag{21}$$

The actual current is expressed as:

$$I_{actual} = V_{GOUTI} \times AMP_I \times CTR \tag{22}$$

Here, PTR is Potential Transformer Ratio, and CTR is Current Transformer Ratio.

The GTAO scaling factor for voltage can be obtained by substituting Equation (3) in Equation (5).

$$V_{actual} = \frac{5}{V_{sf} \times 1000} \times V_{actual} \times AMP_V \times PTR$$

Therefore,

$$V_{sf} = \frac{5 \times AMP_V \times PTR}{1000} \tag{23}$$

Equation (7) can be used to calculate the GTAO scaling factor for voltage. For a PTR equal to 60 and AMP_V of 60, the V_{sf} will be as follows:

$$V_{sf} = \frac{5 \times AMP_V \times PTR}{1000} = \frac{5 \times 60 \times 60}{1000} = 18$$

The GTAO scaling factor for current can be obtained by substituting Equation (4) in Equation (6).

$$I_{actual} = \frac{5}{I_{sf} \times 1000} \times I_{actual} \times AMP_I \times CTR$$

Therefore,

$$I_{sf} = \frac{5 \times AMP_I \times CTR}{1000} \tag{24}$$

Equation (8) can be used to calculate the GTAO scaling factor for the current. For a CTR equal to 120 and AMP_I of 6.4, the I_{sf} will be as follows:

$$I_{sf} = \frac{5 \times AMP_I \times CTR}{1000} = \frac{5 \times 6.4 \times 120}{1000} = 3.84$$

5.2. Obtaining Readings on the Relay Front Panel

After getting the required output from the GTAO card, to show the reading on the relay front panel, i.e., the primary side value, multiply the GTAO output by the current or potential transformer ratio and the scaling factor of the relay. In case of using relay's low-level interface, the formulae are shown below:

$$Relay\ voltage\ reading = V_{GOUTV} \times Relay\ scaling\ factor \times PTR \tag{25}$$

$$Relay\ current\ reading = V_{GOUTI} \times Relay\ scaling\ factor \times CTR \tag{26}$$

Table 3 shows the voltage output and scaling factors obtained for GTAO for our experiment. The current and voltage values shown in Table 3 were the pre-fault values obtained on the relay front panel using the GTAO scales shown in the table. The calculations were verified by the actual readings on the relay.

Table 3. Parameters for GTAO card.

Parameter	Magnitude (RMS)	CTR	PTR	Relay Scale	GTAO Output	GTAO Scale
Current	30 A	120	-	50	0.67 V	30
Voltage	7133 V	-	60	102	1.63 V	30.6

When using CMS 356, the Relay reading would be calculated as the product of GTAO output and the CMS 356 amplification factor and the PT and CT ratio.

5.3. GTFPI Card

The GTFPI card was used to transmit the trip signal from the relay to the circuit breaker in the RSCAD model, as shown in Figure 16. The relay trip signal is transmitted from the relay to the RTDS digital interface panel, as shown in Figures 17 and 18.

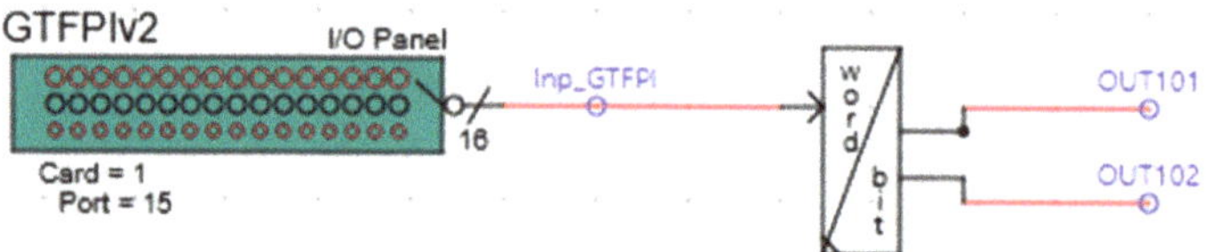

Figure 16. GTFPI element in RSCAD.

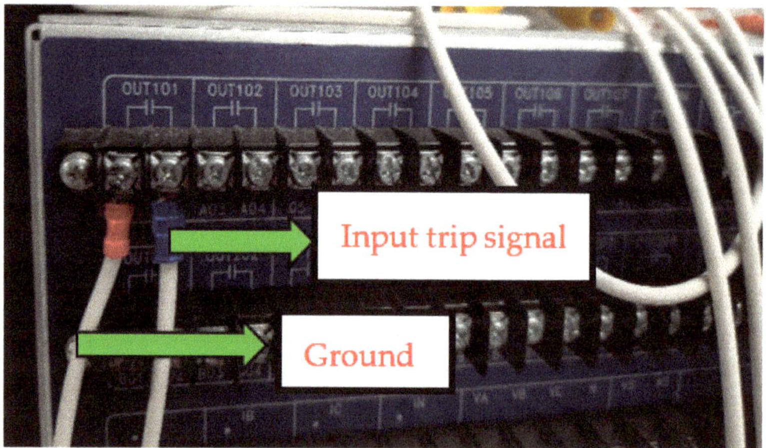

Figure 17. Connection on relay back panel with digital interface panel of RTDS.

In Figure 16, 'OUT 101' and 'OUT102' are the trip signals from R1 and R2, respectively. 'Inp_GTFPI' is the input signal being provided from the GTFPI element. The input signal is a single word that can be converted into multiple logic outputs using a word/bit converter. The outputs of this converter are digital trip signals, i.e., logical outputs. These trip signals are then provided as an input to the digital interface panel provided on the front panel of RTDS, as shown in Figure 18. If there is only one trip signal, the converter is not required as only one port of the digital interface panel is needed. In the case of multiple trip signals, this converter is required as only one GTFPI element can be used in a RSCAD model, and this experiment required more than one port of the digital interface panel. Here, each port was considered as one bit. Two input ports will capture the two trip signals. Hence, two bits were obtained using the converter.

Figure 17 shows the connections done on the relay back panel for providing input signals to the digital interface panel of RTDS. It can be observed from the figure that 'OUT101' terminal of the relay has two connections (i.e., one for trip signal and another for ground). However, in general, the first terminal of the relay was the input signal, and the second terminal was ground, unlike the relay shown in Figure 17.

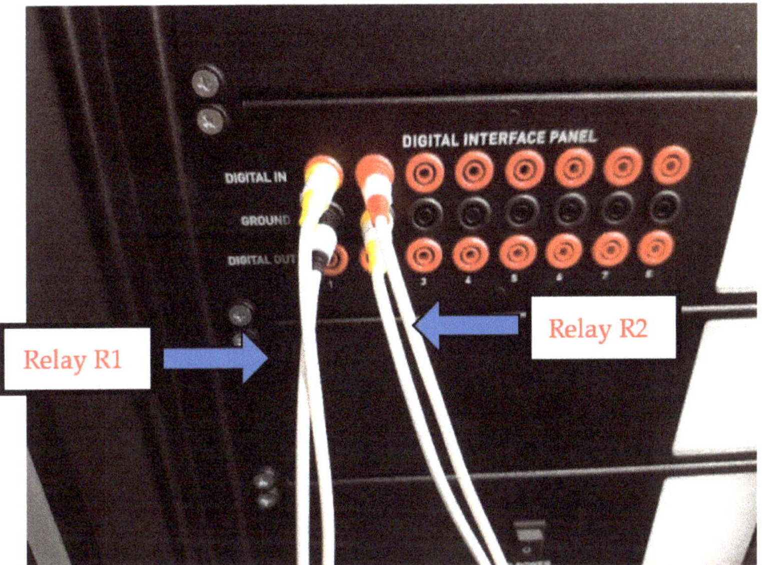

Figure 18. Digital interface panel of RTDS receiving signal from relay.

The connections coming from 'OUT 101' shown in Figure 17 were provided to the digital interface panel, as shown in Figure 18. As mentioned previously, two ports of the panel were used, i.e., one for relay R1 and one for relay R2. The input trip signal coming from relay was connected to the digital input channel on the panel, and the ground signal coming from relay was connected to the ground input on the panel.

5.4. Applying Faults to the System

Figure 19 shows the fault logic developed in RSCAD for the system. 'SW1', 'SW2', and 'SW3' were the switch elements in RSCAD, which denoted phases of the system. Each switch needed to be turned on to introduce the fault on the corresponding phase. For example, to apply a three-phase fault, all three switches must be turned on. Each fault type in RSCAD was assigned a binary value, which had a corresponding decimal value, as shown in Table 4. The three switches shown in Figure 19 were assigned a decimal value of 1, 2, and 4, respectively. The switch positions shown in Figure 19 indicate that a three-phase fault were introduced in the system. Alternatively, a three-phase fault can also be introduced by using a single switch having a decimal value of 7.

Table 4. Selecting fault type in RTDS.

Binary Value	Fault Type	Decimal
001	A-G	1
010	B-G	2
100	C-G	4

Table 4. *Cont.*

Binary Value	Fault Type	Decimal
110	AB-G	6
011	BC-G	3
101	AC-G	5
111	ABC-G	7

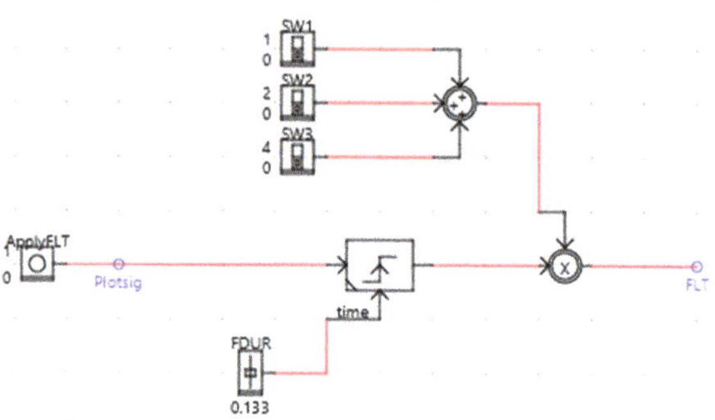

Figure 19. Fault logic.

The fault can be applied to the system using the push button titled 'ApplyFLT' as shown in Figure 19. The duration of the fault can be determined using a slider titled 'FDUR'. The output of this logic system was a control signal 'FLT', which was provided as an input control signal to the fault block in the RSCAD model.

5.5. Circuit Breaker Logic

Figure 20 shows the circuit breaker logic, where 'XCBR1open' and 'XCBR1close' were the push buttons used to open and close the circuit breaker contacts, respectively. The initial status of the breaker was 'closed', which was assigned a logical input of 1. 'OUT101' was the trip signal coming from relay R1. The first OR-gate was assigned for opening the circuit breaker and, hence, had two inputs, which were 'XCBR1open' and 'OUT101'. The output of the first OR-gate served as the input $(\overline{S})$ to the S-R flip flop. The second OR-gate was assigned for closing circuit breaker contacts. Since this experiment was not concerned with the reclosing of the circuit breaker after the occurrence of a fault, the control signal 'PROTIED_Reclose_R1' was assigned a zero value. The output of the second OR-gate served as the second input $(\overline{R})$ to the flip flop. The output (Q) of the flip flop passed through a timer block, which determined the duration of the operation of the circuit braker. The final output of the circuit breaker logic for relay R1, i.e., 'BRK_R1', was provided as an input to the circuit breaker element assigned for R1 in the model. Similar logic was used for relay R2 with its corresponding trip signal and breaker signal.

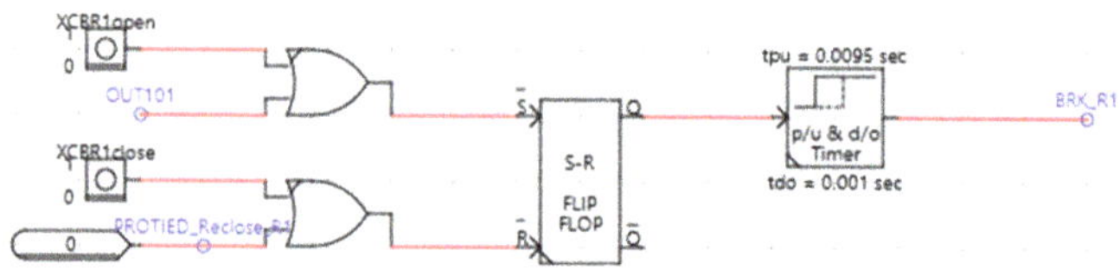

Figure 20. Circuit breaker logic for relay R1.

6. Results

As shown in Figure 1, relay R1 is connected between the source and bus 1, and relay R2 is connected between bus 1 and bus 4. Thus, the phase voltages on bus 1, phase currents from the source, and the phase currents on the sending end of the feeder between bus 1 and bus 4 are of particular interest. The results have been obtained for the system for pre-fault, during fault, and post-fault period. The x-axis in Figures 21–29 (excluding Figure 23) represents time in seconds, and the multiple y axes represent different signals, which have been explained in the previous sections.

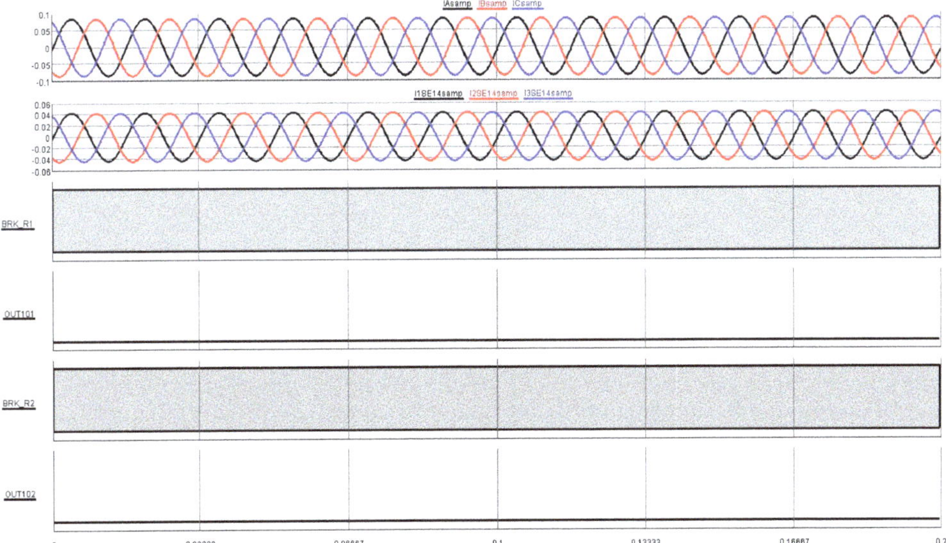

Figure 21. Current waveforms for pre-fault condition.

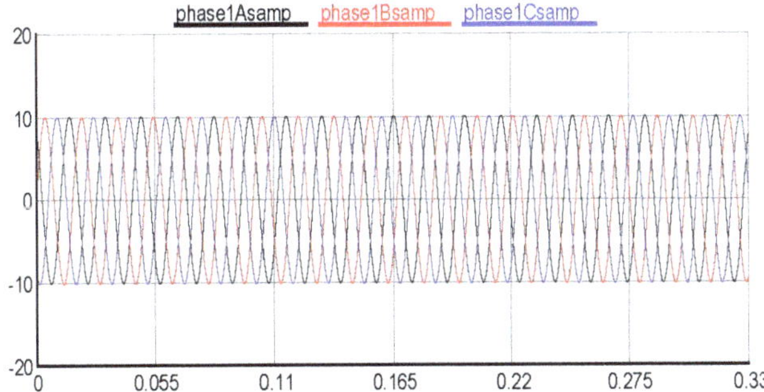

Figure 22. Voltage at bus 1 preceding the fault.

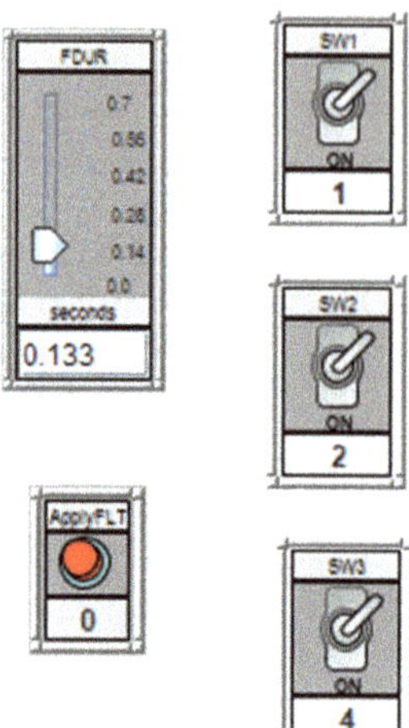

Figure 23. Applying fault to the system in RTDS environment.

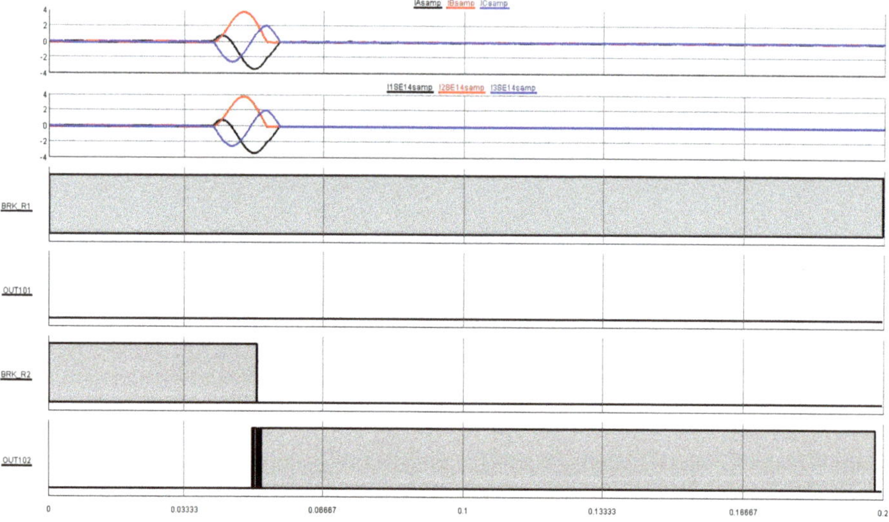

Figure 24. Waveforms for fault condition when R2 operates—case 1.

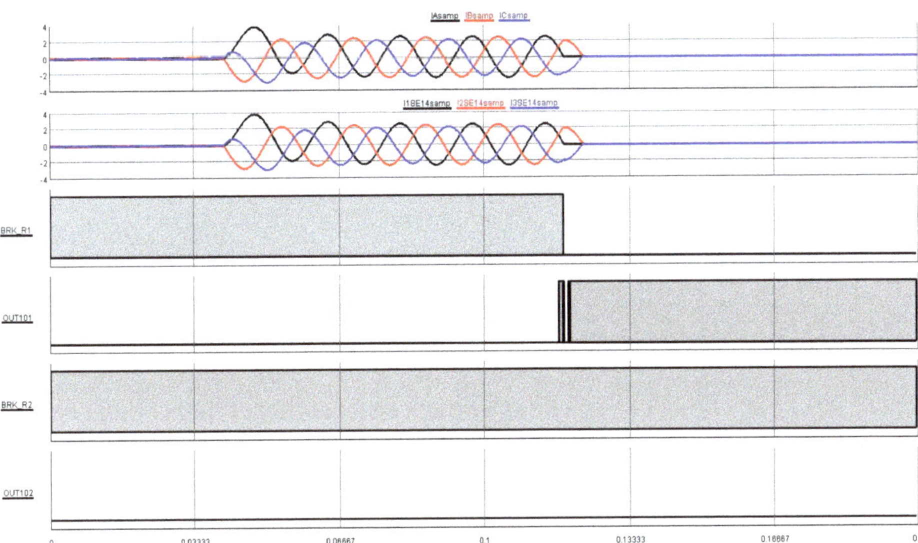

Figure 25. Waveforms for fault condition when R1 operates—case 2.

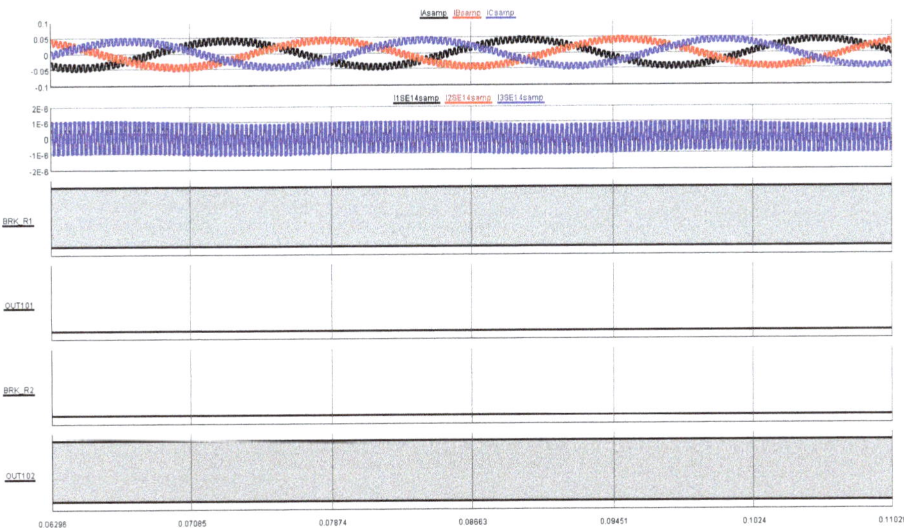

Figure 26. Waveforms for post-fault condition when R2 operates—case 1.

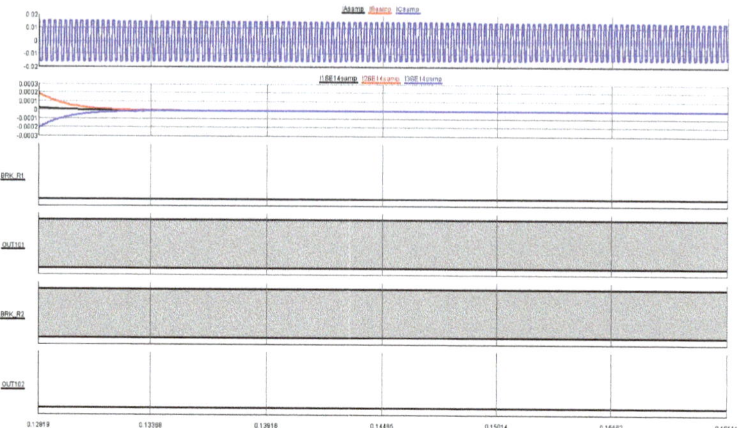

Figure 27. Waveforms for post-fault condition when R1 operates—case 2.

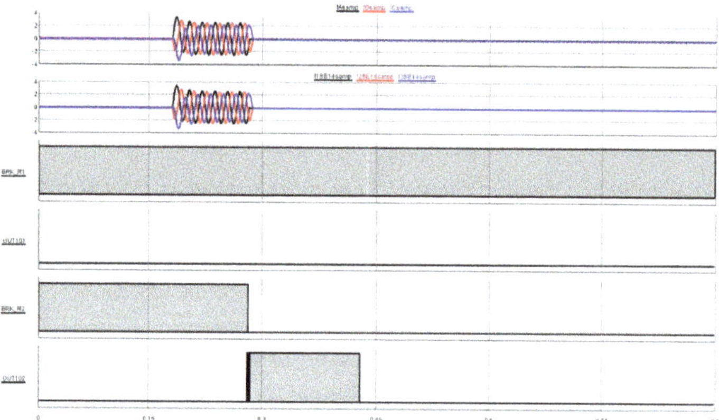

Figure 28. Waveforms when relay R2 operates to clear the fault.

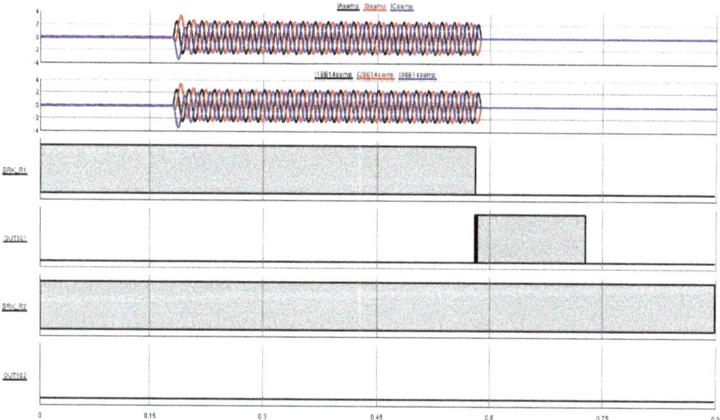

Figure 29. Waveforms when relay R1 operates to clear the fault.

6.1. Pre-Fault Period

Figure 21 depicts the waveforms for the pre-fault condition. The first y-axis shown in Figure 21 shows the sampled phase currents from source. The second y-axis denotes the sampled phase currents between bus 1 and bus 4. The third y-axis shows the status of the circuit breaker for relay R1. The initial status of the signal is 'high', as the circuit breaker is initially closed. The fourth y-axis shows the trip signal of relay R1, whose initial status is 'low' as there is no fault present in the system. Axes 5 and 6 show the circuit breaker status of relay R2 and trip signal of relay R2, respectively.

Figure 22 shows the line-neutral voltage (in kV) at bus 1 during pre-fault period. 'phase1Asamp', 'phase2Asamp', and 'phase3Asamp' are the sampled voltages for phase A, B, and C, respectively, at bus 1.

6.2. Fault Period (Instantaneous and Definite Time Scheme)

A three-phase fault with fault resistance of 0.1 ohms for a duration of 8 cycles has been applied to the system. The fault is applied in the runtime module of RTDS using a red push button shown in Figure 23. The switches shown in Figure 23 are used to select the fault type as explained in Section 5.4.

The waveforms have been obtained for two cases. In case 1, R2 operates to remove the fault. For case 2, R2 is in open loop, and R1 is in closed loop. Case 2 is simulating the scenario where R2 has failed to trip for the fault.

Figure 24 shows the current waveform and trip signal for case 1. The first y-axis in Figure 24 shows the phase current between source and bus 1 during the fault period. It can be observed that the fault current is very high for the fault duration. The second y-axis shows the phase currents between bus 1 and bus 4 during the fault. As R2 operates to remove this fault, the circuit breaker signal of R1 is high and trip signal of R1 is low as shown in third and fourth y-axis, respectively. Conversely, as observed from the sixth y-axis, the trip signal for R2 goes from low to high when the fault occurs, giving the command to the circuit breaker to open its contact. The fifth y-axis shows the breaker signal status for R2 switching from high to low when the trip signal for R2 goes from low to high.

Figure 25 depicts the current waveforms and trip signal for case 2. As the relay R1 operates to remove this fault, the circuit breaker signal of R2 is high, and the trip signal of R2 is low, as shown in fifth and sixth y-axis of Figure 25, respectively. As observed from the fourth y-axis, the trip signal for R1 goes from low to high when a fault occurs, giving the command to the circuit breaker to open its contact. The third y-axis shows the breaker signal status for R1 switching from high to low when the trip signal for R1 goes from low to high. It can be observed from Figures 24 and 25 that when the relay sends the trip command, the circuit breaker opens immediately, but there is a small delay between opening of the circuit breaker and the current dropping to zero, which is characteristic of a real system.

6.3. Post-Fault Period

Figure 26 depicts the post-fault condition for case 1. The high-frequency components in the current waveforms are due to numerical chatter of RTDS simulation. As shown in the second y-axis in Figure 26, the post-fault values for sampled currents are approximately zero as the contacts of the circuit breaker for R2 are opened and does not reclose, as the reclose control signal has been set to zero in the circuit breaker logic. In the first y-axis, the phase source currents are reduced in comparison to the pre-fault condition as the feeder where the circuit breaker for relay R2 was deployed is now open, thus the source is now only supplying load on the remaining feeder. The trip signal of R2 remains high, and correspondingly, the status of circuit breaker of R2 is low because the breaker is still open.

Figure 27 shows the post-fault condition for case 2. In Figure 27, the trip signal of R1 remains high, and correspondingly, the status of the circuit breaker of R1 is low as the breaker is still open. Both phase currents are almost reduced to zero in this scenario as the breaker near the source is open.

6.4. Coordination between R1 and R2 (IDMT Scheme)

Given the operating times of R1 and R2 obtained in Section 4.3, the fault duration and the relay trip logic in this case has been modified from the one chosen for the instantaneous and definite time schemes. The fault duration has been changed to 0.637 s using the fault slider shown in Figure 19. The relay trip logic equation used in this case for both relays is 51PT. In the occurrence of a fault (three-phase fault in this case), R2 is supposed to act as per its operating time. In case of R2 failure, R1 will wait for the time duration specified by the coordination interval (CI) and then will act as per its operating time.

Case 1—When R2 operates

Figure 28 shows the waveforms when relay R2 operates to clear the fault. The first y-axis shows the phase currents between bus 1 and the source, and the second y-axis shows the phase currents between bus 1 and bus 4. When a three-phase fault occurs, the digital trip signal for R2 shown in the sixth y-axis goes from low to high, which in turn causes the circuit breaker signal for R2 on fifth y-axis to switch from high to low. It is observed that after a certain period, the trip signal again switches to a low value even though there is no reclosing performed in this experiment. It is because of the SELogic equations defined for R2 (referred to SEL-351's instruction manual) which causes R2 to deenergize the trip output after a certain period. In the post-fault period, the phase currents between bus 1 and bus 4 shown in the second y-axis are zero as the breaker for R2 remains open. However, as seen in the first y-axis the post-fault currents between source and bus 1 are lower than the pre-fault period because even though breaker between bus 1 and bus 4 are open, the remaining feeder lines between bus 1, 2, and 3 are still operating.

Case 2—When R1 operates

When relay R2 fails to operate, R1 operates after specified CI to clear the fault. Figure 29 depicts the waveforms when R1 operates to clear the fault. It can be observed from Figure 29, the digital trip signal for relay R1 in the fourth y-axis switches from low to high while the trip signal of R2 still remains low as shown in the sixth y-axis. Correspondingly, the circuit breaker signal for R1 goes from high to low, shown in the third y-axis. In the post-fault period, all the phase currents between the source and bus 1 are zero as the breaker for R1 near the source remains open. In fourth y-axis, it can be observed as well that R1 is deenergized after some time, bringing the signal back to the 'low' state.

6.5. Relay Events

The software AcSELerator Quickset is used to retrieve the fault events from the two relays. Figures 30–32 depict the current, voltage, and trip signals, respectively, from R2 during the instantaneous scheme. Figures 33–35 show the current, voltage, and trip signals, respectively, from R1 during definite-time scheme. This section presents waveforms actually recorded by the relays to corroborate with RTDS simulation waveforms.

Figure 30 shows the phase currents for pre-fault, fault, and post-fault between bus1 and bus 4 obtained from relay R2 when deployed in instantaneous scheme. The pre-fault event shown in Figure 30 is similar to the pre-fault period for R2 shown in Figure 21. The fault event obtained from Figure 30 matches the fault period for R2 shown in Figure 24 and similar post-fault event can be seen in Figure 26. These observations indicate that RTDS and relays operate correctly as expected from the model.

The phase voltages seen in relay event for R2 in Figure 31 are similar to the phase voltage waveforms shown in Figure 22 which confirms the voltage waveforms obtained from RTDS.

Figure 32 shows the trip signal generated by R2 as per the specified trip logic equation. When comparing the trip signals in Figure 32 with Figures 21 and 24, it can be noticed that they are similar and thus verified.

Figure 33 shows the phase currents for pre-fault, fault, and post-fault between bus1 and bus 4 obtained from relay R1 when deployed in the definite-time scheme. The pre-fault event shown in Figure 33 aligns with the pre-fault period for R1 shown in Figure 25. The fault event shown in Figure 33 matches the fault period for R1 shown in Figure 25 and similar post-fault event can be seen in Figure 27. This verifies the waveforms obtained for R1 from RTDS.

Figure 34 shows the voltages obtained at bus 1 when R1 is active. When compared with the voltage waveforms shown in Figure 22, it can be observed that they are similar, which verifies the waveforms obtained for R1 from RTDS.

The trip signal generated by R1 under definite-time scheme can be seen in Figure 35. When compared with the digital trip signals shown in Figures 21 and 25, a similar operation of R1 can be observed in pre-fault and fault condition.

Figure 36 shows the phase currents recorded for R2 when deployed under IDMT scheme. When compared with the waveforms shown in Figure 28, the waveforms match, which validates the current waveforms achieved for R2 from RTDS.

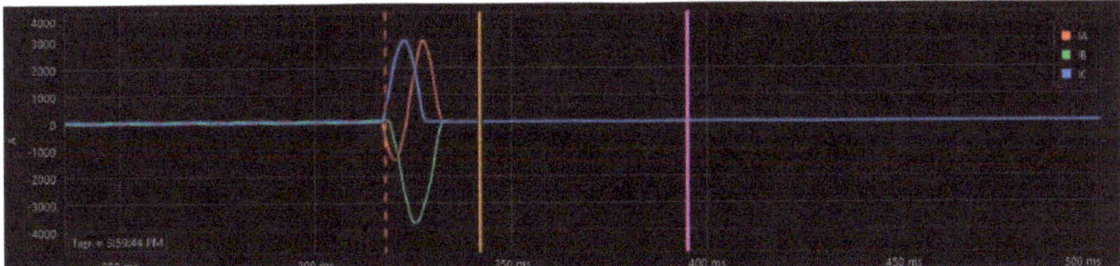

Figure 30. Phase currents recorded by R2 under instantaneous scheme.

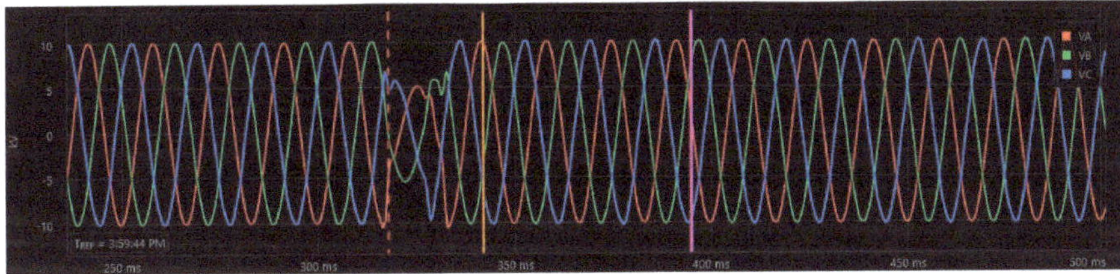

Figure 31. Phase voltages recorded by R2 under instantaneous scheme.

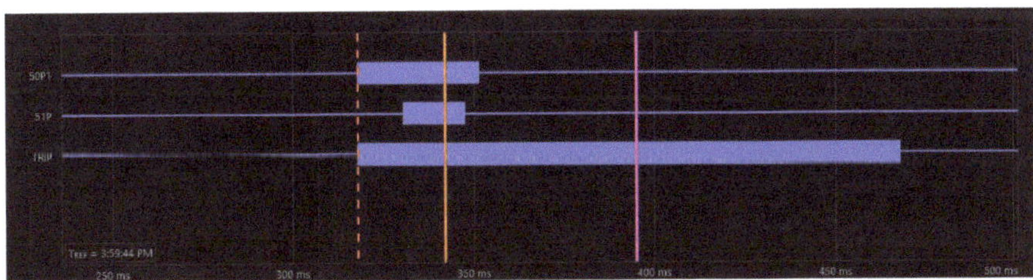

Figure 32. Trip logic and trip signal recorded by R2 under instantaneous scheme.

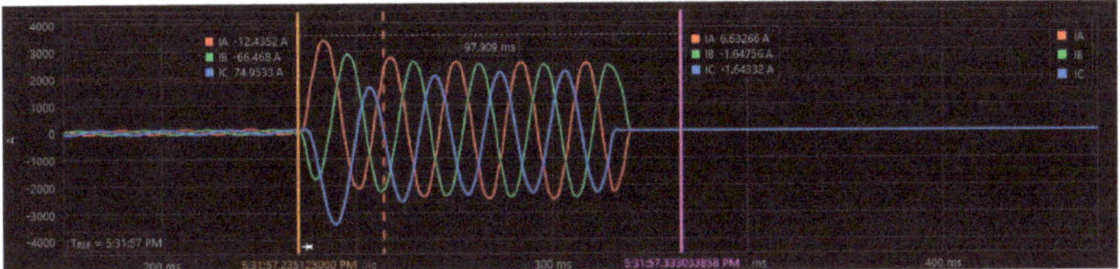

Figure 33. Phase currents recorded by R1 under definite-time scheme.

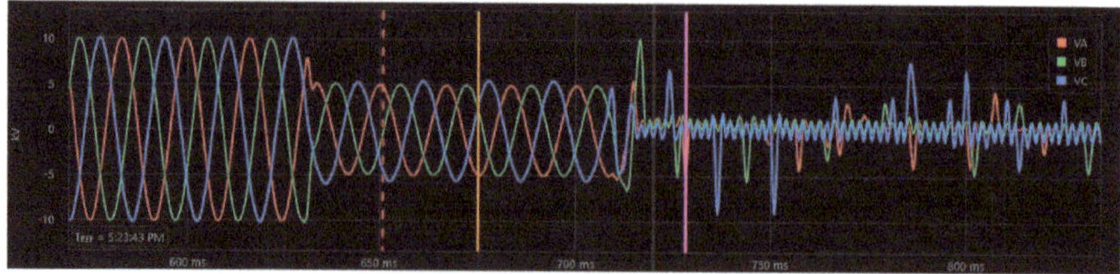

Figure 34. Phase voltages recorded by R1 under definite-time scheme.

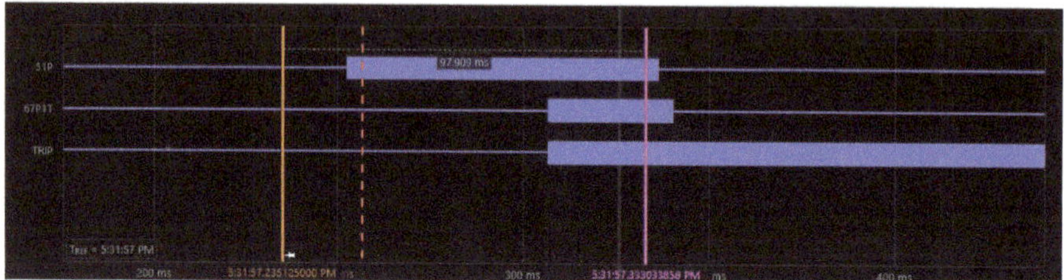

Figure 35. Trip logic and trip signal recorded by R1 under definite-time scheme.

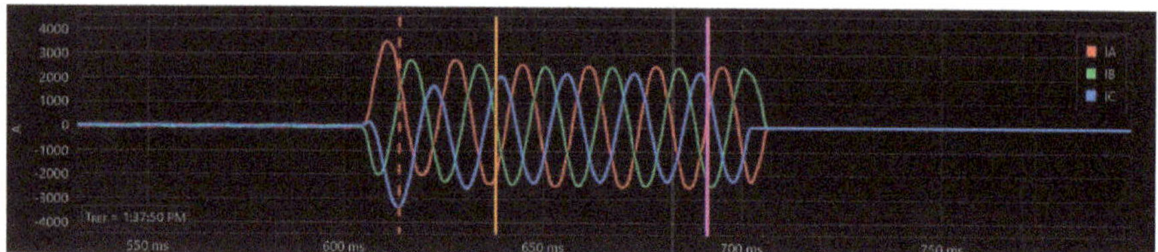

Figure 36. Event obtained for phase currents recorded by R2 under IDMT scheme.

Under IDMT scheme, the relay operates as per the operating time set for it while relay configuration. The relay is supposed to trip after the specified fault duration. The aforesaid functioning can be observed in Figure 37. The similar functioning can also be observed in Figure 28 where R2 operates the clear the fault using IDMT scheme. This validates the waveforms obtained for R2 under IDMT scheme.

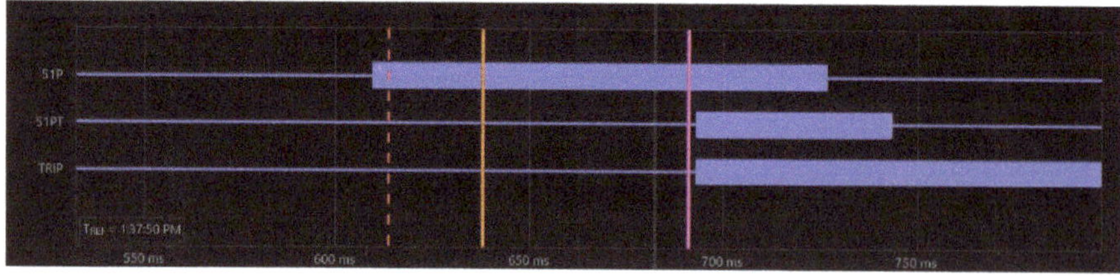

Figure 37. Trip signal generated by R2 under IDMT scheme.

In case when R2 fails to operate, R1 will act as per its operating time calculated in Section 4.3. Figure 38 shows the event for phase currents obtained for R1. When compared with Figure 29, similar current waveforms can be seen for R1, which validates the phase current waveforms obtained for R1 in RTDS.

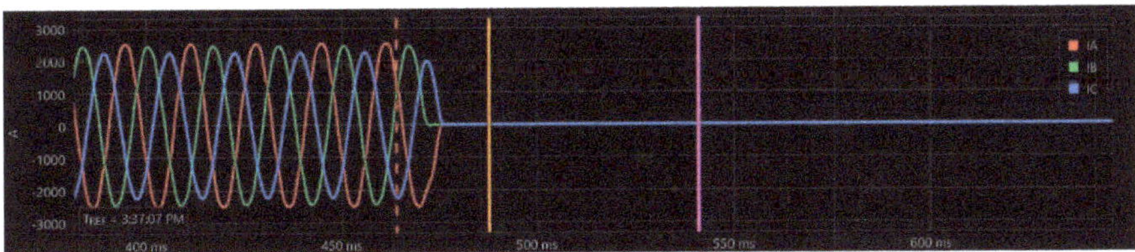

Figure 38. Event obtained for phase currents recorded by R1 under IDMT scheme.

Figure 39 shows the trip signal generated by R1 under IDMT scheme, which when compared with digital trip signal for R1 in Figure 29 verifies the RTDS waveforms.

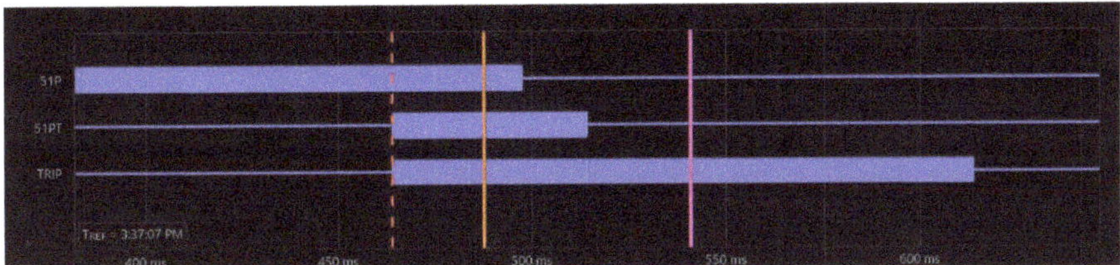

Figure 39. Trip signal generated by R1 under IDMT scheme under IDMT scheme.

7. Conclusions

Hardware-in-the-Loop (HIL) testing plays an increasingly critical role in system design and deployment. There is a lack of literature for providing guidance on how HIL tests can be setup including the connection of various HIL platform components, external power amplifiers, and the equipment under test and how critical parameters, such as scaling factors are calculated.

This paper presents the Hardware-in-the-Loop (HIL) testing done for a five-bus system. This paper thoroughly illustrates the procedure and steps for performing HIL test using RTDS platform, Omicron amplifiers, and SEL 351 protective relays as examples. Different tripping schemes including the Instantaneous overcurrent, Definite time, and Inverse Definite Minimum Time (IDMT) scheme have been implemented. The test results obtained by the testing platform RTDS and those recorded by the relays are compared, and the comparison confirms that the experiments are correct, and the testing platform and setup are suitable for testing protective relays. It is expected that the test setup is generally suitable for testing other equipment, such as power inverters, controllers, and machines, and such transients-based testing provides a realistic environment to better de-sign and test the equipment under test.

The procedure and framework presented in this paper will be useful for designing HIL experiments for research, education, and training purposes. In particular, the paper will provide guidance not only for students at universities, but also for professional engineers that intend to adopt and implement HIL tests.

Author Contributions: Conceptualization, Y.L.; Formal analysis, G.Y. and Y.L.; Funding acquisition, Y.L.; Methodology, G.Y., Y.L. and A.D.B.; Supervision, Y.L.; Validation, G.Y., Y.L. and A.D.B.; Writing—original draft, G.Y. and Y.L.; Writing—review & editing, Y.L. and A.D.B. All authors have read and agreed to the published version of the manuscript.

Funding: This research was funded by the Office of Naval Research, USA, under award number N00014-21-1-2972.

Data Availability Statement: Not applicable.

Acknowledgments: The authors would like to thank Arunprasanth Sakthivel, Heather Meiklejohn and Sachintha Kariyawasam from RTDS Technologies Inc. for providing their valuable guidance towards completing this research.

Conflicts of Interest: The authors declare no conflict of interest.

References

1. Anderson, D.; Zhao, C.; Hauser, C.H.; Venkatasubramanian, V.; Bakken, D.E.; Bose, A. A Virtual Smart Grid. *IEEE Power & Energy Magazine*. February 2012, pp. 49–57. Available online: https://magazine.ieee-pes.org/january-february-2012/a-virtual-smart-grid/ (accessed on 23 October 2022).
2. Nutaro, J. Designing power system simulators for the smart grid: Combining controls, communications, and electro-mechanical dynamics. In Proceedings of the IEEE Power and Energy Society General Meeting, Detroit, MI, USA, 24–28 July 2011. [CrossRef]
3. McLaren, P.; Nayak, O.; Langston, J.; Steurer, M.; Sloderbeck, M.; Meeker, R.; Lin, X.; Yu, M.; Forsyth, P. Testing the 'smarts' in the smart T & D grid. In Proceedings of the IEEE/PES Power Systems Conference and Exposition (PSCE), Phoenix, AZ, USA, 20–23 March 2011; pp. 1–8. [CrossRef]
4. Podmore, R.; Robinson, M.R. The role of simulators for smart grid development. *IEEE Trans. Smart Grid* **2010**, *1*, 205–212. [CrossRef]
5. Montano, F.; Ould-Bachir, T.; David, J.P. An Evaluation of a High-Level Synthesis Approach to the FPGA-Based Submicrosecond Real-Time Simulation of Power Converters. *IEEE Trans. Ind. Electron.* **2018**, *65*, 636–644. [CrossRef]
6. Chen, Y.; Dinavahi, V. Hardware emulation building blocks for real-time simulation of large-scale power grids. *IEEE Trans. Ind. Inform.* **2014**, *10*, 373–381. [CrossRef]
7. Adegbohun, F.R.; Lee, K.Y. Real-time modeling, simulation and analysis of a grid connected PV system with hardware-in-loop protection. In Proceedings of the North American Power Symposium (NAPS), Morgantown, WV, USA, 17–19 September 2017. [CrossRef]
8. Fan, W.; Liao, Y. Fault Location for Distribution Systems with Distributed Generations without Using Source Impedances. In Proceedings of the 51st North American Power Symposium (NAPS), Wichita, KS, USA, 13–15 October 2019. [CrossRef]
9. Asbery, C.; Liao, Y. Fault Identification on Electrical Transmission Lines Using Artificial Neural Networks. Ph.D. Thesis, University of Kentucky, Lexington, KT, USA, 2022; pp. 13–14.
10. Fan, W.; Liao, Y.; Kang, N. Optimal fault location for power distribution systems with distributed generations using synchronized measure-ments. *Int. J. Emerg. Electr. Power Syst.* **2020**, *21*, 20200093.
11. Kuffel, R.; Forsyth, P.; Peters, C. The Role and Importance of Real Time Digital Simulation in the Development and Testing of Power System Control and Protection Equipment. *IFAC-PapersOnLine* **2016**, *49*, 178–182. [CrossRef]
12. Tatcho, P.; Zhou, Y.; Li, H.; Liu, L. A real time digital test bed for a smart grid using RTDS. In Proceedings of the 2nd International Symposium on Power Electronics for Distributed Generation Systems (PEDG 2010), Hefei, China, 16–18 June 2010; pp. 658–661. [CrossRef]
13. Marttila, R.J.; Dick, E.P.; Fischer, D.; Mulkins, C.S. Closed-loop testing with the real-time digital power system simulator. *Electr. Power Syst. Res.* **1996**, *36*, 181–190. [CrossRef]
14. Mishra, P.; Pradhan, A.K.; Bajpai, P. Adaptive Relay Setting for Protection of Distribution System with Solar PV. In Proceedings of the 20th National Power Systems Conference (NPSC 2018), Tiruchirappalli, India, 14–16 December 2018; Volume 2, pp. 1–5. [CrossRef]
15. Afshar, M.; Majidi, M.; Gashteroodkhani, O.A.; Amoli, M.E. Analyzing Performance of Relays for High Impedance Fault (HIF) Detection Using Hardware-In-The-Loop (HIL) Platform. *Electr. Power Syst. Res.* **2022**, *209*, 108027. [CrossRef]
16. Chen, Y.; Wen, M.; Wang, Z.; Yin, X.; Peng, J.; Zhang, R. An improved numerical distance relay based on CCVT transient characteristic matching. *Int. J. Electr. Power Energy Syst.* **2020**, *122*, 106146. [CrossRef]
17. Cho, Y.S.; Lee, C.K.; Jang, G.; Kim, T.K. Design and implementation of a real-time training environment for protective relay. *Int. J. Electr. Power Energy Syst.* **2010**, *32*, 194–209. [CrossRef]
18. So, K.H.; Heo, J.Y.; Kim, C.H.; Aggarwal, R.K.; Kim, J.C.; Jang, G.S. An implementation of current differential relay and directional comparison relay using EMTP MODELS. *Int. J. Electr. Power Energy Syst.* **2006**, *28*, 261–272. [CrossRef]
19. Dias, O.; Tavares, M.C.; Magrin, F. Hardware implementation and performance evaluation of the fast adaptive single-phase auto reclosing algorithm. *Electr. Power Syst. Res.* **2019**, *168*, 169–183. [CrossRef]
20. Ledwaba, T.; Senyane, K.; Van Coller, J. Hardware-In-loop testing of a differential relay used to Protect single/double circuit transmission lines. In Proceedings of the Southern African Universities Power Engineering Conference/Robotics and Mecha-tronics/Pattern Recognition Association of South Africa (SAUPEC/RobMech/PRASA 2019), Bloemfontein, South Africa, 28–30 January 2019; pp. 347–352. [CrossRef]
21. Naveen, P.; Jena, P. Directional Overcurrent Relays Coordination Scheme for Protection of Microgrid. In Proceedings of the IEEE First International Conference on Smart Technologies for Power, Energy and Control (STPEC 2020), Nagpur, India, 25–26 September 2020. [CrossRef]
22. Singh, M.; Telukunta, V.; Srivani, S.G. Enhanced real time coordination of distance and user defined over current relays. *Int. J. Electr. Power Energy Syst.* **2018**, *98*, 430–441. [CrossRef]
23. Langston, J.; Steurer, M.; Schoder, K.; Borraccini, J.; Dalessandro, D.; Rumney, T.; Fikse, T. Power hardware-in-The-loop simulation testing of a flywheel energy storage system for shipboard applications. In Proceedings of the IEEE Electric Ship Technologies Symposium (ESTS), Arlington, VA, USA, 14–17 August 2017; pp. 305–311. [CrossRef]

24. Stifter, M.; Cordova, J.; Kazmi, J.; Arghandeh, R. Real-time simulation and hardware-in-the-loop testbed for distribution synchrophasor applications. *Energies* **2018**, *11*, 876. [CrossRef]
25. Sidwall, K.; Forsyth, P. A Review of Recent Best Practices in the Development of Real-Time Power System Simulators from a Simulator Manufacturer's Perspective. *Energies* **2022**, *15*, 1111. [CrossRef]

Article

An Improved Over-Speed Deloading Control of Wind Power Systems for Primary Frequency Regulation Considering Turbulence Characteristics

Xiaolian Zhang [1,*], Baocong Lin [2], Ke Xu [1], Yangfei Zhang [1], Sipeng Hao [1] and Qi Hu [1]

[1] School of Electric Power Engineering, Nanjing Institute of Technology, Nanjing 211167, China
[2] School of Automation, Nanjing University of Science and Technology, Nanjing 210094, China
* Correspondence: zhangxl530@163.com

Abstract: Wind power systems participating in primary frequency regulation have become a novel trend. In order to solve the problem of the over-speed deloading (OSD) control of wind power systems failing to provide reserved capacity for primary frequency regulation while under turbulent winds, this paper analyzes the influence mechanism of turbulence characteristics on the OSD control and the relationship between the reserve capacity of OSD control and the deloading power coefficient under turbulent wind speeds, while also quantifying the relationship between the turbulence characteristic index and deloading power coefficient. The range of the deloading power coefficient is obtained accordingly, based on which improved OSD control is proposed to dynamically optimize the deloading power coefficient according to the turbulence characteristics, which improves the frequency regulation performance of wind power systems under turbulent wind speed. According to the simulations and experimental results, the improved method proposed in this paper has good effectiveness and superiority in frequency regulation effect and rotor speed performance.

Keywords: wind power system; primary frequency regulation; OSD control; turbulence characteristics

Citation: Zhang, X.; Lin, B.; Xu, K.; Zhang, Y.; Hao, S.; Hu, Q. An Improved Over-Speed Deloading Control of Wind Power Systems for Primary Frequency Regulation Considering Turbulence Characteristics. *Energies* **2023**, *16*, 2813. https://doi.org/10.3390/en16062813

Academic Editors: Francesco Castellani and Abu-Siada Ahmed

Received: 4 January 2023
Revised: 7 March 2023
Accepted: 15 March 2023
Published: 17 March 2023

1. Introduction

With the proposal of "carbon peak" and "carbon neutrality", renewable energy sources such as wind and solar energy have been vigorously developed in China, but the high proportion of wind power results in some brand-new problems. Nowadays, mainstream wind turbines use converters to connect to the power grid, causing the uncoupled relationship between the rotor speed of wind turbines and the power grid's frequency, which leads to decreased inertia of the power grid with a high wind power penetration rate, reducing the frequency stability of the system [1]. Therefore, some research about frequency regulation by wind power systems has been carried out.

At present, the methods of wind power systems participating in frequency regulation can be divided into rotor kinetic energy control, power reserve control and wind power-energy storage hybrid control [1,2]. The rotor kinetic energy control, such as the droop control and inertia control, can maintain the frequency adjustment for a short time but will eventually lead to a second drop in the system frequency [3,4]. The power reserve control, such as the over-speed deloading (OSD) control and variable pitch control, can preset reverse capacity by part-load operation. The wind power-energy storage hybrid control can provide enough power for system frequency adjustment by the energy storage system, but this type of control needs to add storage units which increases the cost of construction and maintenance [5,6]. The comparison of these methods is shown in Table 1.

Table 1. The comparison of different control methods.

The Control Methods		Work Principle	Advantage	Disadvantage
Rotor kinetic energy control	Droop control	Using the kinetic energy of wind turbine rotor	Simple, easy to carry out	Leading to a second drop in system frequency
	Inertia control	Using the kinetic energy of wind turbine rotor	Simple, easy to carry out	Leading to a second drop in system frequency
Power reserve control	The over-speed deloading control	Presetting reverse capacity by part-load operation	Easy to control, can provide more power	Reducing the wind power efficiency
	Variable pitch control	Presetting reverse capacity by part-load operation	Can provide more power	Reducing the wind power efficiency
Wind power-energy storage hybrid control	Cooperative control	Power support by both systems	Better performance than other methods	More complex and more expensive

The OSD control has the advantages of not needing to add equipment and being easy to control [7,8], so it is widely applied in wind power systems. References [9–11] apply OSD control to a single wind power system and wind farm, improving the performance of the wind power system participating in power grid frequency regulation to a certain extent. However, the research above is all conducted under constant wind speed, and whether the same results would be obtained under turbulence wind speed needs further study.

Although some control methods consider the change in wind speed, it is mainly a rough classification of wind speed conditions. For example, wind speed is divided into three levels and control parameters are set separately according to these levels. It has also been studied that the wind speed is divided into three levels, and the pitch control and OSD control are used to coordinate at different levels to achieve frequency regulation [9,10]. The reference [12] considered the impact of wind power and load power fluctuations on system frequency and smoothed the output power of wind turbines to improve frequency stability. However, these studies are relatively rough when considering wind speed characteristics, ignoring the turbulence characteristics, such as average wind speed and turbulence intensity.

The actual wind speed is mainly turbulent wind speed with strong randomness and fluctuations, which will lead to difficulties or failures in wind power control [13,14]. Additionally, fluctuating wind power under turbulent wind speed will also cause system frequency changes [15,16]; therefore, it is necessary to study the influence of turbulent wind speed on the frequency regulation performance of wind power and improve the control effect under turbulent wind speed.

All in all, the existing control strategies mainly focus on constant wind speed, but the dynamic response of wind power systems under turbulent wind speed is different from that of constant wind speed [13,14], which weakens the effect of the traditional control strategy. In addition, different turbulence characteristics also have different influences on the control effect, and the existing research has not studied these issues.

In the previous research, the author analyzed the influence of turbulence characteristics on system frequency and proposed an improved frequency controller [17], but this method requires additional power control blocks. Additionally, through further research, it was found that the OSD control could not provide preset reserve capacity under turbulent wind speed, which weakened its frequency regulation ability under turbulent wind speed.

Therefore, further study has been conducted based on the author's previous research in this paper to obtain an improved OSD control. The main contributions of this paper are as follows:

1. the influence mechanism of turbulent wind speed on the OSD control is analyzed;
2. the relationship between the reserve capacity of OSD control and deloading power coefficient under turbulent wind speed is analyzed, and the quantitative relationship between the turbulence characteristics and the deloading power coefficient is obtained;
3. based on the relationships above, the range of the deloading power coefficient of the OSD control is obtained, and a reasonable algorithm for setting the coefficient is designed, therefore the OSD control is improved to respond to turbulence characteristics.

Through simulations and experiments based on the experimental platform, it is verified that the improved strategy can effectively reserve enough power for frequency regulation. Compared with existing methods, the improved strategy has a better frequency regulation effect and rotor speed performance.

2. The Model and Control Strategy

2.1. The Model of Wind Power Participating in Primary Frequency Regulation

The wind power system is a combination of mechanical, electrical and control equipment, converting wind energy into electric power [18]; it usually consists of wind turbines, a transmission system, a variable pitch system, an electrical system and a control system. In this paper, according to [19], a model for wind power systems participating in primary frequency regulation of the power grid is established, including wind turbine models, maximum power point tracking (MPPT) controllers, frequency controllers, and transmission systems, as shown in Figure 1.

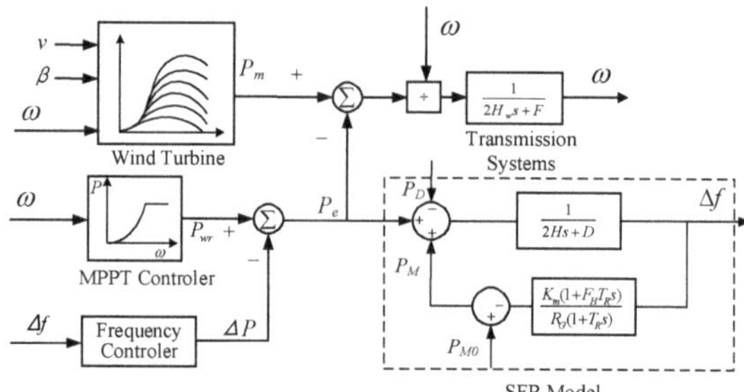

Figure 1. The model of wind power participating in primary frequency regulation of power grid.

According to Bertz's Law, the mechanical power of wind turbine captured from wind energy P_m is:

$$\begin{cases} P_m = 0.5\rho\pi R^2 v^3 C_p(\lambda,\beta) \\ C_p = 0.5(116/\lambda - 0.4\beta - 5)e^{-21/\lambda_1} + 0.0068\lambda \\ \lambda_1 = [1/(\lambda + 0.08\beta) - 0.035/(\beta^3 + 1)]^{-1} \end{cases} \tag{1}$$

where ρ is the air density; R is the radius of the wind turbine; v is the wind speed; C_p is the power coefficient, β is the pitch angle, λ is the tip speed ratio, it is defined as $\lambda = \omega R/v$, ω is rotor speed for wind turbines.

The mechanical power absorbed by the wind turbine is transferred to the generator through the transmission system. Here, the transmission system model is shown as follows:

$$\omega = \frac{1}{2H_w s + F}(T_m - T_e) \tag{2}$$

where H_w is the inertia coefficient of the transmission system, and F is the friction coefficient of the transmission system. The detail of Equation (2) can be seen in [19].

The MPPT control of the wind turbines uses the power curve feedback method, the active power reference $P_{\max}$ can be defined by the equation below [20].

$$\begin{cases} P_{\max} = K_{opt}\omega^3 \\ K_{opt} = \dfrac{0.5\rho\pi R^5 C_p^{\max}}{\lambda_{opt}^3} \end{cases} \tag{3}$$

The power grid is modeled by a low-order system frequency response (SFR) model shown in Figure 1. This model can be used to estimate the frequency changes under power disturbance [18]. In Figure 1, P_e is the output power of the wind turbine; P_M is a traditional synchronous generator output power, P_{M0} is the initial value, P_D is load power, Δf is power grid frequency deviation, H is inertia constant, D is damping factor, K_m is power gain factor of the synchronous generator, F_H is the fraction of total power generated by the high-pressure turbine, T_R is reheat time constant, R_G is the adjustment factor of the synchronous generator.

2.2. OSD Control

When the wind turbine operates at the maximum power point, there is no reserve capacity for frequency regulation, but when the wind turbine operates at a point with a larger rotor speed, the output power of the wind turbine can be reduced, and the reserve capacity can be obtained, as the P_{del} shown in Figure 2. This method is called the over-speed deloading (OSD) control. The OSD control is generally adopted to realize the power reduction operation of the wind turbine [2].

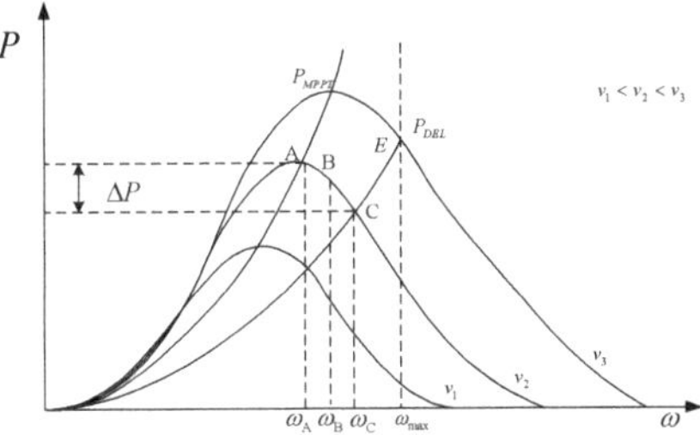

Figure 2. Working principle of OSD control.

The OSD control tracks deloading power curve P_{del}; the relationship between P_{del} and rotor speed ω can be expressed as follows:

$$\begin{cases} P_{del} = K_d K_{opt}\omega^3 \\ K_d = 1 - d\% \end{cases} \tag{4}$$

where K_d is the deloading power coefficient, and $d\%$ is the deloading coefficient.

The working principle of OSD control is shown in Figure 2. The MPPT curve shifts to the right, from the optimal working point A to the suboptimal working point C, at which time the wind power output decreases, reserving certain power. When the system frequency decreases, the rotor speed decreases from ω_C to ω_B, the MPPT curve shifts to the left, and the working point shifts from point C to point B so that the wind turbine output increases in response to the system frequency change.

3. Analysis of Difficulties and Mechanism of OSD Control under Turbulent Wind Speed

Wind is caused by atmospheric movement. Due to the complex surface topography and the change in air temperature, pressure, or humidity, the atmosphere's movement is a random and turbulent motion, which makes the wind speed fluctuate constantly, and these are the causes of turbulence. Therefore, these factors will influence the values of wind speed v, as well as the air density ρ; the wind power P_m will also be influenced according to Equation (1). It can be seen that wind speed fluctuates constantly, this is one cause of frequency deviation; therefore, wind speed fluctuation should be addressed to guarantee the stable operation of the system.

Under constant wind speed, when the OSD control is used, the wind turbine can work stably at the deloading operating point and provide the reserve capacity stably. While under turbulent wind speed, due to the fluctuation of wind speed and the influence of moment of inertia, the wind turbine cannot work at the deloading operating point in time and provide stable reserve capacity. In addition, it will even reduce the reserve capacity and weaken the frequency regulation effect. Therefore, the control strategies based on constant wind speed or steady-state will have new problems.

The example shown in Figure 3 shows that the system response of turbulent wind speed is different from that of constant wind speed with the same control parameters, the difference in dynamic performance and the influence of wind turbulence on the control are also verified.

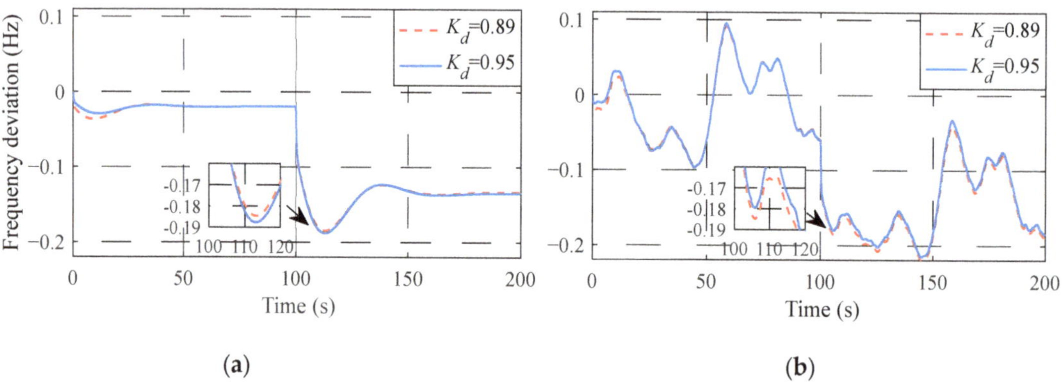

Figure 3. Frequency regulation effect of OSD control at different wind conditions. (**a**) Constant wind speed, (**b**) turbulent wind speed.

It is found that the slope of the optimal power curve is reduced when the wind turbine uses OSD control (such as P_{del} shown in Figure 2). Therefore, the difference between mechanical torque and electromagnetic torque is increased, which improves the performance of the wind power system in tracking the maximum power point under turbulent wind speed [12], then the available reserve capacity will be reduced due to the improved wind energy capture efficiency. That means K_d will change the dynamic performance of the wind power system, and then affect the reserve capacity and frequency regulation effect of the wind power system. The elaboration is shown as follows.

Under constant wind speed, when the load suddenly increases at 100 s, the smaller the K_d is, the better the lowest frequency and then the better the frequency regulation effect will be, as shown in Figure 3a. However, under turbulent wind speed, the result is just the opposite. The smaller the K_d is, the more severe the frequency falls, as shown in Figure 3b.

Figure 4 shows the wind power output of the example shown in Figure 3b. It can be seen that when load suddenly increases at 100s, the wind power with K_d of 0.89 is lower than that of 0.95 and the MPPT control.

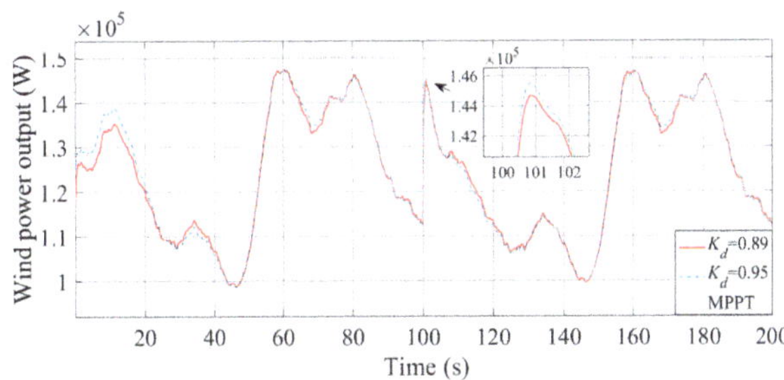

Figure 4. Wind power output under turbulent wind speed.

Table 2 shows the average wind energy capture efficiency, the average frequency deviation and the reserve capacity factor [21] under different strategies in Figure 4. The average wind energy capture efficiency is defined by Equation (5), and the average frequency deviation is defined by Equation (6).

Table 2. The average wind energy capture efficiency and reserve capacity factor under different strategies.

	$K_d = 0.89$	$K_d = 0.95$	**MPPT**
The average wind energy capture efficiency	0.4577	0.4575	0.4570
Reserve capacity factor k_{res}	1.03351	1.03353	1.03419
Average frequency deviation (Hz)	0.09732	0.09596	0.09576

The reserve capacity factor represents the reserve capacity of wind turbines. The higher the value, the lager the reserve capacity. The definition can be given by Equation (7) of Section 4.

$$C_{pavg} = \frac{\sum\limits_{i=1}^{n} C_{pi}}{n} \tag{5}$$

where C_{pi} is the power coefficient of the ith sample, n is the total number of samples.

$$|\Delta f|_{avg} = \frac{\sum\limits_{i=1}^{n} |\Delta f_i|}{n} \tag{6}$$

where Δf_i is the frequency deviation of the ith sample, n is the total number of samples, and the sampling period is 0.04 s.

It can be known from Table 2 that when K_d is 0.89, the wind energy capture efficiency is the highest, but the reserve capacity factor is the smallest, so it is difficult to provide enough power to support frequency adjustment in case of load fluctuation. The wind energy capture efficiency of MPPT control is the lowest, while the reserve capacity factor is the highest. It is verified that under turbulent wind speed, it will have a worse control effect with the same K_d as the constant wind speed.

So, the OSD control under turbulent wind speed does not play its function of reserving power for frequency regulation, and the frequency regulation effect is even worse than that of MPPT control without reserving power, which is the difficulty of OSD control applied to turbulent wind speed.

4. Improved OSD Control Considering Turbulence Characteristics

To solve the difficulties above, the influence of K_d on frequency regulation and the variation law of K_d under different turbulence characteristics are first analyzed, then an improved OSD control strategy for turbulent wind speed is proposed.

4.1. Analysis of the Influence of K_d on Frequency Regulation

4.1.1. Relationship between K_d and Frequency Deviation under Turbulent Wind Speed

In this paper, a typical turbulent wind speed with turbulence intensity level A in IEC-614000-1 standard is used, and the average frequency deviation of the power grid corresponding to different K_d is obtained through simulation.

The variation range of K_d is from 0.75 to 1.25, the statistical results are shown in Figure 5. It can be seen from Figure 5 that under turbulent wind speed, a different K_d will have different frequency control effects, and there is a K_d to maximize the average frequency deviation.

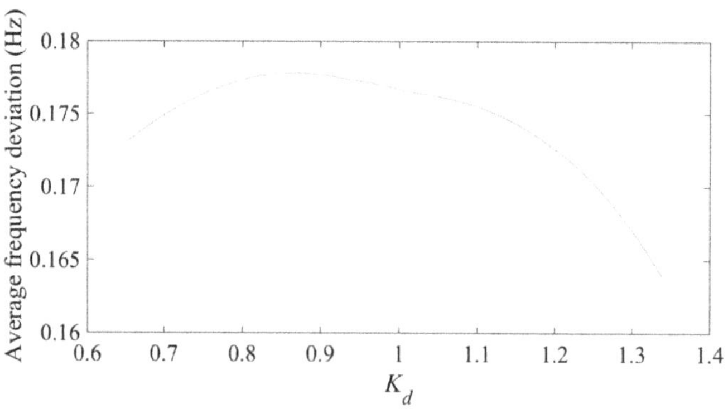

Figure 5. Relationship between average frequency deviation and K_d.

It is found that the K_d with the maximum average frequency deviation improves the tracking performance of the wind turbine, and maximizes its wind energy capture efficiency, thereby reducing the reserve capacity of the wind turbine, leading to the maximum average frequency deviation.

In this paper, the relationship between reserve capacity and K_d is further analyzed quantitatively. Here, the reserve capacity factor in reference [21] is used to define the reserve capacity of the wind turbine, as shown in Equation (7):

$$k_{\text{res}} = \frac{P_{wn} - P_w}{P_{wn} - P_{w1}} \tag{7}$$

where, P_{wn} is the rated power of the wind power system; P_w is the current wind power output; P_{w1} is the wind power at the lowest rotor speed.

To evaluate the reserve capacity at the same turbulent wind speed as the example shown in Figure 5, this paper calculates the average reserve capacity factor corresponding to different K_d, and the results are shown in Figure 6. It can be seen that the trend of average reserve capacity factor is opposite to the average frequency deviation shown in Figure 5, and there is a K_d to minimize the reserve capacity. The K_d is defined as K_{d_minres}. The result in Figure 6 verifies the influence mechanism mentioned in Section 3.

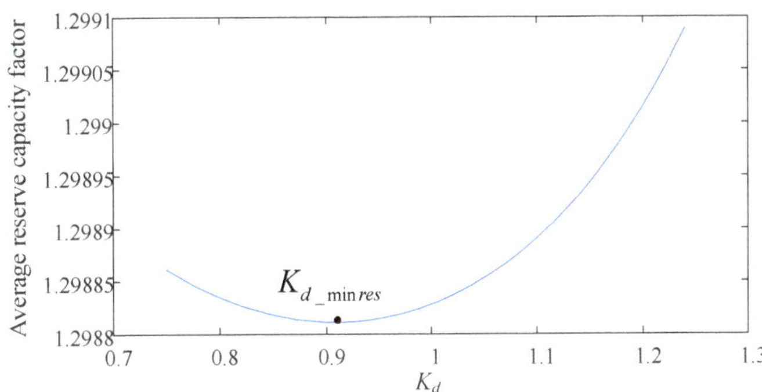

Figure 6. Relationship between average reserve capacity factor and K_d.

4.1.2. Mathematical Relationship between K_{d_minres} and Turbulence Characteristics

There is a K_{d_minres} to minimize the reserve capacity, and it is found that K_{d_minres} will change under different wind conditions; therefore, to analyze the relationship between K_{d_minres} and turbulence characteristics, a large number of wind speed samples are produced according to the given turbulence characteristic index, and then the traversal algorithm is used to obtain the K_{d_minres} of different turbulence characteristic indexes, then the change rule of K_{d_minres} under various turbulence characteristics can be achieved.

The average wind speed and turbulence intensity are selected as the turbulence characteristic indexes, the average wind speed ranges from 6.2 to 8.2 m/s with an interval of 0.2 m/s. According to the A, B and C three turbulence levels corresponding to IEC-614000-1 standard [22], the turbulence intensity value range is 0.18~0.32, and the interval is 0.02. Based on this, 70 wind speed samples with 10 kinds of average wind speeds and seven kinds of turbulence intensities can be obtained.

In each wind speed sample, the reserve capacity of the wind turbine is analyzed when the value of K_d is from 0.6 to 1.4, and the minimum reserve capacity of K_{d_minres} under each wind speed is achieved; the statistical results are shown in Figure 7.

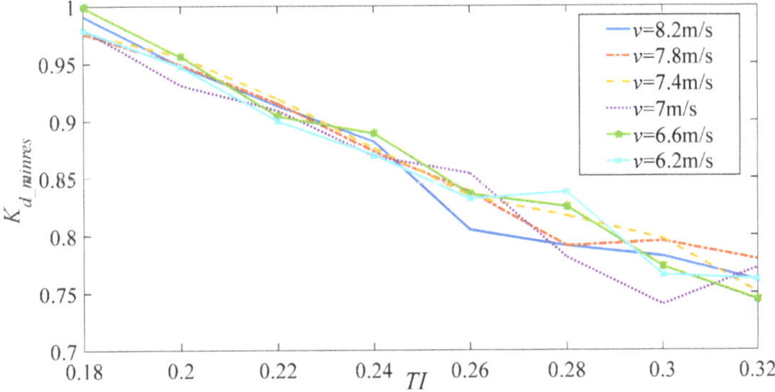

Figure 7. Relationship between K_{d_minres} and turbulence characteristic index.

It can be seen from Figure 7 that the K_{d_minres}-TI curves of different average wind speeds are relatively concentrated, when the average wind speed is constant, K_{d_minres} decreases with the increase of TI, and the change amplitude is obvious. Therefore, the impact of average wind speed to the K_{d_minres} is ignored, focusing on the relationship between K_{d_minres} and turbulence intensity TI. We obtained the average K_{d_minres}-TI curve, then used the ordinary least squares method to fit the K_{d_minres}-*TI* curve. The fitting expression is as follows.

$$K_{d\_minres} = -1.6452 \cdot TI + 1.2718 \tag{8}$$

It can be seen from Figure 6 that, when K_d is set as K_{d_minres}, the average reserve capacity of the wind turbine is the lowest, and it is not good for frequency regulation. To make the OSD control effective, the selection of K_d need to avoid K_{d_minres}.

If K_d is greater than K_{d_minres}, in order to obtain more reserve capacity than MPPT control, K_d should be greater than one, and if it is changed to under-speed deloading control, in this case, deceleration will occur, taking the wind power system out of service. A large number of studies show that when using under-speed deloading control, the rotor speed will be at a low level, then stalling and taking the wind power system out of service are likely to occur [2]. Thus, the reasonable K_d should be less than K_{d_minres}.

4.1.3. The Range of K_d

According to the above analysis K_d should be less than K_{d_minres}. However, the smaller K_d is, the higher the rotor speed will be; there is an upper limit for the rotor speed of the wind turbine. When K_d is too small, the rotor speed will approach the upper limit, then MPPT control cannot be realized normally and it may easily cause damage to the rotor.

According to the deloading operation curve shown in Figure 2, point E is the upper limit of the rotor speed. At point E, the K_d and wind speed v will satisfy the following relationship.

$$K_d \cdot K_{opt} \cdot \omega_{max}^3 = 0.5\rho\pi R^2 v^3 C_p(\omega_{max}R/v, \beta) \tag{9}$$

while

$$K_d = \frac{0.5\rho\pi R^2 v^3 C_p(\omega_{max}R/v, \beta)}{K_{opt} \cdot \omega_{max}^3} \tag{10}$$

The relationship is shown in Figure 8. Because the equation is too complex, in order to simplify the control strategy, this paper uses the ordinary least square method to fit the curve, and the fitting expression is shown in Equation (11).

$$K_{dmin} = 0.1116 \cdot v - 0.6355 \tag{11}$$

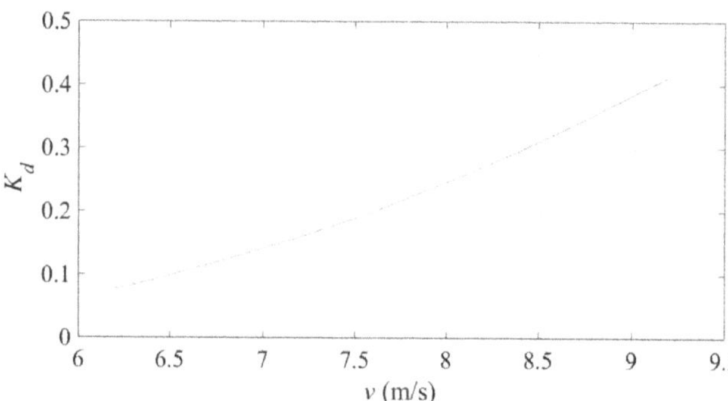

Figure 8. K_d-v curve when the rotor speed reaches the upper limit.

Therefore, in order to make sure the rotor speed does not exceed the limit and the OSD control can be normally used, the value of K_d should be greater than K_{dmin}.

In conclusion, to make the OSD control operate effectively under turbulent wind speed, K_d should be set in the range of $K_{dmin} < K_d < K_{d\_minres}$.

4.2. Improved OSD Control Considering Turbulence Characteristics

The above analysis gives a reasonable range of K_d. When K_d is set to K_{d_minres}, the wind turbine will provide the minimum reserve capacity and cannot provide enough power for frequency regulation; when K_d is set to K_{dmin}, the wind turbine will basically operate at the upper limit rotor speed, losing MPPT ability and greatly reducing the wind energy capture efficiency. Therefore, it is reasonable to set K_d as the midpoint of the range, which is shown in Equation (12):

$$K_d = \frac{K_{dmin} + K_{d\_minres}}{2} = 0.058 \cdot \bar{v} - 0.8226 \cdot TI + 0.31815 \tag{12}$$

where the average wind speed $\bar{v}$ is used to characterize the turbulent wind speed.

In this paper, Equation (12) is used to periodically adjust the K_d. The control block diagram is shown in Figure 9.

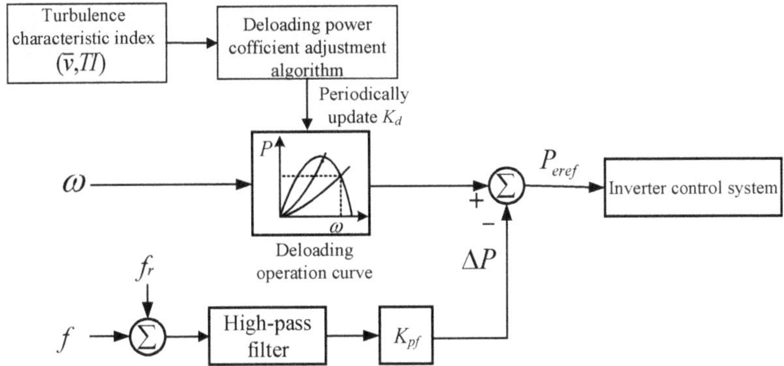

Figure 9. Block diagram of the improved OSD control.

The updating steps of K_d are as follows:

Step 1: initialize. Set the update cycle T of the deloading power coefficient, then put the number k of cycles into 0;

Step 2: let $k = k + 1$, and use the wind speed prediction to obtain the average wind speed $\bar{v}(k)$ and turbulence intensity $TI(k)$ of the wind speed during current cycle;

Step 3: calculate the $K_d(k)$ of the current cycle with Equation (12);

Step 4: judge whether the current cycle is over. If it is over, return to step 2.

5. Example Analysis

5.1. Simulation Analysis

In order to verify the superiority and effectiveness of the improved strategy in this paper, the method with fixed K_d (K_d is 0.9), the variable deloading power coefficient method proposed in [23] (hereafter referred to as variable K_d) and the improved strategy in this paper are compared.

1. Simulation parameters

A turbulent wind speed with average wind speed of 8 m/s and turbulence intensity of level A is used for simulation, and the load suddenly increases at 100 s.

The simulation parameters are shown in Table 3. The rated power of the wind power system in the simulation is 0.6 MW, and the rated wind speed is 12 m/s. The wind speed used in this paper mainly runs between 5 m/s and 12 m/s. The wind speed profile used in simulation is shown in Figure 10.

Table 3. Simulation parameters.

Wind turbine	R/m	20
	H_w/s	5.54
	λ_{opt}	5.8
	C_p^{max}	0.4603
The Grid	H/s	4
	D	1
	K_m	0.95
	F_H	0.3
	T_R/s	8.0
	R_G	0.05
Controller	K_{pf}	2×10^5

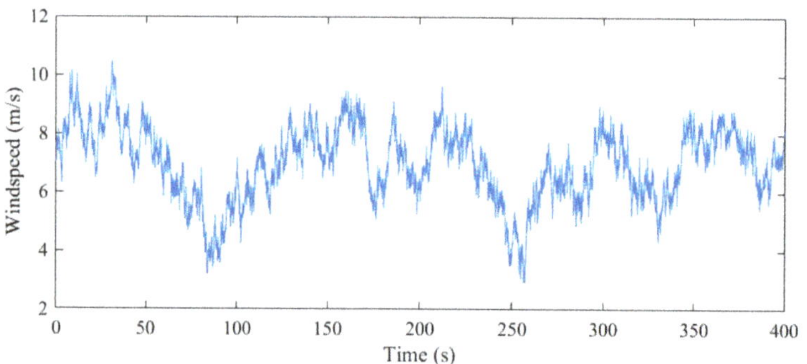

Figure 10. The wind speed used in simulation.

2. Analysis of simulation results

Figures 11–15, respectively, show the frequency deviation of the grid, wind power output, the operation curve of wind turbine before and after load change (95 s–120 s), and the K_d of different strategies.

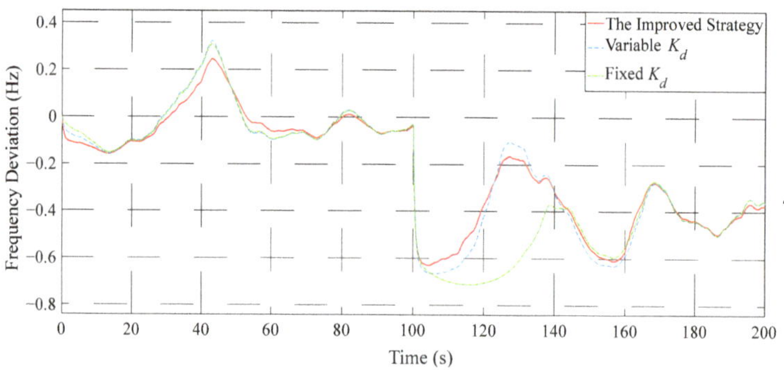

Figure 11. Frequency deviation.

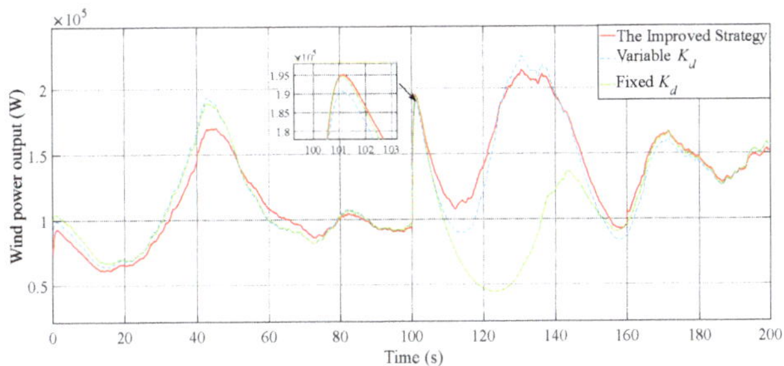

Figure 12. Wind power output.

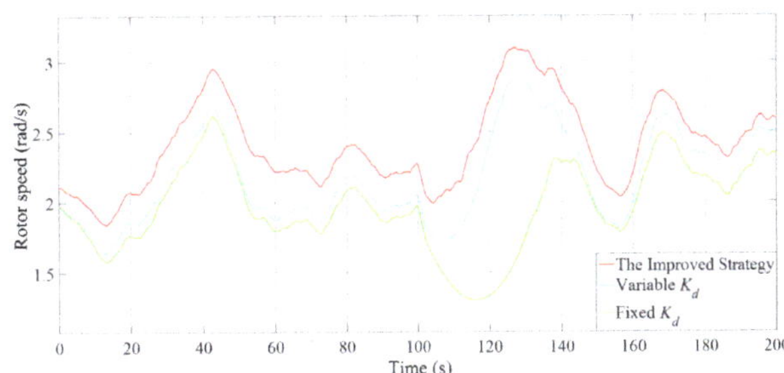

Figure 13. Rotor speed.

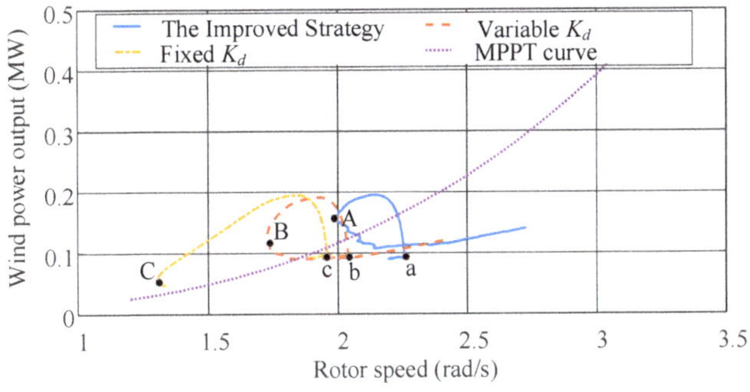

Figure 14. The operation curve before and after load fluctuation.

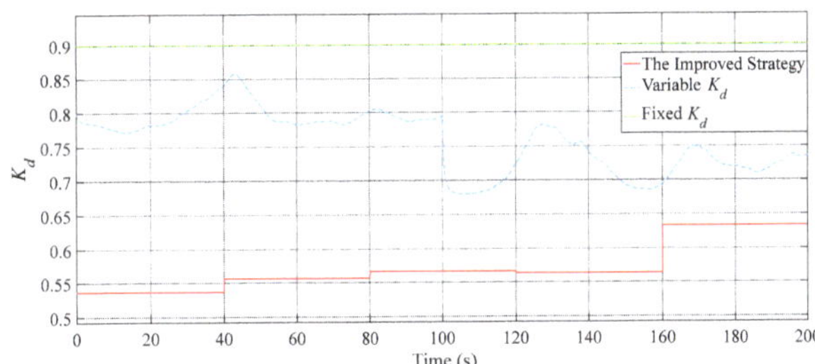

Figure 15. K_d of different methods.

It can be seen from Figure 11 that after sudden load increasing at 100 s, the frequency of the three strategies decreases, but the decline of the improved strategy in this paper is significantly less than the other two strategies. And when the frequency support using the reserve capacity is completed, the improved strategy in this paper can recover the tracking of wind speed more quickly. However, the method with variable K_d and the one with fixed K_d have a longer recovery process due to the large drop in rotor speed. It can also be seen from Figure 12 that the improved strategy in this paper provides greater output power when the load suddenly increases at 100 s.

It can be seen from Figure 14 that a, b and c are the operating points of the three strategies at 100 s, A, B and C are the times when the rotor speed stops falling and begins to recover. When the load suddenly increases at 100 s, the operating points of the three strategies rise rapidly. After a short period of power support, the speed starts to drop, and the operating point moves to the left. It can be seen that the rotor speed drop of the improved strategy in this paper is obviously smaller than that of the other two strategies, and it can recover faster. However, the rotor speed of the fixed K_d method has a long time drop, which leads to continuous low output power that is not conducive to the frequency adjustment of the system. At the same time, the rotor stall problem is prone to occur, and there is a risk of making the wind power system go out of service.

Table 4 shows the average frequency deviation of the three strategies and the improvement of the improved strategy compared with other strategies.

Table 4. Average frequency deviation of different strategies.

	The Improved Strategy	Variable K_d	Fixed K_d
Average frequency deviation (Hz)	0.2510	0.2625	0.2994
Improvement of the improved strategy	——	4.58%	19.28%

The average wind energy capture efficiency of the three strategies is compared in Table 5.

Table 5. Average wind energy capture efficiency of different strategies.

	The Improved Strategy	Variable K_d	Fixed K_d
Average wind energy capture efficiency	0.4321	0.4389	0.4084

The wind energy capture efficiency of the improved strategy is higher than that of the method with fixed K_d and lower than that of variable K_d. When the load fluctuation occurs in the control with fixed K_d, the participation of the wind power system in frequency regulation leads to a serious drop in the rotor speed, so the wind energy capture efficiency is low. However, the rotor speed of the improved strategy in this paper is at a high level, so there is no serious rotor speed drop.

Due to the lower K_d than that of variable K_d, the wind energy capture efficiency of the improved strategy in this paper is lower than that of variable K_d, but the improved strategy obtains a better frequency regulation effect.

Wind energy capture efficiency and frequency regulation are always in conflict because the wind power system will give up tracking the maximum power point when it is participating in frequency regulation. The performance of frequency regulation becomes a more important goal. In the standard of GB/T 19963-2011 "Technical Regulations for Wind Farm Access to Power System", it is required that wind power systems should provide power to support the power system frequency. Additionally, the proportion of wind power systems in the power system will be larger and larger and the traditional synchronous generator will reduce, so wind power systems will play an important role in frequency regulation. In this case, the improved strategy is reasonable and valuable.

From the above analysis, it can be seen that the improved strategy in this paper can provide better reserve capacity, and the frequency regulation effect is much better. Compared with the control with fixed K_d, the frequency deviation can be increased by 19.28%, and the rotor speed can be well maintained to avoid the secondary frequency drop caused by rotor stall and rotor speed drop.

5.2. Experimental Analysis

In order to verify the effectiveness of the improved strategy, it is verified on the frequency regulation experimental platform with a wind power system. As shown in Figure 16, the experimental platform consists of seven parts: a synchronous generator simulator, wind turbine simulator, rectifier inverter for wind turbine simulator, synchronous drive inverters, synchronous machine dragging frequency converter, convergence cabinet and feed-back load. The experimental platform can simulate the frequency response of the power grid with wind turbines.

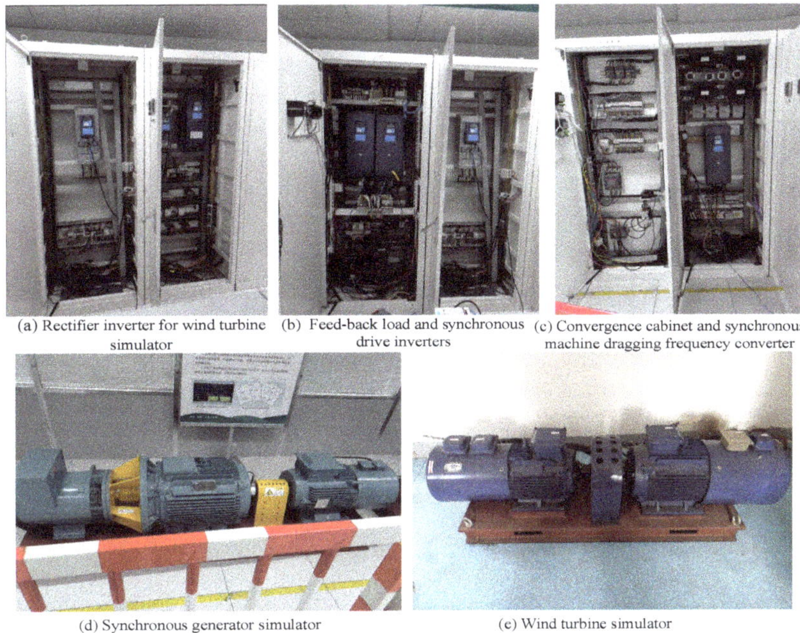

(a) Rectifier inverter for wind turbine simulator   (b) Feed-back load and synchronous drive inverters   (c) Convergence cabinet and synchronous machine dragging frequency converter

(d) Synchronous generator simulator   (e) Wind turbine simulator

Figure 16. Experimental platform.

Table 6 gives the main parameters of the experimental platform. Through the power scaling and inertia compensation algorithm [24], the wind turbine simulator in the experimental platform realizes the dynamic characteristics of the large capacity, large inertia wind power system, the frequency response characteristics of the actual power grid on the small capacity and low inertia experimental platform, which can be used to verify the control performance of the primary frequency regulation control strategy of the wind turbine. The wind turbine parameters simulated by the wind turbine simulator are the same as those of the simulation model.

A turbulent wind sequence with an average wind speed of 8 m/s and turbulence intensity level A is adopted for experimental analysis which is shown in Figure 17, and the load is set as in Figure 18. The improved strategy and the traditional deloading control with fixed K_d (K_d is 0.9) are compared by the experiment. The experimental results are shown in Figures 19–21.

Table 6. The main parameters of the experimental platform.

Parameters	The Value
Rated power of the wind turbine simulator	15 kW
Starting generation speed of the wind turbine simulator	0.48 rad/s
Rated power of the synchronous generator	25 kW
Inertia time constants of the synchronous generator	5.96 s
Rated speed of the synchronous generator	1800 r/min
Droop coefficient of the synchronous generator	0.05
The rated frequency of the grid	60 Hz
Rated power of feed-back loads	37 kVA
Control cycle of PLC	40 ms

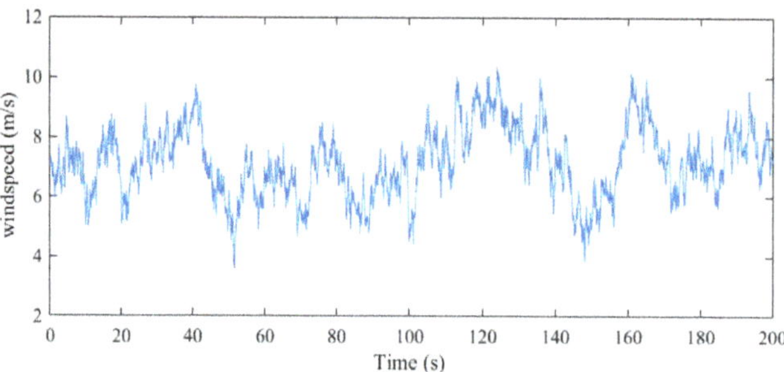

Figure 17. The wind speed used in experimental analysis.

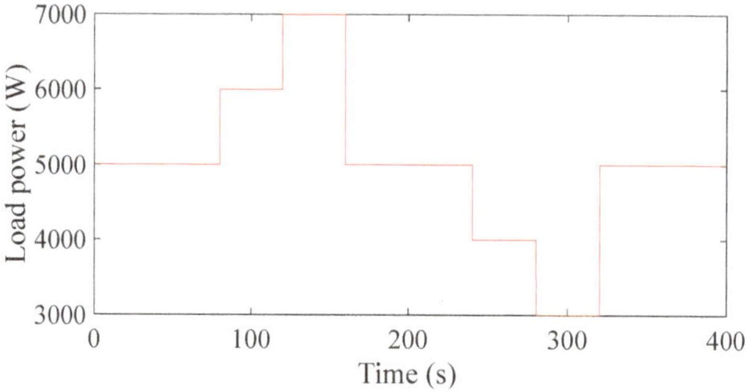

Figure 18. The load power.

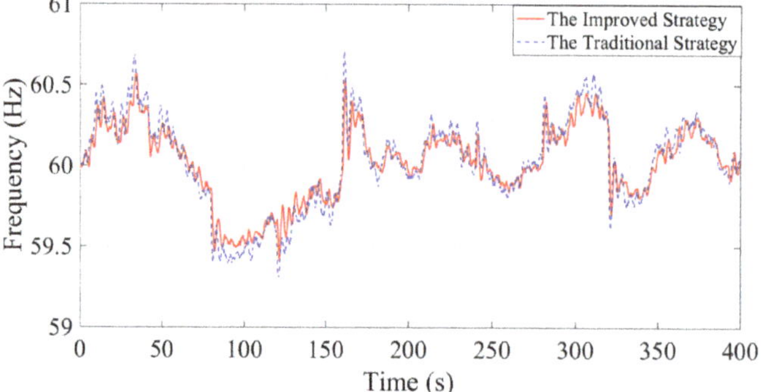

Figure 19. Frequency regulation effect.

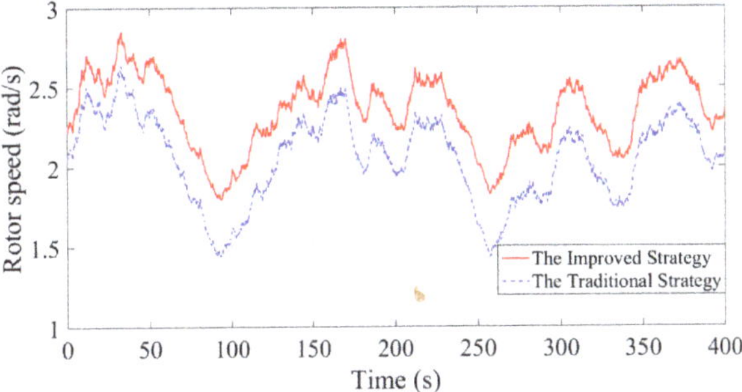

Figure 20. Rotor speed of the experiment.

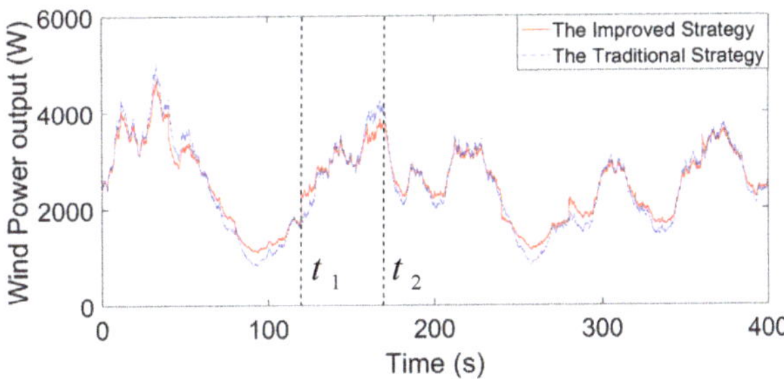

Figure 21. Wind power output of the experiment.

It can be seen from Figure 19 that the frequency fluctuation of the improved strategy in this paper is significantly smaller than that of the traditional strategy in the case of sudden load changes, and the frequency deviation of the improved strategy has increased by 8.3%.

Figure 20 shows the rotor speed of the wind turbine. Compared with the traditional strategy, the rotor speed of the improved strategy in this paper is at a higher level, and the rotor speed fluctuation is small, which can effectively mitigate the fatigue damage of the rotor and improve the stall problem caused by the low rotor speed.

The wind power output in Figure 21 shows that the power fluctuation of the improved strategy in this paper is smaller than that of the traditional strategy, which also improves the frequency stability to a certain extent. In addition, wind power can be increased or decreased faster and better in the case of sudden load changes. When the load suddenly increases at time t_1, the increased power of the improved strategy is higher than that of the traditional strategy. When the load suddenly decreases at time t_2, the output power of the improved strategy can better respond to frequency changes and be reduced faster.

6. Conclusions

This paper aims at improving the performance of OSD control under turbulent wind speed, proposing an improved OSD control strategy, which improves the effect of wind power participating in power grid frequency regulation under turbulent wind speed. The main contributions and conclusions of this paper are as follows.

1. The mechanism of poor control effect of OSD control under turbulent wind speed is revealed; the OSD control has the effect of improving the output power of wind power system, thus reducing the reserved power and weakening the frequency regulation effect.

2. The influence of deloading power coefficient on reserve capacity and frequency deviation is analyzed. It is found that there is an extreme value that minimizes the reserve capacity of the wind power system, and the extreme value will change with the turbulence characteristics of turbulent wind speed.

3. The mathematical relationship between the deloading power coefficient corresponding to the minimum reserve capacity and the turbulence characteristic index is obtained, and the reasonable range of the deloading power coefficient under different wind conditions is proposed accordingly.

4. An algorithm for optimizing deloading power coefficient based on turbulence characteristics is proposed, which can make wind turbines effectively respond to different wind conditions, provide reserve capacity stably and quickly, and improve the frequency regulation effect under turbulent wind speed.

Compared with the existing methods, the simulation and experimental results based on the experimental platform show that the improved control strategy proposed in this paper has better performance in decreasing frequency deviation, degree of frequency drop, expediting rotor speed recovery and avoiding the secondary frequency drop, and that the improvement in frequency deviation can reach 19.28%.

Author Contributions: X.Z. conceived the study and wrote the paper; B.L. analyzed the problems and designed the method and simulation examples; K.X. performed the simulation examples and help to wrote the paper; Y.Z. and S.H. analyzed the data; Q.H. help to wrote the paper. All authors have read and agreed to the published version of the manuscript.

Funding: This work was supported by National Natural Science Foundation of China (52107098).

Data Availability Statement: Not applicable.

Conflicts of Interest: The authors declare no conflict of interest.

References

1. Lu, L.; Saborío-Romano, O.; Cutululis, N.A. Frequency Control in Power Systems with Large Share of Wind Energy. *Energies* **2022**, *15*, 1922. [CrossRef]
2. Liu, H.; Wang, P.; Zhao, T.; Fan, Z.; Pan, H. A Group-Based Droop Control Strategy Considering Pitch Angle Protection to Deloaded Wind Farms. *Energies* **2022**, *15*, 2722. [CrossRef]
3. Ding, L.; Yin, S.; Wang, T.X.; Liu, J. Active Rotor Speed Protection Strategy for DFIG-based Wind Turbines with Inertia Control. *Autom. Electr. Power Syst.* **2015**, *39*, 29–34.
4. Chen, Y.; Wang, G.; Shi, Q.; Fu, L.J.; Jiang, W.T.; Huang, H. A New Coordinated Virtual Inertia Control Strategy for Wind Farms. *Autom. Electr. Power Syst.* **2015**, *39*, 27–33.
5. Yan, X.; Cui, S.; Chang, W. Primary Frequency Regulation Control Strategy of Doubly-Fed Induction Generator Considering Supercapacitor SOC Feedback Adaptive Adjustment. *Trans. China Electrotech. Soc.* **2021**, *36*, 1027–1039. [CrossRef]
6. Yan, X.; Cui, S.; Wang, D.; Wang, Y.; Li, T. Inertia and Primary Frequency Regulation Strategy of Doubly-fed Wind Turbine Based on Super-capacitor Energy Storage Control. *Autom. Electr. Power Syst.* **2020**, *44*, 111–129.
7. De Almeida, R.G.; Pecas Lopes, J.A. Participation of Doubly Fed Induction Wind Generators in System Frequency Regulation. *IEEE Trans. Power Syst.* **2007**, *22*, 944–950. [CrossRef]
8. Ghosh, S.; Kamalasadan, S.; Senroy, N.; Enslin, J. Doubly Fed Induction Generator (DFIG)-Based Wind Farm Control Framework for Primary Frequency and Inertial Response Application. *IEEE Trans. Power Syst.* **2016**, *31*, 1861–1871. [CrossRef]
9. Hu, J.X.; Xu, G.Y.; Bi, T.S.; Cui, H.B. A strategy of frequency control for deloaded wind turbine generator based on coordination between rotor speed and pitch angle. *Power Syst. Technol.* **2019**, *43*, 3656–3662. [CrossRef]
10. Boyle, J.; Littler, T.; Foley, A.M. Frequency Regulation and Operating Reserve Techniques for Variable Speed Wind Turbines. In Proceedings of the 2021 IEEE PowerTech, Madrid, Spain, 28 June–2 July 2021; pp. 1–5.
11. Wang, R.; Gao, L.; Shen, J.; Wang, Q.; Deng, Y.; Li, H. Frequency Regulation Strategy with Participation of Variable-speed Wind Turbines Power System with High Wind Power Penetration. *Autom. Electr. Power Syst.* **2019**, *43*, 101–110.
12. Cui, S.; Yan, X.; Wang, Y.; Li, R.; Li, T. A Smooth Adjustment Method for Primary Frequency of Doubly-fed Wind Generators Considering Random Fluctuation Characteristics of Source-load Power. *Proc. CSEE* **2021**, *41*, 143–154. [CrossRef]
13. Zhang, X.; Zhang, Y.; Hao, S.; Wu, L.; Wei, W. An improved maximum power point tracking method based on decreasing torque gain for large scale wind turbines at low wind sites. *Electr. Power Syst. Res.* **2019**, *176*, 105942. [CrossRef]
14. Zhang, X.; Huang, C.; Hao, S.; Chen, F.; Zhai, J. An Improved Adaptive-Torque-Gain MPPT Control for Direct-Driven PMSG Wind Turbines Considering Wind Farm Turbulences. *Energies* **2016**, *9*, 977. [CrossRef]
15. Ye, H.; Pei, W.; Qi, Z. Analytical Modeling of Inertial and Droop Responses from a Wind Farm for Short-Term Frequency Regulation in Power Systems. *IEEE Trans. Power Syst.* **2016**, *31*, 3414–3423. [CrossRef]
16. Wang, H.; Yang, J.; Chen, Z.; Ge, W.; Ma, Y.; Xing, Z.; Yang, L. Model Predictive Control of PMSG-Based Wind Turbines for Frequency Regulation in an Isolated Grid. *IEEE Trans. Ind. Appl.* **2018**, *54*, 3077–3089. [CrossRef]

17. Zhang, X.; Lin, B.; Yin, M.; Hao, S.; Zhang, C.; Zuo, T. A Novel Frequency Control Method for Smart Buildings with Wind Power System Integration Considering Turbulence Characteristics. *IEEE Trans. Ind. Appl.* **2022**, *59*, 15–25. [CrossRef]
18. Li, S.; Huang, Y.; Wang, L.; Lei, X.; Tang, H.; Deng, C. Modeling Primary Frequency Regulation Auxiliary Control System of Doubly Fed Induction Generator Based on Rotor Speed Control. *Proc. CSEE* **2017**, *37*, 7077–7086, 7422. [CrossRef]
19. Yao, Q.; Liu, J.; Hu, Y.; Meng, H. Equivalent modeling and simulation for primary regulation of wind farm with asynchronous variable-speed wind turbine. *Autom. Electr. Power Syst.* **2019**, *43*, 185–192. [CrossRef]
20. Huang, C.; Li, F.; Jin, Z. Maximum Power Point Tracking Strategy for Large-Scale Wind Generation Systems Considering Wind Turbine Dynamics. *IEEE Trans. Ind. Electron.* **2015**, *62*, 2530–2539. [CrossRef]
21. Shi, Q.; Wang, G.; Li, H. Coordinated Virtual Inertia Control Strategy of Multiple Wind Turbines in Wind Farms Considering Frequency Regulation Capability. *Power Syst. Technol.* **2019**, *43*, 4005–4017.
22. Committee, I.E. *IEC 61400-1: Wind Turbines Part 1: Design Requirements*; International Electrotechnical Commission: London, UK, 2005.
23. Wang, Y.; Meng, J.; Zhang, X.; Xu, L. Control of PMSG-Based Wind Turbines for System Inertial Response and Power Oscillation Damping. *IEEE Trans. Sustain. Energy* **2015**, *6*, 565–574. [CrossRef]
24. Yin, M.; Li, W.; Chung, C.Y.; Chen, Z.; Zou, Y. Inertia compensation scheme of WTS considering time delay for emulating large-inertia turbines. *IET Renew. Power Gener.* **2017**, *11*, 529–538. [CrossRef]

MDPI

St. Alban-Anlage 66

4052 Basel

Switzerland

www.mdpi.com

Energies Editorial Office

E-mail: energies@mdpi.com

www.mdpi.com/journal/energies

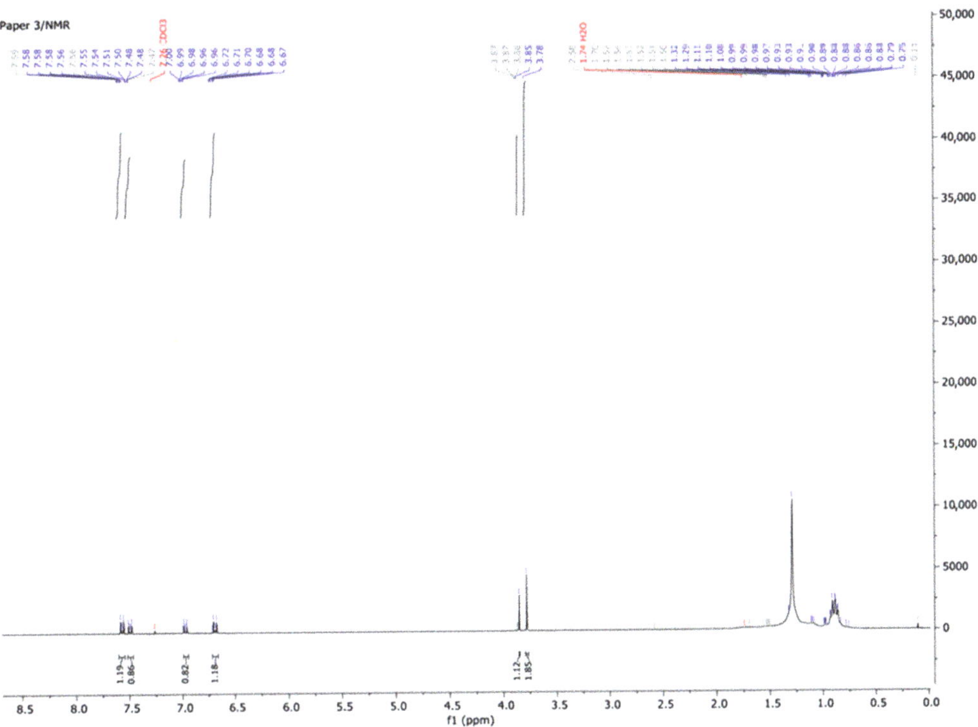

Figure 10. ^{1}H NMR of synthesised coupling compound.

4. Conclusions

The synthesis of the materials started with a low-cost carbon source, BT, which was then modified through a two-step synthesis. The synthesis resulted in the production of amorphous AC, BTSA, which was used as a precursor for the synthesis of oxidised BTHM. The materials were characterized using SEM, HR-TEM, XRD, and Raman and FTIR spectroscopy to determine their structure as well as chemical composition. The Raman active D- and G-bands at ~1340 cm^{-1} and ~1590 cm^{-1} from the planar graphite arising from sp^3 hybridised carbons consist of defects and sp^2 hybridized carbon, respectively. We observed an increase in intensity after each stage of synthesis, while the Raman shift remained the same. Each stage of modification changed the size, structure, and morphology of the material as well as functional groups available on the surface (active sites) of the materials. The raw and modified materials were then used in batch adsorption studies. The kinetic analysis of the Cu^{2+} adsorption to BTHM was found to exhibit two-phase pseudo-first-order behaviour (compartmentalized fast and slow adsorption) dominated by mass transfer to BTHM. It was further determined that BTHM exhibited increased adsorption capacity with the optimum pH at 6 with a 136.1 mg/g capacity. The application of carbonaceous materials for water treatment is, therefore, a promising venture and the modification of these materials and their capacity for Cu^{2+} removal from wastewater need to be studied. Furthermore, the catalytic ability of the materials to promote the synthesis of the coupling reaction, which is important in organic chemistry, will be studied and further catalytic studies will be conducted to evaluate the selectivity, sensitivity, and recyclability of the catalysts. The Cu@BTHM material's catalytic ability was confirmed using coupling reactions where an iodobenzene was coupled with a phenol in a classic organic coupling reaction, which successfully yielded the coupled compound. This verified that the Culaden adsorbent is reusable as a catalyst and secondary waste generation can be avoided in this manner.

Supplementary Materials: The following supporting information can be downloaded at: https://www.mdpi.com/article/10.3390/min14030302/s1, Figure S1. FTIR before adsorption for Raw BT, BTSA and, BTHM; Figure S2. Thermogravimetric analysis of BTSA and BTHM.

Author Contributions: Conceptualization, I.C.A., K.P. and A.M.; methodology, I.C.A., M.B. and A.M.; software, I.C.A., H.G.B. and A.M.; validation, I.C.A., H.G.B., M.B., K.P. and A.M.; formal analysis, I.C.A., M.B., K.P., H.G.B. and A.M.; investigation, I.C.A. and A.M.; resources, A.M.; data curation, A.M.; writing—original draft preparation, I.C.A., K.P., H.G.B. and A.M.; writing—review and editing, H.G.B. and A.M.; visualization, I.C.A., H.G.B. and A.M.; supervision, A.M.; project administration, A.M.; funding acquisition, A.M. All authors have read and agreed to the published version of the manuscript.

Funding: This research received no external funding.

Data Availability Statement: The raw data supporting the conclusions of this article will be made available by the authors on request.

Conflicts of Interest: The authors declare no conflicts of interest.

References

1. Afroze, Q. How Does Mining Affect the Environment. Available online: https://federalmetals.ca/how-does-copper-mining-affect-the-environment/ (accessed on 14 November 2023).
2. Liu, Y.; Wang, H.; Cui, Y.; Chen, N. Removal of Copper Ions from Wastewater: A Review. *Int. J. Environ. Res. Public Health* **2023**, *20*, 3885. [CrossRef]
3. Tchounwou, P.B.; Yedjou, C.G.; Patlolla, A.K.; Sutton, D.J. Heavy Metal Toxicity and the Environment. *Exp. Suppl.* **2012**, *101*, 133–164. [CrossRef] [PubMed]
4. Singh, R.; Gautam, N.; Mishra, A.; Gupta, R. Heavy metals and living systems: An overview. *Indian J. Pharmacol.* **2011**, *43*, 246. [CrossRef] [PubMed]
5. Dong, D.; van Oers, L.; Tukker, A.; van der Voet, E. Assessing the future environmental impacts of copper production in China: Implications of the energy transition. *J. Clean. Prod.* **2020**, *274*, 122825. [CrossRef]
6. Xu, Y.; Weinberg, G.; Liu, X.; Timpe, O.; Schlögl, R.; Su, D.S. Nanoarchitecturing of Activated Carbon: Facile Strategy for Chemical Functionalization of the Surface of Activated Carbon. *Adv. Funct. Mater.* **2008**, *18*, 3613–3619. [CrossRef]
7. Kobya, M.; Demirbas, E.; Senturk, E.; Ince, M. Adsorption of heavy metal ions from aqueous solutions by activated carbon prepared from apricot stone. *Bioresour. Technol.* **2005**, *96*, 1518–1521. [CrossRef]
8. El Fakir, L.; Flayou, M.; Dahchour, A.; Sebbahi, S.; Kifani-Sahban, F.; El Hajjaji, S. Adsorptive removal of copper(II) from aqueous solutions on phosphates: Equilibrium, kinetics, and thermodynamics. *Desalin. Water Treat.* **2015**, *43*, 1–10. [CrossRef]
9. Bobrowska, D.M.; Olejnik, P.; Echegoyen, L.; Plonska-Brzezinska, M.E. Onion-Like Carbon Nanostructures: An Overview of Bio-Applications. *Curr. Med. Chem.* **2019**, *26*, 6896–6914. [CrossRef]
10. Ren, X.; Li, J.; Tan, X.; Wang, X. Comparative study of graphene oxide, activated carbon and carbon nanotubes as adsorbents for copper decontamination. *Dalt. Trans.* **2013**, *42*, 5266. [CrossRef]
11. Deng, J.; You, Y.; Sahajwalla, V.; Joshi, R.K. Transforming waste into carbon-based nanomaterials. *Carbon N. Y.* **2016**, *96*, 105–115. [CrossRef]
12. McDonough, J.K.; Gogotsi, Y. Carbon Onions: Synthesis and Electrochemical Applications. *Interface Mag.* **2013**, *22*, 61–66. [CrossRef]
13. Eatemadi, A.; Daraee, H.; Karimkhanloo, H.; Kouhi, M.; Zarghami, N.; Akbarzadeh, A.; Abasi, M.; Hanifehpour, Y.; Joo, S.W. Carbon nanotubes: Properties, synthesis, purification, and medical applications. *Nanoscale Res. Lett.* **2014**, *9*, 393. [CrossRef] [PubMed]
14. Zhang, B.-T.; Zheng, X.; Li, H.-F.; Lin, J.-M. Application of carbon-based nanomaterials in sample preparation: A review. *Anal. Chim. Acta* **2013**, *784*, 1–17. [CrossRef] [PubMed]
15. Meng, X.; Deng, D. Trash to Treasure: Waste Eggshells Used as Reactor and Template for Synthesis of Co 9 S 8 Nanorod Arrays on Carbon Fibers for Energy Storage. *Chem. Mater.* **2016**, *28*, 3897–3904. [CrossRef]
16. Akhavan, O.; Bijanzad, K.; Mirsepah, A. Synthesis of graphene from natural and industrial carbonaceous wastes. *RSC Adv.* **2014**, *4*, 20441. [CrossRef]
17. Yang, X.; Wan, Y.; Zheng, Y.; He, F.; Yu, Z.; Huang, J.; Wang, H.; Ok, Y.S.; Jiang, Y.; Gao, B. Surface functional groups of carbon-based adsorbents and their roles in the removal of heavy metals from aqueous solutions: A critical review. *Chem. Eng. J.* **2019**, *366*, 608–621. [CrossRef]